AF389339

Optimization and Communication in UAV Networks

Optimization and Communication in UAV Networks

Editors

Christelle Caillouet
Nathalie Mitton

MDPI • Basel • Beijing • Wuhan • Barcelona • Belgrade • Manchester • Tokyo • Cluj • Tianjin

Editors
Christelle Caillouet
Université Côte d'Azur
France

Nathalie Mitton
Inria
France

Editorial Office
MDPI
St. Alban-Anlage 66
4052 Basel, Switzerland

This is a reprint of articles from the Special Issue published online in the open access journal *Sensors* (ISSN 1424-8220) (available at: https://www.mdpi.com/journal/sensors/special_issues/UAV_net).

For citation purposes, cite each article independently as indicated on the article page online and as indicated below:

LastName, A.A.; LastName, B.B.; LastName, C.C. Article Title. *Journal Name* **Year**, *Article Number*, Page Range.

ISBN 978-3-03943-310-0 (Hbk)
ISBN 978-3-03943-311-7 (PDF)

Contents

About the Editors

Christelle Caillouet has served as Associate Professor at University of Côte d'Azur since September 2011 and is a member of the joint team Coati between the I3S (CNRS, University of Nice-Sophia Antipolis) laboratory and Inria. She completed her MSc in Optimization and Game Theory at University Paris VI in 2006, and Ph.D. in Computer Science at University of Nice Sophia Antipolis in 2009. Her research interests are focused on optimization, linear programming, and algorithmics as applied to telecommunication networks (such as wireless mesh networks and backhaul networks) and biological networks.

Nathalie Mitton received her M.Sc. and Ph.D. degrees in Computer Science from INSA Lyon in 2003 and 2006, respectively. She received her Habilitation à diriger des recherches (HDR) in 2011 from Université Lille 1. She is currently an Inria Full Researcher since 2006, and has served as Scientific Head of the Inria FUN team from 2012, which is focused on small computing devices like electronic tags and sensor networks. Her research interests focus on self-organization from PHY to routing for wireless constrained networks. She has published her research in more than 30 international revues and more than 100 international conferences. She is involved in the setup of the FIT IoT LAB platform (http://fit-equipex.fr/,https://www.iot-lab.info), the H2020 CyberSANE and VESSEDIA projects, and in several program and organization committees such as Infocom 2019 and 2020, PerCom 2019 and 2020, DCOSS 2019, Adhocnow 2015–2019, ICC (since 2015), Globecom (since 2017), Pe-Wasun 2017, and VTC (since 2016). She also supervises numerous Ph.D. students and engineers.

Preface to "Optimization and Communication in UAV Networks"

Nowadays, Unmanned Aerial Vehicles (UAVs) have received growing popularity in the Internet-of-Things (IoT) which often deploys many sensors in a relatively wide region. Current trends focus on deployment of a single UAV or a swarm of it to generally map an area, perform surveillance, monitoring or rescue operations, collect data from ground sensors or various communicating devices, provide additional computing services close to data producers, etc. Applications are very diverse and call for different features or requirements. But UAV remain low-power battery powered devices that in addition to their mission, must fly and communicate. Thanks to wireless communications, they participate to mobile dynamic networks composed of UAV and ground sensors and thus many challenges have to be addressed to make UAV very efficient. And behind any UAV application, hides an optimization problem. There is still a criterion or multiple ones to optimize such as flying time, energy consumption, number of UAV, quantity of data to send/receive, etc.

This book, which is a Special Issue of the Sensors journal, deals with the wedding of optimization and communication in UAV networks. We hope you will enjoy it. We wish you a pleasant reading.

Christelle Caillouet, Nathalie Mitton
Editors

Editorial

Optimization and Communication in UAV Networks

Christelle Caillouet [1,2,*,†] and **Nathalie Mitton** [1,*,†]

1   Inria, 40 Avenue Halley, 59650 Villeneuve d'Ascq, France
2   Department INFO of IUT Nice Côte d'Azur, Université Côte d'Azur, CNRS, I3S, 2004 Route des Lucioles, 06902 Sophia Antipolis, France
*   Correspondence: christelle.caillouet@univ-cotedazur.fr (C.C.); nathalie.mitton@inria.fr (N.M.)
†   These authors contributed equally to this work.

Received: 2 September 2020; Accepted: 3 September 2020; Published: 4 September 2020

Abstract: Nowadays, Unmanned Aerial Vehicles (UAVs) have received growing popularity in the Internet-of-Things (IoT) which often deploys many sensors in a relatively wide region. Current trends focus on deployment of a single UAV or a swarm of it to generally map an area, perform surveillance, monitoring or rescue operations, collect data from ground sensors or various communicating devices, provide additional computing services close to data producers, etc. Applications are very diverse and call for different features or requirements. But UAV remain low-power battery powered devices that in addition to their mission, must fly and communicate. Thanks to wireless communications, they participate to mobile dynamic networks composed of UAV and ground sensors and thus many challenges have to be addressed to make UAV very efficient. And behind any UAV application, hides an optimization problem. There is still a criterion or multiple ones to optimize such as flying time, energy consumption, number of UAV, quantity of data to send/receive, etc

Keywords: UAV; drones; wireless; self-organization; optimization; swarm; communication; algorithms

---

1. Introduction

With new technological advances, UAVs are becoming a reality and are attracting more and more attention. UAVs or drones are flying devices that can be remotely controlled or, more recently, completely autonomous. They can be used alone or as a fleet, and in a large set of applications: from rescue operations to event coverage going through servicing other networks such as sensor networks for replacing, recharging, or data offloading. They are hardware-constrained since they cannot be too heavy and rely on batteries. Depending on their use (alone or in a swarm) and the targeted applications, they must evolve differently and meet different requirements (energy preservation, delay of covering an area, coverage, limited number of devices, etc.) with limited resources (energy, speed, etc.). Yet, their use still raises a large set of new exciting challenges, in terms of trajectory optimization, positioning, when they are used alone or in cooperation, coordination, and communication when they evolve in a swarm, just to name a few. This Special Issue was calling for any new original submissions that deal with UAV or UAV swarm optimization or communication aspects. Among the numerous submissions, only twelve of them have been selected after a rigorous selection process. The main themes that arise from them are: (i) ground data collection from the air, (ii) control of UAV swarm UAV-based Mobile Edge Computing and (iii) application-driven UAV based measurements.

In the following, we sum up the contributions of the papers published to this Special Issue for each category to then conclude by drawing future challenges and still open issues.

2. Ground Data Collection from the Air and Path Planning

It is becoming more and more common to imagine having data sensed from ground wireless sensors collected by UAV to alleviate wireless peer-to-peer communications between ground sensors and reduce their energy consumption. However, such a paradigm raises a set of new challenges such as how to prioritize the sensors to visit, how to optimize the time to collect all data by visiting all devices, etc. This is an exciting optimization problem. Works [1–4] propose different approaches to address this issue with different perspectives. Different criteria are considered to plan the trajectory of the UAV and different functions are optimized.

Reference [1] proposes to visit the nodes in a given order and for a variable time that depend on a node priority, while in [2], the authors aim to maximize the data collection utility by jointly optimizing the communication scheduling and trajectory of each UAV. The data collection utility is determined by the amount and value of the collected data and a novel trajectory planning algorithm is designed to maximize it. The author of [3] focuses on the problem of minimizing the mission completion time (flying time and hovering time) for a multi-UAV system in a monitoring scenario while ensuring that the information of each sensor is collected. As for [4], the authors aim to improve the secrecy performance of cellular-enabled unmanned aerial vehicle communication networks through an aerial cooperative jamming scheme.

3. Control of UAV Swarm

When more than a drone is required, a swarm of UAVs can be deployed. Although bringing more performances in terms of coverage and connectivity, new optimization challenges pop up due to the difficulty to control and scale such swarms both in a distributed or centralized way. References [5–8] tackle these numerous challenges going from connectivity maintenance to swarm control.

Reference [5] studies the different factors that may impact the accuracy and efficiency of an unmanned aerial vehicle (UAV) swarm coordination. The authors propose a mathematical data model to demonstrate the fundamental properties of antenna arrays and study the performance of the data collection system framework. Numerical examples and practical measurements are provided to demonstrate the feasibility of the proposed data collection system framework using an iterative-MUSIC algorithm and benchmark theoretical expectations.

Reference [6] deals with multi-UAV systems where the UAV autonomy is much smaller than the time to complete their mission. The authors thus introduce a UAV replacement procedure as a way to guarantee ground users' connectivity over time, formulating the practical UAV replacements problem in moderately large multi-UAV swarms and proves it to be an NP-hard problem in which an optimal solution has exponential complexity. Reference [7] focuses on the maintenance formation with time-varying shape of a swarm proposing a virtual leader approach while [8] investigates a stochastic model of the UAV Swarm system with multiplicative noises.

4. UAV Enabled Mobile Edge Computing

The potential offered by the abundance of sensors, actuators, and communications in the Internet of Things (IoT) era is hindered by the limited computational capacity of local nodes. However, the latter do not necessarily always have the capacity to offload data to an edge server. In such a case, mobile edge servers can go to them thanks to the deployment of UAV-assisted Multi-access Edge Computing systems, which raises new challenging optimization and networking issues as addressed in [9,10].

Reference [9] proposes to provide an Unmanned Aerial Vehicle (UAV)-assisted Multi-access Edge Computing (MEC) system based on a usage-based pricing policy for allowing the exploitation of the servers' computing resources while the authors of [10] introduce the DRUID-NET perspective, aiming to adapt to a rapidly varying demand by applying different tools from Automata and Graph theory, Machine Learning, Modern Control Theory, and Network Theory combined.

5. Application-Driven UAV Based Measurements

In such cases, the application that has asked for UAV deployment comes with very specific constraints and requirements and calls for specific optimization models. Reference [11,12] gives two such examples dedicated respectively for three-dimensional measurements and surveillance.

For instance, Reference [11] aims to provide a comparative analysis of the precision of ground geodetic data versus the three-dimensional measurements from unmanned aerial vehicles (UAV), while establishing the impact of herbaceous vegetation on the UAV 3D model. A constraint to take into account in this application is the fact that herbaceous vegetation can impede the establishment of the anthropogenic roughness of the surface and deteriorates the identification of minor surfaces.

Reference [12] focuses on UAV cooperative surveillance networks and introduces the use of complex field network coding (CFNC) for this application. According to whether there is a direct communication link between any source drone and the destination, the information transfer mechanism at the downlink is set to one of two modes, either mixed or relay transmission, and two corresponding irregular topology structures for CFNC-based networks are proposed. Theoretical analysis and simulation results with an additive white Gaussian noise (AWGN) channel show that the CFNC obtains a throughput as high as $1/2$ symbol per source per channel use. Results show that CFNC applied to the proposed irregular structures under the two transmission modes can achieve better reliability due to full diversity gain as compared to that based on the regular structure.

6. Conclusions

As you can notice, challenges in UAV networks are huge, numerous and heterogeneous. They are concerning different aspects of the deployment of drones, from path trajectory to connectivity maintenance going through energy management. Much of them have been addressed with optimization tools but there remain a lot of open issues and research directions. Contributions presented in this special issue are only a first step to pave the way towards even more exciting investigations.

Funding: This research received no external funding.

Conflicts of Interest: The authors declare no conflict of interest.

References

1. Ma, X.; Liu, T.; Liu, S.; Kacimi, R.; Dhaou, R. Priority-Based Data Collection for UAV-Aided Mobile Sensor Network. *Sensors* **2020**, *20*, 3034. [CrossRef]
2. Qin, Z.; Dong, C.; Wang, H.; Li, A.; Dai, H.; Sun, W.; Xu, Z. Trajectory Planning for Data Collection of Energy-Constrained Heterogeneous UAVs. *Sensors* **2019**, *19*, 4884. [CrossRef] [PubMed]
3. Qin, Z.; Li, A.; Dong, C.; Dai, H.; Xu, Z. Completion Time Minimization for Multi-UAV Information Collection via Trajectory Planning. *Sensors* **2019**, *19*, 4032. [CrossRef] [PubMed]
4. Sun, H.; Duo, B.; Wang, Z.; Lin, X.; Gao, C. Aerial Cooperative Jamming for Cellular-Enabled UAV Secure Communication Network: Joint Trajectory and Power Control Design. *Sensors* **2019**, *19*, 4440. [CrossRef] [PubMed]
5. Chen, Z.; Yeh, S.; Chamberland, J.F.; Huff, G.H. A Sensor-Driven Analysis of Distributed Direction Finding Systems Based on UAV Swarms. *Sensors* **2019**, *19*, 2659. [CrossRef] [PubMed]
6. Sanchez-Aguero, V.; Valera, F.; Vidal, I.; Tipantuna, C.; Hesselbach, X. Energy-Aware Management in Multi-UAV Deployments: Modelling and Strategies. *Sensors* **2020**, *20*, 2791. [CrossRef] [PubMed]
7. Cordeiro, T.F.K.; Ishihara, J.Y.; Ferreira, H.C. A Decentralized Low-Chattering Sliding Mode Formation Flight Controller for a Swarm of UAVs. *Sensors* **2020**, *20*, 3094. [CrossRef] [PubMed]
8. Zhao, H.; Wu, S.; Wen, Y.; Liu, W.; Wu, X. Modeling and Flight Experiments for Swarms of High Dynamic UAVs: A Stochastic Configuration Control System with Multiplicative Noises. *Sensors* **2019**, *19*, 3278. [CrossRef] [PubMed]

9. Mitsis, G.; Tsiropoulou, E.E.; Papavassiliou, S. Data Offloading in UAV-Assisted Multi-Access Edge Computing Systems: A Resource-Based Pricing and User Risk-Awareness Approach. *Sensors* **2020**, *20*, 2434. [CrossRef] [PubMed]
10. Dechouniotis, D.; Athanasopoulos, N.; Leivadeas, A.; Mitton, N.; Jungers, R.; Papavassiliou, S. Edge Computing Resource Allocation for Dynamic Networks: The DRUID-NET Vision and Perspective. *Sensors* **2020**, *20*, 2191. [CrossRef] [PubMed]
11. Cesnulevicius, A.; Bautrėnas, A.; Bevainis, L.; Ovodas, D. A Comparison of the Influence of Vegetation Cover on the Precision of an UAV 3D Model and Ground Measurement Data for Archaeological Investigations: A Case Study of the Lepelionys Mound, Middle Lithuania. *Sensors* **2019**, *19*, 5303. [CrossRef] [PubMed]
12. Xue, R.; Han, L.; Chai, H. Complex Field Network Coding for Multi-Source Multi-Relay Single-Destination UAV Cooperative Surveillance Networks. *Sensors* **2020**, *20*, 1542. [CrossRef] [PubMed]

Article

Priority-Based Data Collection for UAV-Aided Mobile Sensor Network

Xiaoyan Ma [1], Tianyi Liu [2],*, Song Liu [1], Rahim Kacimi [3] and Riadh Dhaou [4]

[1]   College of Architecture and Urban Planning, Tongji University, Shanghai 200092, China;
      xiaoyan.ma@enseeiht.fr (X.M.); liusong5@tongji.edu.cn (S.L.)
[2]   School of Aerospace Engineering and Applied Mechanics, Tongji University, Shanghai 200092, China
[3]   IRIT-UPS, University of Toulouse, 31062 Toulouse, France; rahim.kacimi@irit.fr
[4]   IRIT-ENSEEIHT, University of Toulouse, 31071 Toulouse, France; riadh.dhaou@enseeiht.fr
*   Correspondence: tianyi.liu@tongji.edu.cn

Received: 31 March 2020; Accepted: 25 May 2020; Published: 27 May 2020

Abstract: In this work, we study data collection in multiple unmanned aerial vehicle (UAV)-aided mobile wireless sensor networks (WSNs). The network topology is changing due to the mobility of the UAVs and the sensor nodes, so the design of efficient data collection protocols is a major concern. We address such high dynamic network and propose two mechanisms: prioritized-based contact-duration frame selection mechanism (PCdFS), and prioritized-based multiple contact-duration frame selection mechanisms (PMCdFS) to build collision-free scheduling and balance the nodes between the multi-UAV respectively. Based on the two mechanisms, we proposed a Balance algorithm to conduct the collision-free communication between the mobile nodes and the multi-UAVs. Two key design ideas for a Balance algorithm are: (a) no need of higher priority for those nodes that have lower transmission rate between them and the UAV and (b) improve the communication opportunity for those nodes that have shorter contact duration with the UAVs. We demonstrate the performance of proposed algorithms through extensive simulations, and real experiments. These experiments using 15 mobile nodes at a path with 10 intersections and 1 island, present that network fairness is efficiently enhanced. We also confirm the applicability of proposed algorithms in a challenging and realistic scenario through numerous experiments on a path at Tongji campus in Shanghai, China.

Keywords: wireless sensor networks; multiple unmanned aerial vehicles; mobile nodes; data collection; collision-free

1. Introduction

Unmanned Aerial Vehicle-aided wireless sensor networks (UAV-aided WSN) have gained more and more interest due to their many applications in monitoring, surveillance, and exploring in healthcare, agriculture, industry, and military [1–5]. Among UAVs' applications, one of the key functions is the data collection [6–11]. These works focus on deterministic topology where the nodes are deployed statically, and the locations of the sensors are known. The data collection issues addressed on dynamic topology, which are usually used in applications such as maritime detection, traffic surveillance, and wilderness rescuing where the targets are moving and no static sensors are deployed in advance, are seldom covered.

The main difference between the static network and mobile network are: the transmission opportunities for nodes that are within the coverage of the UAV are different. In static case, all covered nodes are static, the relative velocity (v_r) between the nodes and the UAV are the same. Thus, the contact durations (CD) between them with the UAV depend on the relative distance (d_r) between them ($CD = \frac{d_r}{v_r}$, see [12,13] for more details). The relative distances almost have no difference if the UAV flies at a higher altitude. However, in mobile case, when the nodes move at different velocities, the

CD are different greatly even the relative distance is the same. Intuitively, the shorter the CD between them, the smaller the opportunities for the mobile node to communicate with the UAV. When the CD is very short, the mobile node may have no opportunity to communicate with the UAV if no attention is paid on the CD between it with the UAV. Thus, a contact-duration-based data collection algorithm should be designed for such context despite a large array of existing data collection algorithms (see Section 2. related works) on UAV-aided static WSNs.

The impact factors of the CD between mobile nodes and the UAV include two aspects: (a) the relative distance between the sensor and the UAV, and (b) the relative velocity between them. Priority-based Frame Selection (PFS) [14,15] is a one-hop mechanism based on the relative distance according to which the nodes are divided into different priority groups. Communications are conducted from higher to lower priorities. A multi-hop highest velocity opportunistic algorithm which is based on relative velocity between mobile nodes and the UAV is proposed in [16]. The ones that have higher velocity have longer CD with the UAV, therefore were selected as forwarded nodes. In our previous work [12,13], we studied the data collection maximization issues in single UAV enabled mobile WSN where the pre-defined path is a straight path without comparison with existing works and real experiments. The curve path and multi-UAVs aspects are also not covered in the previous work. Thus, a large room for enhancing the network performance still exists.

In this work, we focus on multi-UAV aided mobile WSN, Figure 1, where the nodes are deployed on mobile bicycles and move along a pre-defined curve path. Considering that, in the context of the nodes move along a path, two UAVs are enough to cover all mobile nodes when (as in Figure 1) UAV_1 take-off from the original point of the path and fly along the path, UAV_2 take-off from the end-point of the path and fly along the path. Data collection issues in such contexts contain two aspects. End-to-end data collection is a very complex problem. In this paper, we focus on the access link. As the literature, on this kind of link between the sensors and the UAV [6,7], still does not propose efficient solutions. The access link suffers from the synchronization problem due to the high dynamic network, the coordination between the mobile nodes and the multi-UAVs. Providing the opportunity of communication to the nodes that have a very short duration with the UAVs reduces the congestion risk. On the other hand, extensive literature can be referred to, on the second link, on the backhaul link, between the UAVs and gateways [17]. The second link is also challenging on several levels such as the data security, the security of UAVs, and the dimensioning of the backhaul. In our previous work [18,19], we focused on the backahul link with the satellite system. The proposed algorithms on mobile mules, in [18,19] are applicable for UAV-aided sensor networks. Moreover, because that the collected data (considering the value of data and distinguish the data collected from each sensor) could be stored in SD cards embedded on the UAV, thus, in this work, we focus on the access link. The data collection optimization objectives in such context include two aspects: (i) maximizing the number of collected packets, and (ii) maximizing the number of nodes that successfully send at least one packet during the collection period. Our main purpose is to jointly maximize the two aspects through formulating the dynamic parameters. Our main contributions are summarized as follows:

- We study the impact of dynamic parameters, including the speed and flying height of UAV, the sensor speed, network size, and different priority areas. We mathematically formulate the data collection issue into the optimization with the objective of maximizing the number of collected packets and the number of sensors that successfully send packets to the UAVs.
- Based on the dynamic parameters, we adopt a time-discrete mechanism and propose a prioritized-based multiple contact-duration frame selection algorithm (PMCdFS). PMCdFS algorithm is used for the balancing between the nodes (that are within the range of multi-UAVs at the same time) and multi-UAVs.
- We improve the contact duration mechanism in our previous work (see [12,13] for more details) with the Prioritized Frame Selection (PFS) mechanism (see [14,15] for more details) and propose a prioritized-based contact-duration frame selection algorithm (PCdFS). PCdFS algorithm is a

one-hop and slotted mechanism which is used to allocate the time-slot for the nodes that covered only by one of the UAVs.

- We propose a Balance algorithm to solve the collision between the nodes and UAVs so as to optimize the aforementioned data collection performance.
- Through extensive simulations, and real experiments, we examine the effectiveness of the proposed algorithms, and compare it with existing algorithm under different configurations.

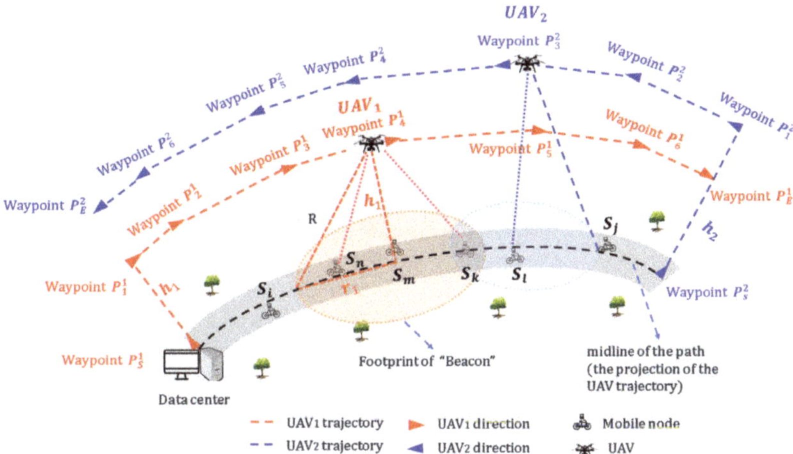

Figure 1. An illustration of the unmanned aerial vehicle (UAV)-aided data collection for a mobile wireless sensor network. The exemplar trajectory of the UAV_1 is shown as: Waypoint $P_S^1 \rightarrow$ Waypoint $P_1^1 \rightarrow$ Waypoint $P_2^1 \rightarrow$ Waypoint $P_3^1 \rightarrow$ Waypoint $P_4^1 \rightarrow$ Waypoint $P_5^1 \rightarrow$ Waypoint $P_6^1 \rightarrow$ Waypoint P_E^1.

The remainder of this paper is organized as follows: in the next section, we discuss previous related work. Section 3 presents the system model and the problems formulated. Section 4 present the proposed algorithms. Section 5 evaluated the proposed algorithms through extensive simulations and real experiments. Section 6 concludes this paper and gives some future work suggestions.

2. Related Works

There exists an extensive array of research on data collection in UAV-aided WSN with different objectives ranging from completion time minimization [20], power controlling [21], trajectory distance minimizing [22] to energy consumption minimization [23,24]. We classify these existing data collection algorithms by two criteria: (i) Static or mobile nodes, and (ii) sensors are deployed along a path or deployed within an interesting area. In (i), algorithms are differentiated by whether the sensors mobile or not because the dynamic parameters brought by the movement of nodes in the network structure have a much greater impact on the system performance. In (ii), algorithms are differentiated by whether the nodes deployed along a path or not. The nodes deployed along a given path [12,13,25,26] so the UAV trajectory planning has very little impact on the network performance.

(i) Data collection algorithms addressed on mobile nodes. There are many works on studying how to collect data from WSN. The authors in [4,9,27–29] review these works. According to the [4,9,27–29], most of these algorithms only based on the mobile sink or only focused on mobile sensors. In our previous works [12,13,16], we studied how to use UAV to collect data from mobile nodes based on an assumption that both the nodes and the UAV move along a straight path with constant speeds. The case where both the UAV and the nodes move in a curved path is not considered. Numerous researches have been done on statically deployed networks [6,7,11,14,15,20,24,25,30–40].

(ii) Most of the aforementioned data collection algorithms can also be classified according to the deployed status of the nodes. Authors in [12,13,16,25,34] studied how to use UAV to collect data from nodes that deployed along a straight path. Especially in [25], the nodes deployed on a straight line, and the UAV flies over this line to collect data from nodes. In such context, the trajectory of the UAV is dependent on the path (or line) and has a light impact on the performance if the path is long enough. For instance, in [25], the authors aim to minimize the flight time through jointly optimizing the transmit power of nodes, the UAV speed and the transmission intervals. For the case that nodes are deployed within the area of interest, one of the main issues is to plan the UAV's trajectory so as to enhance the network performance. Numerous research has been done on the UAV trajectory planning issues [6,7,20,24,30–40]. These works are different from the optimization method and objective function because of different scenarios. They are mainly classified into two types: single-UAV trajectory planning [6,7,24,30–34] and multi-UAV trajectory planning [20,35–40].

The first is the single-UAV trajectory planning. Authors in [33] use a UAV for the mobile edge computing system. They minimize the maximum delay of all ground users through jointly optimizing the offloading ratio, the users' scheduling variables, and UAV's trajectory. While, in [24], the authors aim to minimize the maximum energy consumption by optimizing the trajectory of a rotary-wing UAV. The authors utilize a UAV to collect data from IoT devices with each has limited buffer size and target data upload deadline [6]. In this study, the data should be transmitted before it loses its meaning or becomes irrelevant. To maximize the number of served IoT devices, they jointly optimize the radio resource allocation and the UAV's trajectory.

The second is the multi-UAV trajectory planning. Multi-UAVs were used as mobile base stations to provide service for ground users in [38]. They aim to maximize the minimum throughput of ground users by optimizing the trajectory for each UAV. Scholars in [20] employ multi-UAVs to collect data from nodes. Through jointly optimizing the trajectories of UAVs, wake-up association and scheduling for sensors, they minimize the maximum mission completion time of all UAVs. The authors studied a multiple casting network utilizing the UAV to send files to all ground users [37]. They aim to minimize the mission completion time of the UAVs through designing the UAV's trajectory. Meanwhile, the proposed algorithms guarantee that each ground user can successfully recover the file. In urban applications, the authors proposed a risk-aware trajectory planning algorithm [36] for multi-UAVs. Under the same test scenarios, authors in [39] aim to minimize the mission time by planning the trajectory of each UAV. The scholars exploit the nested Markov chains to analyze the probability for successful data transmission [40]. They propose a sense-and-send mechanism [40] for real-time sensing missions, and a multi-UAVs enabled Q-learning algorithm for decentralized UAV trajectory planning.

In other cases. The authors in [11] use a single UAV to collect data from harsh terrains. Due to the large scale of the detection area, the network has a high demand for power. They adopted a rechargeable mechanism to extend the lifetime of the UAV so as to enhance the collection period. The PFS mechanism in [14,15] is based on the nodes' positions for the data collection in single-UAV aided static sensor networks. The nodes are divided into different priority groups according to two steps: (i). increasing group and decreasing group (Figure 2). The nodes within the decreasing group was given higher priority than the ones within the increasing group. (ii). For each group in (i), the nodes were divided into sub-groups according to which power level does it belong to. The sets of nodes within "power level 1" in the increasing group and in the decreasing group are denoted by $\mathbb{S}_{a,I}^1$ and $\mathbb{S}_{a,D}^1$, respectively. The priority values for nodes within $\mathbb{S}_{a,I}^1$ and $\mathbb{S}_{a,D}^1$ are denoted by $P_{a,I}^1$ and $P_{a,D}^1$, respectively. The authors give high priority to those nodes that are within high power level (Figure 2), and applied opposite actions to the increasing and decreasing groups: (a) in the increasing group, the nodes within high power level was given high priority value; (b) in the decreasing group, the nodes within lower power level were given high priority. After these actions, almost all nodes at the best channel conditions have been considered.

Table 1 presents the key focuses and the key difference of our proposed algorithms from existing algorithms. Although a lot of research has been done on data collection, there is still room to enhance the network performance through balancing the dynamic parameters in the first link in mobile sensor networks.

Table 1. Summary of related works.

Ref.	Sensor Status	N_{uav}	Descriptions
[6]	Static deployed	1	Through UAV trajectory planning to achieve timely data collection from IoT devices where the data has deadlines and needs to be sent before the data loses its meaning or becomes irrelevant.
[7]	Static deployed	1	Considering the age of information, characterized by the data uploading time and the time elapsed since the UAV leaves a node, when designing the UAV trajectory.
[11]	Static deployed	1	To extend the lifetime of the network through charging for the UAV in the air.
[14,15]	Static deployed	1	The authors divided the interesting area into different priority groups, and the data communication conducted from higher to lower priorities (PFS mechanism). Based on PFS, the authors proposed MAC protocols for UAV-aided WSN.
[24]	Static deployed	1	the authors through optimizing the trajectory of a rotary-wing UAV to collect data with an objective of minimizing the maximum energy consumption of all devices.
[25]	Static deployed	1	To minimize the flight time, and jointly optimize the transmit power of nodes, the UAV speed and the transmission intervals.
[31]	Static deployed	1	To minimize the energy consumption of the system through optimizing the UAV's trajectory and devices' transmission schedule, while ensuring the reliability of data collection and required 3D positioning performance.
[32]	Static deployed	1	To maximize the minimum average data collection rate from all nodes subject to a prescribed reliability constraint for each node by jointly optimizing the UAV communication scheduling and three-dimensional trajectory.
[33]	Static deployed	1	To minimize the maximum delay of all ground users through jointly optimizing the offloading ratio, the users' scheduling variables, and UAV's trajectory.
[34]	Static deployed	1	To maximize the minimum received energy of ground users by optimizing the trajectory of the UAV. They first presented the globally optimal one-dimensional (1D) trajectory solution to the minimum received energy maximization problem.
[20]	Static deployed	Multiple	Minimize the maximum mission completion time through jointly optimize the wake-up scheduling and association for sensors, the UAV trajectory, while ensuring that each node can successfully upload the targeting amount of data with a given energy budget.
[35]	Static deployed	Multiple	To maximize the data collection utility by jointly optimizing the communication scheduling and trajectory for all UAVs.
[36]	Static deployed	Multiple	The authors proposed a risk-aware trajectory planning algorithm for multi-UAVs for urban applications.
[37]	Static deployed	Multiple	To minimize the mission completion time of the UAVs through designing the UAV's trajectory, and meanwhile, they guaranteed that each ground user can successfully recover the file.
[38]	Static deployed	Multiple	To maximize the minimum throughput of ground users through optimizing the trajectory for each UAV.
[39]	Static deployed	Multiple	To minimize the mission time by planning the trajectory of each UAV, while satisfying the time requirements.
[40]	Static deployed	Multiple	Use nested Markov chains to analyze the probability for successful data transmission, and propose a sense-and-send mechanism for real-time sensing missions, and a multi-UAVs based Q-learning algorithm for decentralized UAV trajectory planning.
this paper	Mobile	Multiple	Collect data from mobile nodes through balancing the different contact durations between mobile nodes, and multi-UAVs.

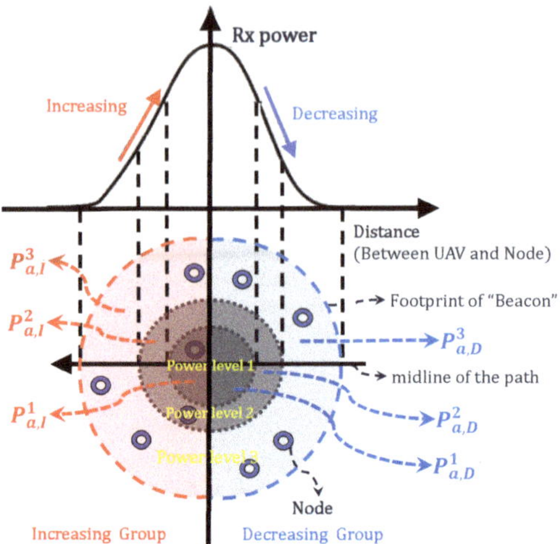

Figure 2. The Priority Frame Selection (PFS) mechanism.

3. System Model and Problem Formulation

3.1. System Model

This paper considers a UAV-assisted mobile sensor network which has N mobile bicycles with each equipped a sensor, and M UAVs with each equipped a sensor (as illustrated in Figure 1, where $M = 2$). $\mathbb{S} = \{S_1, S_2, \cdots, S_N\}$ is the set of mobile sensors. N nodes move along a pre-defined path (path length is denoted as L) with each has a speed v_i. The UAV U_i is dispatched to collect data from mobile sensors at a given height h_i and speed v_u^i along a predefined trajectory (Figure 1).

The trajectory consists of a few line segments that contain the waypoint start and waypoint end (e.g., in Figure 1, waypoint P_S^i and waypoint P_E^i in the trajectory of UAV_i, $i = 1, 2$), and k intermediate waypoints (e.g., in Figure 1, waypoint P_1^i, waypoint P_2^i, waypoint P_3^i, waypoint P_4^i, waypoint P_5^i and waypoint P_6^i in the trajectory of UAV_i, $i = 1, 2$). Let $\mathbb{P}_i = \{P_S^i, P_1^i, P_2^i, \cdots, P_k^i, P_E^i\}$ denote the set of all waypoints of UAV_i. The coordinates for each waypoint P_m^i is denoted by $P_m^i(x_m^i, y_m^i, h_i)$. The UAV's flight time between any two waypoints P_m^i and P_n^i is given by,

$$\lambda_{m,n}^i = \frac{\| P_m^i - P_n^i \|}{v_u^i}, \quad P_m^i, P_n^i \in \mathbb{P}_i. \tag{1}$$

The collection period of the UAV_i is the duration from waypoint P_1^i to the waypoint P_E^i. It is denoted by T_i,

$$T_i = \Sigma_{m=1}^{k-1} \lambda_{m,m+1}^i + \lambda_{k,E}^i. \tag{2}$$

The trajectory length for UAV_i is,

$$L_i = \Sigma_{m=1}^{k-1} \| P_{m+1}^i - P_m^i \| + \| P_E^i - P_k^i \|. \tag{3}$$

Generally, in a given path, the coordinates (x-axis and y-axis) of the waypoints for the UAVs are the same except the height (z-axis). For instance, the point (x_m^i, y_m^i, h_j) is one of the waypoints for UAV_j (i.e., $P_m^j(x_m^i, y_m^i, h_j) \in \mathbb{P}_j$) if $P_m^i(x_m^i, y_m^i, h_i) \in \mathbb{P}_i$. Thus, we have $L_i = L_j$. Intuitively, the straighter the pre-defined path, the smaller the ΔL ($\Delta L = |L - L_i|$). The larger the number of waypoints, the smaller the ΔL. Major notations used in this work are defined in Table 2.

Table 2. Major notations used in this article.

Parameters	Descriptions
N	Network size;
M	The number of UAVs;
UAV_i	The ith UAV;
$\mathbb{S}$	The sensors set;
$\mathbb{S}_{kB}^{i}$	The sensors set that within the range of UAV_i in t_k;
$\mathbb{S}_{kB}^{i,o}$	The sensors set that only within the range of UAV_i in t_k;
$\mathbb{S}_{kB}^{i,j}$	The sensors set that within the range of both UAV_i and UAV_j in t_k;
$\mathbb{U}$	The UAVs set;
T_i	The collection period of the UAV_i;
h_i	The fly height of the UAV_i;
L	The path length;
L_i	The length of the trajectory of UAV_i;
N_{ts}	The number of time slots;
$\mathbb{T}$	The set of time-slots;
α	The duration of one time-slot;
P_m^i, P_S^i, P_E^i	The "mth", the "start", and the "end" way points of the UAV_i respectively;
$\mathbb{P}_i$	The set of waypoints for UAV_i;
$\lambda_{m,n}^i$	The UAV's flight time between any two waypoints P_m^i and P_n^i of UAV_i;
$\mathbb{F}$	The set of nodes that send at least one packet in collection period;
t_B	The duration between adjacent two "Beacon";
$N_{ts,a}(i,j,k)$	A matrix where value is "0" and "1". $N_{ts,a}(i,j,k) = 1$ implies in t_k, the UAV_i will communicate with S_j, and othwise it is "0";
σ_{ijk}	Boolean function. $\sigma_{ijk} = 1$ implies that the UAV_i successfully collect data from S_j in t_k;
$N_{t,a}^i$	The number of time slots that sensor S_i ($S_i \in \mathbb{S}$) was allocated in time T;
N_p	The total number of collected packets;
N_{node}	The number of nodes that successfully send at least one packets.

To well present the impact of the dynamic parameters on the system, we using homogeneous UAVs ($v_u^i = v$) to reduce the influence brought by UAVs' speeds. Accordingly, the collecting period is denoted by T, and $T = T_i$.

3.2. Discrete Time Mechanism

Considering the waypoint selection and beacon sending, we introduce a discrete-time mechanism where the collecting period T is divided into N_{ts} time-slots with each lasting α time units, $N_{ts} = \left\lfloor \frac{T}{\alpha} \right\rfloor$, where $\lfloor \cdot \rfloor$ is the rounding down function. It is assumed that the time-slots are indexed as $1, 2, \cdots, N_{ts}$, and $\mathbb{T} = \{t_1, t_2, \cdots, t_{N_{ts}}\}$ (Figure 3). It is worth note that, in each time slot, a sensor could communicate only with one UAV. For example, in t_k, S_i communicate with UAV_m, and S_j communicate with UAV_n ($i \neq j$ and $m \neq n$).

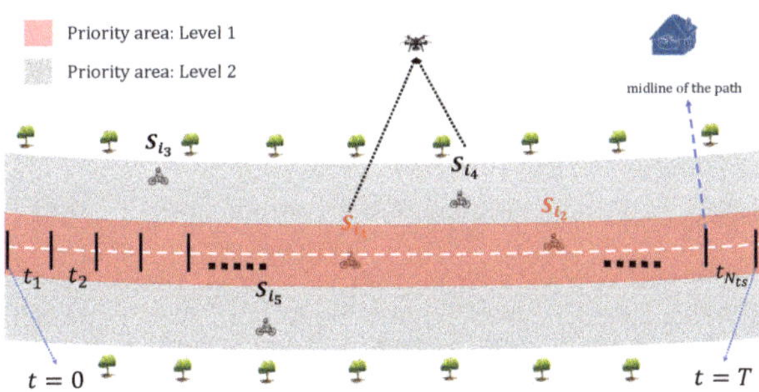

Figure 3. An illustration of studied scenario.

From Figure 1, the nodes that are covered by the UAV_i and deployed nearly complete to communicate with the UAV_j. For instance, S_m, S_n, S_k in Figure 1 complete to communicate with the UAV_1. Meanwhile, there are more than one UAV within the range of one node. For example, S_k in Figure 1 with the range of both UAV_1 and UAV_2. The S_k should choose one from them to send packets. Hence, how to balance the communication between nodes and the UAVs so as to maximize the data collection is a challenging task.

3.3. Data Collection Protocols Using UAV

In this paper, we present a distributed method for the data collection issues in UAV-aided mobile sensor networks as follows. The collection period T is divided into N_{ts} time slots. At the beginning of every time slot (Figure 4), UAV sends a beacon message to tell the mobile nodes that UAV is coming. The beacon includes the UAV's information, e.g., the aerial height, speed, etc. The new comers send a JOIN message which includes the sensors' information to the UAV to update the network topology. The UAV judges whether the nodes are within its range or not according to these messages. Then, it calculates the contact duration, the relative distance, and the potential time slots for each node that successfully sends the JOIN message. According to the time slot allocation algorithms that we proposed in Section 4, the UAV provides scheduling for the covered sensors, and broadcasts them a scheduling message which contains the assignment of the time-slots. Having received the scheduling message, every sensor transmits its data in its own time slots.

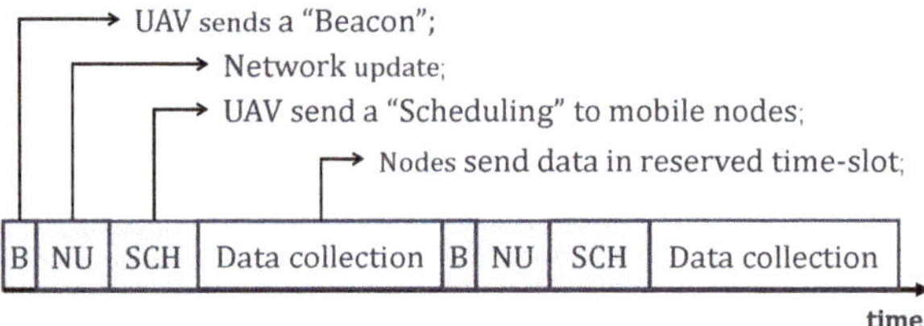

Figure 4. The procedure of allocating.

3.3.1. Collecting Packets

Allocating the N_{ts} time slots to individual mobile sensors under the proposed mechanism is equivalent to maximizing the usage of time slots. Let

$$N_{ts,a}(i,j,k) = \begin{cases} 1 & UAV_i communicate with S_j in t_k, \\ 0 & otherwise. \end{cases}$$

The data collection maximization problem is to maximize the number of collected packets, N_p,

$$N_p = \sum_{i=1}^{M} \sum_{j=1}^{N} \sum_{k=1}^{N_{ts}} N_{ts,a}(i,j,k) \cdot \sigma_{ijk} \cdot D_r \cdot \alpha . \tag{4}$$

where D_r is the transmission rate, and

$$\sigma_{ijk} = \begin{cases} 1 & successfully transmission, \\ 0 & otherwise. \end{cases}$$

Our objective is to balance the communication between the two UAVs and N mobile nodes to maximize the overall data collection utility. Therefore, the optimization problem can be formulated as,

$$\mathscr{P}_1 \;:\; \max_{S_j \in \mathbb{S}, t_k \in \mathbb{T}} \{N_p\}, \tag{5}$$

$$s.t. \quad \sum_{k=1}^{N_{ts}} N_{ts,a}(i,j,k) \le N_{ts} \,, \forall i,j \,, \tag{6}$$

$$\sum_{j=1}^{N} N_{ts,a}(i,j,k) \le N \,, \forall i,k \,, \tag{7}$$

$$\sum_{i=1}^{M} N_{ts,a}(i,j,k) \le M \,, \forall j,k \,, \tag{8}$$

$$\sum_{i=1}^{M} \sum_{j=1}^{N} \sum_{k=1}^{N_{ts}} N_{ts,a}(i,j,k) \le M \cdot N_{ts} \,, \forall i,j \,. \tag{9}$$

Constraints (6)–(8) imply that, in a given time-slot, a UAV chooses only one node to collect data, and one node selects only one UAV to send data. Constraint (9) ensures that, in a given time-slot, no more than two communications happen between UAVs and mobile nodes.

3.3.2. The Number of Nodes that Successfully Send Packets to the UAV

During the communication between the UAVs and mobile nodes, the sensors transmission state contains: have no opportunity to send packets, have an opportunity to send but fail to transmit, and successfully send data to the UAVs. The larger number of nodes (N_{node}) that successfully transmit packets, the higher the system performance. Thus, to enhance the number of nodes that successfully send data to the UAVs is one of the key points in designing data collection algorithms.

Let matrix $I_{M \times N \times N_{ts}}$ is given by,

$$I_{ijk} = N_{ts,a}(i,j,k) \cdot \sigma_{ijk} \cdot i, \qquad UAV_i \in \mathbb{U}, S_j \in \mathbb{S} \,and\, t_k \in \mathbb{T}.$$

The elements in matrix I are the node ID. Then, we can obtain the number of nodes that successfully transmit at least one packet,

$$N_{node} \triangleq Hist(I). \tag{10}$$

where "*Hist*" is used to calculated the number of different elements in the I matrix. The N_{node} maximization problem can be regarded as the formulated problem,

$$\mathscr{P}_2 \;:\; \max_{S_j \in \mathbb{S}, t_k \in \mathbb{T}} \{N_{node}\}, \tag{11}$$

$$s.t. \quad \sum_{k=1}^{N_{ts}} N_{ts,a}(i,j,k) \le N_{ts} \,, \forall i,j \,, \tag{12}$$

$$\sum_{j=1}^{N} N_{ts,a}(i,j,k) \le N \,, \forall i,k \,, \tag{13}$$

$$\sum_{i=1}^{M} N_{ts,a}(i,j,k) \le M \,, \forall j,k \,, \tag{14}$$

$$\sum_{i=1}^{M} \sum_{j=1}^{N} \sum_{k=1}^{N_{ts}} N_{ts,a}(i,j,k) \le M \cdot N_{ts} \,, \forall i,j \,. \tag{15}$$

When $i = 1$ (single-UAV enabled sensor network), it is a classical NP-hard problem that we have studied in [12,13]. When $i = 2$ (multi-UAV enabled sensor network), this problem is also an NP-hard combinatorial maximization problem [41]: under the given conditions, its objective is to select items which have unique weight and value to maximize the total value.

4. Proposed Algorithms

In this section, we study how to balance the communication between multi-UAVs and mobile nodes, and we propose a balance mechanism. For the two cases, multiple nodes within the range of both two UAVs and multiple nodes only with the range of only one UAV, we propose two algorithms: PCdFS (Section 4.2) and PMCdFS (Section 4.3) algorithms.

4.1. Balance Algorithm between UAVs and Mobile Nodes

In a given time slot t_k ($t_k \in \mathbb{T}$), there are multiple nodes within the range of the UAV. The nodes that are potentially for UAV_1 and UAV_2 are denoted by $\mathbb{S}_{kB}^1$ and $\mathbb{S}_{kB}^2$ respectively. When $\mathbb{S}_{kB}^1 \cap \mathbb{S}_{kB}^2 = \varnothing$, there is no node within the range of the UAV_1 and UAV_2 at the same time. In this case, we propose PCdFS mechanism (see Section 4.2 for more details) to balance the communications between $\mathbb{S}_{kB}^1$ and UAV_1, $\mathbb{S}_{kB}^2$ and UAV_2 respectively. When $\mathbb{S}_{kB}^1 \cap \mathbb{S}_{kB}^2 \neq \varnothing$, and $\mathbb{S}_{kB}^{1,2} \triangleq \mathbb{S}_{kB}^1 \cap \mathbb{S}_{kB}^2$. Then,

$$\mathbb{S}_{kB}^{1,o} \triangleq \mathbb{S}_{kB}^1 - \mathbb{S}_{kB}^{1,2} , \tag{16}$$

$$\mathbb{S}_{kB}^{2,o} \triangleq \mathbb{S}_{kB}^2 - \mathbb{S}_{kB}^{1,2} , \tag{17}$$

$\mathbb{S}_{kB}^{1,o}$ and $\mathbb{S}_{kB}^{2,o}$ denote the sensors set only within the range of the UAV_1 and UAV_2 respectively. We use the PCdFS mechanism to balance the communications between $\mathbb{S}_{kB}^{1,o}$ and UAV_1, $\mathbb{S}_{kB}^{2,o}$ and UAV_2 respectively. For the nodes within $\mathbb{S}_{kB}^{1,2}$, we proposed the PMCdFS algorithm to balance between $| \mathbb{S}_{kB}^{1,2} |$ mobile nodes and multi-UAVs. The Balance algorithm is detailed in Algorithm 1.

Algorithm 1 Balance Algorithm.

Input: Initial deployed information of nodes and UAVs
Output: N_p, N_{node}
1:   $N_p = N_{node} = 0$, $k = 1$, $T_{now} = 0$;
2:   **Step 1. Synchronization;**
3:   UAV sends k-th 'Beacon' message;
4:   Network update, obtain the $\mathbb{S}_{k_B}^1$ and $\mathbb{S}_{k_B}^2$;
5:   **Step 2. Data Communication;**
6:   **while** $T_{now} < T$ **do**
7:      Let $\mathbb{S}_{kB}^{1,2} \triangleq \mathbb{S}_{kB}^1 \cap \mathbb{S}_{kB}^2$, $\mathbb{S}_{kB}^{1,o} \triangleq \mathbb{S}_{kB}^1 - \mathbb{S}_{kB}^{1,2}$, and $\mathbb{S}_{kB}^{2,o} \triangleq \mathbb{S}_{kB}^2 - \mathbb{S}_{kB}^{1,2}$;
8:      **if** $\mathbb{S}_{kB}^{1,2} = \varnothing$ **then**
9:          Apply PCdFS mechanism (Algorithm 2) to balance the communication between $\mathbb{S}_{kB}^{1,o}$ and UAV_1, $\mathbb{S}_{kB}^{2,o}$ and UAV_2 respectively;
10:     **else**
11:         Apply PMCdFS algorithm (Algorithm 3) for the balancing between mobile nodes in $\mathbb{S}_{kB}^{1,2}$ and multi-UAVs, and obtain $\mathbb{S}_{k_B}^1$ and $\mathbb{S}_{k_B}^2$ through PMCdFS algorithm;
12:         Apply PCdFS mechanism (Algorithm 2) to balance the communication between $\mathbb{S}_{kB}^{1,o}$ and UAV_1, $\mathbb{S}_{kB}^{2,o}$ and UAV_2 respectively;
13:     **end if**
14:     Update T_{now}, k, N_p and N_{node};
15: **end while**
16: **return** N_p and N_{node};

4.2. Priority-Based Contact-Duration Frame Selection Mechanism

In the PCdFS mechanism, the priority areas division includes two steps: (i) divide the nodes into different groups according to their power level. For example, the nodes are divided into two groups and three groups in Figures 3 and 5, respectively. In Figure 3, S_{i_1} and S_{i_2} are within the same priority area (level 2), S_{i_3}, S_{i_4} and S_{i_5} are within level 2. If we take more priority levels into account, e.g., 3 levels as in Figure 5, S_{i_1} and S_{i_2} belong to level 1, S_{i_4} is in level 2, S_{i_3} and S_{i_5} are in level 3. The more levels, the more detailed group. (ii) For each group, the nodes are given different priority according to their contact duration with the UAV. The ones that have short CD with the UAV are given higher priority values. In PCdFS, different nodes are given different priority values except the case that more than one node have the same CD with the UAV. In PCdFS, it makes the nodes facing a connection lose with the UAV highly concerned. In addition, PCdFS provides the nodes within a higher power level to send data exactly at the moment of their good channel condition so as to reduce the packet's loss. The PCdFS algorithm is detailed in Algorithm 2.

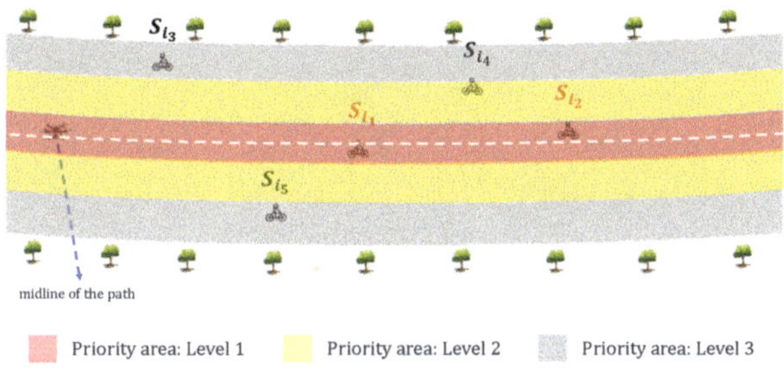

Figure 5. Priority areas.

Algorithm 2 Prioritized-based contact-duration frame selection mechanism (PCdFS) Algorithm.

Input: Initial deployed information of nodes and UAVs, $\mathbb{S}^1_{k_B}$, $\mathbb{S}^2_{k_B}$, N_p, N_{node}.
Output: N_p, N_{node}

1: **for** $\forall S_i \in \mathbb{S}^1_{k_B}$, $\forall S_j \in \mathbb{S}^2_{k_B}$ **do**
2:　　Make a judgement for sensor S_i and S_j: which priority area does them in;
3:　　Calculate the *contact duration* between S_i and the UAV_1, S_j and the UAV_2, respectively;
4: **end for**
5: For UAV_1 (and UAV_2), t_k allocated to the one (e.g., S_{i_k}, and $S_{i_k} \in \mathbb{S}^1_{k_B}$ for UAV_1, and S_{j_l}, $S_{j_l} \in \mathbb{S}^2_{k_B}$ for UAV_2) which is within the higher priority area; When more than one node within the same high priority area, t_k allocated to the one (e.g., S_{i_k} for UAV_1, and S_{j_l} for UAV_2) which has the shorter contact duration with the UAV.
6: In t_k, S_{i_k} and S_{j_l} send packets to UAV_1 and UAV_2 respectively;
7: Update N_p, N_{node};
8: **return** N_p and N_{node};

4.3. Priority-Based Multiple-Contact-Duration Frame Selection Mechanism

The PMCdFS algorithm is used to balance the communications between the UAVs and nodes when these nodes are within the range of the multi-UAVs at the same time. Intuitively, the longer the CD between the nodes and the UAV, the higher the opportunity to send packets to the UAV. Thus, it increases the transmission opportunity of the node if it was arranged to the UAV which has a longer CD between it and the UAV. The PMCdFS is detailed in Algorithm 3. Through the PMCdFS algorithm, we obtain the sensors set in which all nodes only compete to communicate with a single UAV (UAV_1 or UAV_2). Then, we apply the PCdFS algorithm to conduct the communication among them. The proposed algorithms are summarized in Table 3.

Algorithm 3 Prioritized-based multiple contact-duration frame selection mechanisms (PMCdFS) algorithm.

Input: Initial deployed information of nodes and UAVs, $\mathbb{S}^{1,2}_{kB}$, $\mathbb{S}^{1}_{kB}$, $\mathbb{S}^{2}_{kB}$.
Output: $\mathbb{S}^{1}_{kB}$ *and* $\mathbb{S}^{2}_{kB}$
  1: **for** $\forall S_i \in \mathbb{S}^{1,2}_{kB}$ **do**
  2:      Calculate the *contact duration* between S_i and the UAV_1 (denoted as $T_{i,1}$), the UAV_2 (denoted as $T_{i,2}$), respectively;
  3:      **if** $T_{i,1} < T_{i,2}$ **then**
  4:          $\mathbb{S}^{2}_{kB} = \mathbb{S}^{2}_{kB} \cup \{S_i\}$;
  5:      **else**
  6:          $\mathbb{S}^{1}_{kB} = \mathbb{S}^{1}_{kB} \cup \{S_i\}$;
  7:      **end if**
  8: **end for**
  9: **return** $\mathbb{S}^{1}_{kB}$ and $\mathbb{S}^{2}_{kB}$;

Table 3. Proposed algorithms.

Algorithms	Descriptions
Balance mechanism	It is specially used to balance the communications between multiple nodes and multiple UAVs for the system.
PCdFS mechanism	It is used to build the "scheduling" between the nodes (that only with the range of a single UAV) and the UAV.
PMCdFS mechanism	It is specially used to balance the communications between the nodes (these nodes are within the range of multiple UAVs at the same time) and the UAV.

In the following, we will evaluate the proposed algorithms through different configurations, and compare our proposed algorithms with the existing algorithm (PFS).

5. Implementation and Evaluation

We implement the algorithms in both simulations and real experiments as following.

5.1. Simulations

We conduct the simulations in MATLAB/Simulink where the UAV fly (5 min) along a path (the path is 10 m wide). The simulated priority groups are {2, 3, 4, 5} groups. The other simulation parameters are presented in Table 4, the final results are given by the mean of 30 simulation runs. Considering that, the PFS mechanism is proposed and examined based on a single-UAV sensor network. To compare it to the proposed algorithm, we use $M = 1$ in the simulations in Sections 5.1.1–5.1.4. In Section 5.1.5, we compare our proposed algorithms when using single UAV and multiple UAVs. All the simulations are summarized in Table 5.

Table 4. Simulation parameters.

Parameter	Value	Parameter	Value
network size	[5, 200]	path width	10 m
fly time	5 min	fly height	[5, 95] m
UAV speed	[5, 25] ms^{-1}	sensor speeds	[0, 10] ms^{-1}
# priority groups	[2, 5]	packet size	127 Bytes
Data bit rate	250 kbps	inter-beacon duration	2 s to 60 s
receiving threshold	−70 dBm	sensing threshold	−80 dBm
transmission range of the UAV and the node	100 m		

Table 5. Summary of simulations.

Section	Parameters	N_{uav}	Descriptions
Section 5.1.1. Impact of priority level changes.	$N = 200$, $h = 15$ m, $v = 10$ ms^{-1}, IBD $= 2$ s, $v_i \in [0, 10]$ ms^{-1}, $N_{pl} \in \{2,3,4,5\}$.	1	Study the impact of priority levels on the network performance.
Section 5.1.2. Varying beacon intervals.	$N = 200$, $h \in [5, 95]$ m, $v \in [5, 25]$ ms^{-1}, IBD $\in [2, 60]$ s, $v_i \in [0, 10]$ ms^{-1}, $N_{pl} = 2$.	1	Study the impact of different synchronization frequency on the network performance.
Section 5.1.3. Impact of UAV's parameters changes.	$N = 200$, $h = 15$ m, $v = 10$ ms^{-1}, IBD $= 2$ s, $v_i \in [0, 10]$ ms^{-1}, $N_{pl} = 2$.	1	Study the impact of fly height and speeds on the network performance.
Section 5.1.4. Scalability.	$N \in [5, 200]$, $h = 15$ m, $v = 10$ ms^{-1}, IBD $= 2$ s, $v_i \in [0, 10]$ ms^{-1}, $N_{pl} = 2$.	1	Study the impact of the network size on the network performance.
Section 5.1.5. Comparison between Multi-UAVs and Single-UAV.	$N = 200$, $h = 15$ m, $v = 10$ ms^{-1}, IBD $= 2$ s, $v_i \in [0, 10]$ ms^{-1}, $N_{pl} = 2$.	$\{1, 2\}$	Compare our proposed algorithms when using one UAV and two UAVs.

5.1.1. Impact of Priority Level Changes

Figure 6 presents the impact of varying the number of priority groups. The more priority groups, the smaller number of collected packets. The number of collected packets is much improved at two priority groups division as compared to five priority groups division. That is because the nodes in lower priority groups may have changed their state when it was their turn to send packets. The introduction of contact duration provides high priority to them so as to overcome a part of this issue, thus more packets were collected in PCdFS algorithm.

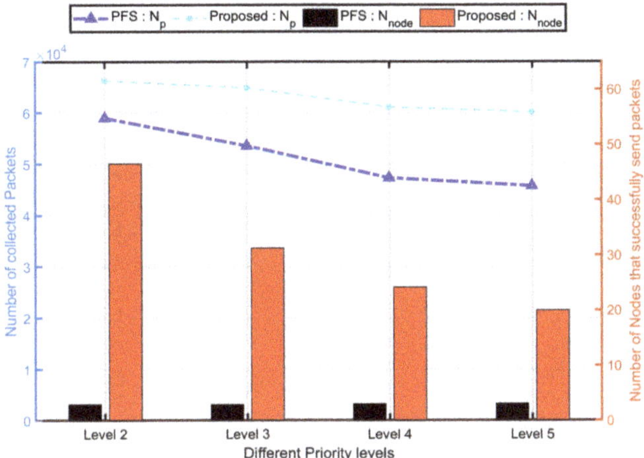

Figure 6. Impact of priority area change. In these simulations, the proposed algorithm is the combination of proposed Balance and prioritized-based contact-duration frame selection mechanism (PCdFS) algorithms.

It also can be concluded that at a larger number of priority groups, a smaller number of nodes were within the highest priority group. Then, the smaller number of nodes have opportunities to send packets, which is unfair for the network. The number of nodes that successfully sent at least one packet in the proposed Priority-based Contact-duration Frame Selection mechanism was 16.2 times larger than in the PFS mechanism which is because the dynamic parameters are concerned in the proposed algorithm.

In the following, in both simulations and real experiments, the number of priority groups is fixed at 2.

5.1.2. Varying Beacon Intervals

Figure 7 shows that both N_p and N_{node} were much improved when the inter-beacon duration at 2 s. Indeed, the longer the beacon intervals, the smaller the number of beacons sent. Thus, the number of network synchronizations is reduced so that nodes were seldom detected during collecting. No node will be detected if no beacon is sent.

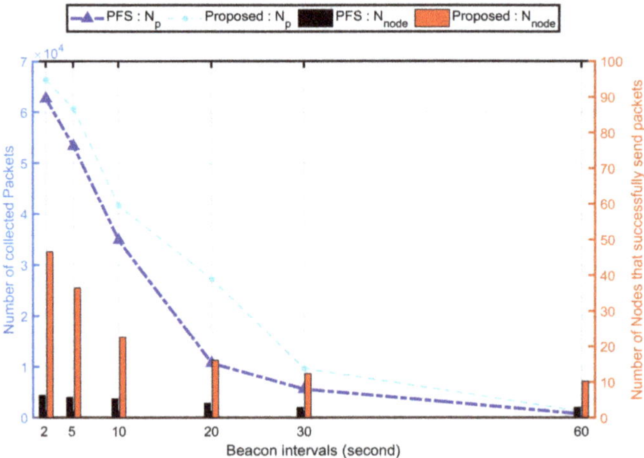

Figure 7. The impact of inter-beacon duration on network performance. In these simulations, the proposed algorithm is the combination of proposed Balance and PCdFS algorithms.

5.1.3. Impact of UAV's Parameters Changes

Figure 8 shows the impact of the total number of collected packets for varying the UAV speed and fly height. The network achieves the optimal ($N_{node} = 46.5$ of 30 simulations) when the fly height is 15 m (Figure 8a). In this simulation, the UAV speed is 10 ms^{-1}, and the size is 200 with nodes speeds vary from 1 ms^{-1} to 10 ms^{-1}. Due to using fixed Dr, the flight height had very slight impact on both N_p and N_{node}. The contact duration which was given by the relative distance between the nodes and the UAV was highly affected by the fly height. Hence, the PCdFS algorithm presents a difference from the PFS mechanism when the fly height is 95 m. Compared to N_p, the N_{node} was affected much when the fly height is larger than 75 m. There is clearly a difference between the two mechanisms when the gap between different fly heights exceeds 50 m.

The change of the UAV speed has a huge impact on both the total number of collected packets and the number of nodes that successfully send packets to the UAV Figure 8b. When the gap between the UAV speed and the maxi speed of all nodes is very small, the network performance is optimal. In this studied scenario, the maxi speed for all nodes is 10 ms^{-1}, thus, the performance is optimal when the UAV speed is 10 ms^{-1}. When $V_{uav} > 10$ ms^{-1}, the higher the V_{uav}, the bigger gap between the UAV speed and the nodes' speeds, the shorter contact duration between them, then, the less opportunities for nodes to communicate with the UAV. Then, the smaller number of packets sent to the UAV, the more it was unfair for the network.

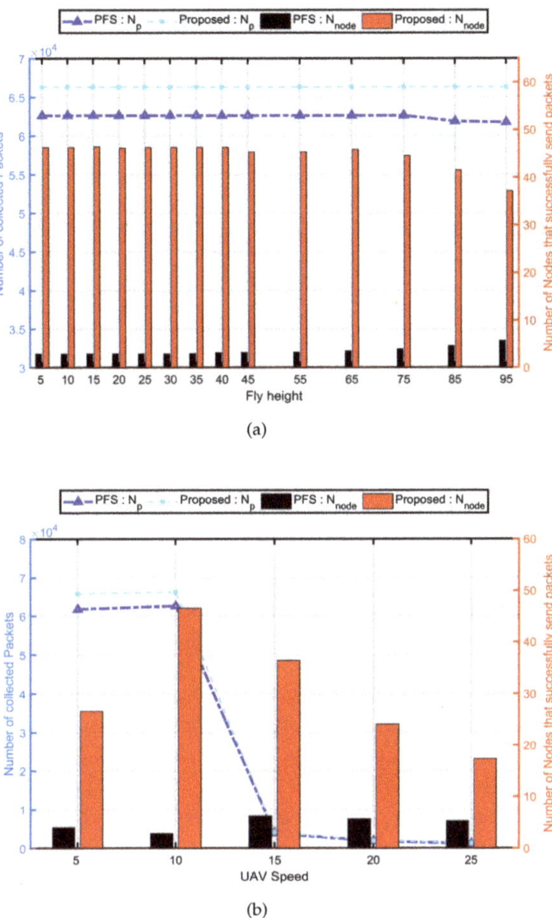

Figure 8. Network performance for varying UAV' parameters: flight height and speed. In these simulations, the proposed algorithm is the combination of proposed Balance and PCdFS algorithms. (**a**) The number of collected packets for the network for varying fly height of the UAV. The number of nodes that successfully send packet to the UAV in the same scenario. (**b**) The number of collected packets for the network for varying UAV' speed. The number of nodes that successfully send packet to the UAV for varying the speed of the UAV.

5.1.4. Scalability

Figure 9 shows the impact of the network size on system performance. In this study, the flight height is 15 m and UAV's speed is 10 ms^{-1} and the size vary from 5 to 200 with nodes' speeds vary from 1 ms^{-1} to 10 ms^{-1}.

The larger the network size, the larger number of nodes has the opportunity to communicate with the UAV, thus, the larger number of packets were sent to the UAV. When the size was larger than 30, each time-slot has successful communication, thus, the number of collected packets in the PFS mechanism keeps steady. It keeps increasing in the PCdFS algorithm until it reaches the transmission upper bound of the collection time. The N_{node} increased steadily in the proposed algorithm. The N_{node} when $N = 200$ in the proposed algorithm is 11.34 times larger than when $N = 5$ while it is almost the same in the PFS mechanism. Hence, the proposed algorithm shows high scalability in terms of sensors density.

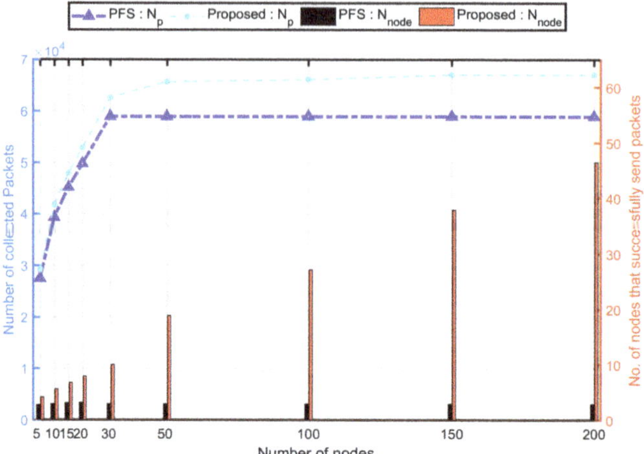

Figure 9. Evaluation of proposed algorithm (the combination of proposed Balance and PCdFS algorithms) on network size.

5.1.5. Comparison between Multi-UAVs and Single-UAV

Figure 10 presents the impact of proposed algorithms on the network size. "Alg1/UAV_1" simulate the combination of proposed Balance and PCdFS algorithms on the UAV_1 which takes-off from the original point of the path, while "Alg1/UAV_2" simulate the same combination algorithms on the UAV_2 which takes-off from the endpoint (the midline of the path) of the path. UAV_1 fly in the same direction as the nodes while UAV_2 fly in the opposite direction. Intuitively, the average contact duration between the UAV_1 and the nodes is longer than the average value between UAV_2 and the nodes. Thus, the communication conducted in the UAV_1 case works better than in UAV_2. There is no doubt that the multi-UAVs work better than single UAVs in data collection issues because of more opportunity provided for mobile nodes.

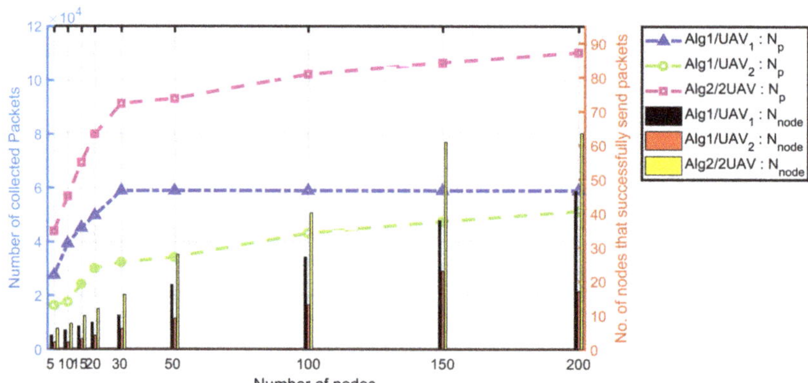

Figure 10. The impact of UAV_1, UAV_2 and multi-UAVs of proposed algorithm on network size. In these simulations, "Alg1" is the combination of proposed Balance and PCdFS algorithms, "Alg2" is the combination of proposed prioritized-based multiple contact-duration frame selection mechanisms (PMCdFS), PCdFS, and Balance algorithms.

5.2. Real Experiment

5.2.1. Set Up

We study a path in Tongji University (Jiading Campus) as in Figure 11a. It is 5 meters wide and 1200 m long, with several intersections and 1 island (Figure 11a). In these experiments, the UAV equips a Pixhawk autopilot system [42,43] (as shown in Figure 11b) so as to fly along a predefined path at a given height. The UAV controlled through a ground station (Figure 12) where the flight height, speed and the packet transmission are controlled. We implement 15 bicycles move along the path with each equips a Pixhawk to simulate the communications based on proposed algorithms (Figure 11c). These nodes start with a random distance from the original point (point A in Figure 11a). Their locations and speeds are expressed in the NED coordinate system, as presented in Figure 11a.

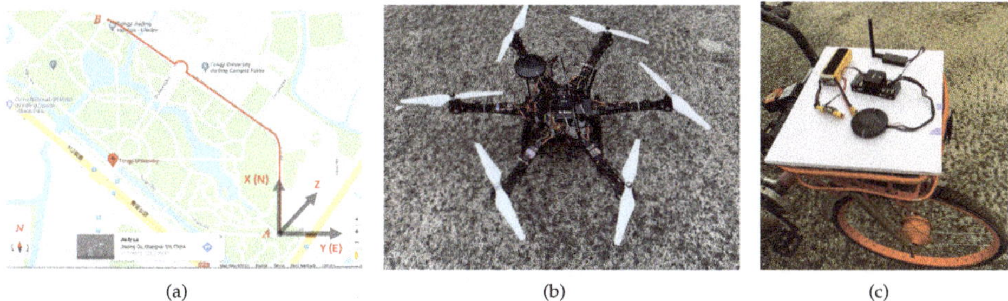

(a) (b) (c)

Figure 11. Presentation of the studied path and hardware in experiments. (**a**) Experiments path in Tongji University-Jiading Campus. (**b**) The UAV employed with a Pixhawk autopilot system. (**c**) The Pixhawk autopilot system deployed on a bicycle.

Figure 12. A screen shot from ground control station.

Pixhawk has built-in MAVLINK protocol [44], the protocol No.24 (GPS_RAW_INT) [44] is used as the "beacon" packet (including the speed and location of the UAV) for the UAV, whose interval can be configured (e.g., in the following experiments, the beacon intervals is set at 2 s). For mobile nodes, the protocol No.24 (GPS_RAW_INT) is used as the "update" packet (including the speed and location of the mobile node). We modified and reused the protocol No.36 (SERVO_OUTPUT_RAW) [44] as the "scheduling" packet (which stores the sensor ID and time-slot ID for the collision-free communication between nodes and the UAV) for the UAV. Each MAVLINK packet contains a system ID field so we can use it to identify the sender. The pixhawk also has a log system so the GPS information, as well as the received packet number and time, is stored in the on-board SD card.

Figure 13 presents the movements for UAV_1 (only one UAV is used in real experiments) where the fly height is 15 m, with control speeds of 5 ms^{-1} (Figure 13a) and 3 ms^{-1} (Figure 13b) according to Pixhawk. In the studied experiments, the UAV flew at 15 m and 30 m. Figure 14 is an example to present the instantaneous speeds and trajectories for 5 nodes (Node 1 to Node 5) according to the Pixhawk.

To make the UAV fly along this path, we set four waypoints along the path as shown in Figure 12. In the experiments, the UAV start from $Point_1$ to achieve its given speed (it is 5 ms^{-1} in Figure 12) to $Point_2$, $Point_3$ and the ending point ($Point_4$). In Pixhawk autopilot system, the UAV will hover on the waypoint and ending point for 2 s. That is why in the Figure 13a, the UAV speed is lower than 5 ms^{-1} at P_2 and P_3. In Figure 13b, both the height and instantaneous speed of UAV have a shock between the $Point_3$ and $Point_4$ because of the influence of wind. The wind has an impact on the dynamic parameters so as to affect the relative velocity between the mobile node and the UAV, the network performance affected accordingly. However, it cannot be control during experiments.

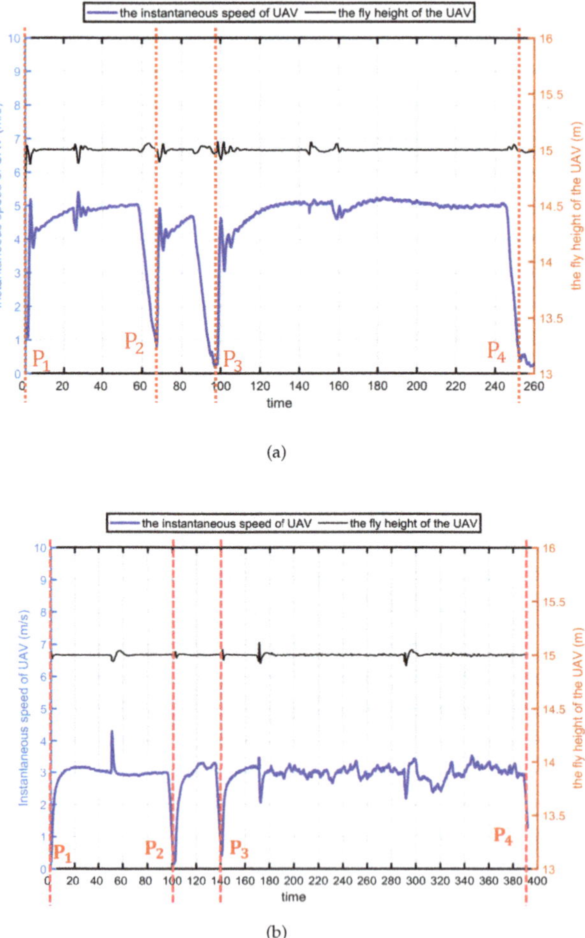

Figure 13. Presentation of the movements for UAV when it fly at 15 m with 3 ms^{-1} and 5 ms^{-1} in ground station. (**a**) The movements for UAV flying at 15 m, and its speed is 5 ms^{-1} in ground control station. (**b**) The movements for UAV flying at 15 m, and its speed is 3 ms^{-1} in ground control station.

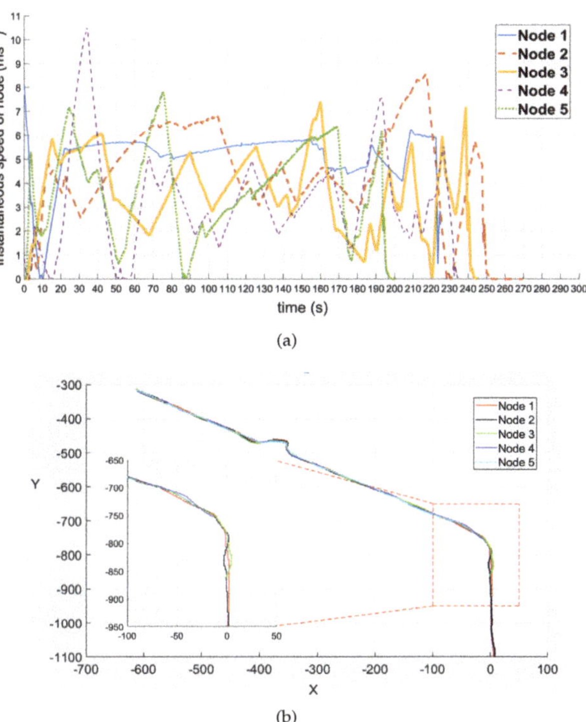

Figure 14. The movements for five nodes. (**a**) Instantaneous speeds of five nodes over time. (**b**) Trajectories of five nodes.

5.2.2. Results

Figures 15 and 16 show the experiments results under the proposed algorithms, the combination of the Balance algorithm and the PCdFS algorithm. From Figure 15, the number of collected packets in simulation is almost two times larger than in the experiments because of the impacts of hardware and environments are not considered in simulations. The flying height has a significant impact on the number of collected packets in experiments, especially when N_{node} is steady between different heights. The higher the height, the larger number of nodes in both PFS and proposed algorithms. The number of collected packets of the proposed algorithm in size 15, $h = 30$ m is more than twice than in $h = 15$ m. The system performance increase as the size increase. The larger the network, the more nodes have opportunities to send packets, the more packets were collected. The number of collected packets in the proposed algorithm (when h is 15 m) is 1.2 times larger than in the PFS algorithm.

From Figure 16, it can be found that the UAV's speed has little impact on data collection in real experiments. This is because the UAV's speed is set at 3 ms^{-1} and 5 ms^{-1} because of the battery constrictions and the campus constrictions. The nodes' speeds are between 2 ms^{-1} and 5 ms^{-1} also (Figure 14). Thus, the relative velocity between the UAV and mobile nodes is very small. The number of collected packets presented in Figure 16 keep the same conclusions as in simulations in Section 5.1.3 where the UAV fly at 5 ms^{-1} and 10 ms^{-1}.

The flight height almost has no influence on the number of nodes that successfully transmit packets to the UAV, as presented in both Figures 15 and 16, which are the same as in Section 5.1.3.

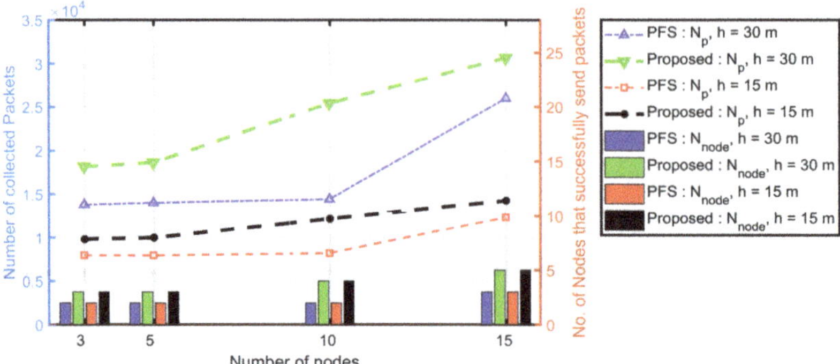

Figure 15. The impact of network size, and flying height over the system performance. In these experiments, the beacon interval is fixed at 2 s according to the simulation results in Figure 7.

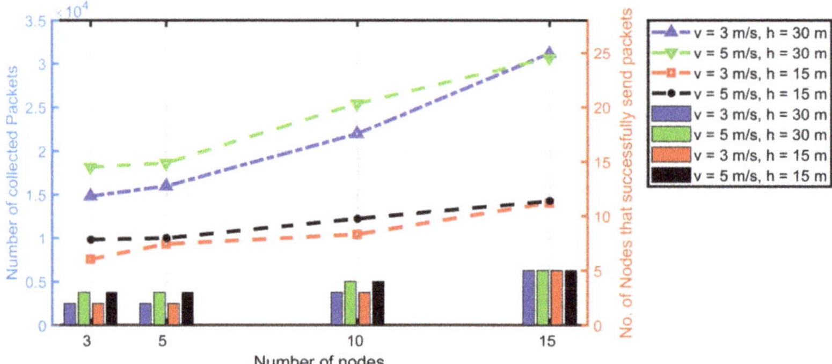

Figure 16. The impact of network size, and UAV's speed over the system performance. In these experiments, the beacon interval is fixed at 2 s. All the results are based on the combination of Balance and PCdFS algorithms.

5.3. Discussions

According to the aforementioned simulations, the beacon interval and the UAV speed have a huge impact on network performance. The shorter the beacon interval, the better the system performance. The UAV speed is constrained by the node speed. The smaller the relative velocity between them, the higher the network performance. It keeps the same conclusions as in the real experiment. In real experiments, the data collection is well conducted when the UAV speed is set at $5\,\text{ms}^{-1}$ which is very close to the average speed of mobile nodes. Compare to the other dynamic parameters, the number of priority levels has a steady impact on data collection in the simulations. From the movements of the nodes in Figure 14, it can be seen that the difference between the trajectories of nodes is very small because the road width is 5 m and the road length is 1200 m.

Compare Figure 13a,b, it also can be found that, the fly time in $3\,\text{ms}^{-1}$ is 1.56 times as in $5\,\text{ms}^{-1}$ while the speed increase by 66.67% (from $3\,\text{ms}^{-1}$ to $5\,\text{ms}^{-1}$). In the studied scenario, there are very small differences between the trajectories when UAV fly at $3\,\text{ms}^{-1}$ and $5\,\text{ms}^{-1}$ because the UAV follow the same path which width is very short compared to its length. Thus, the fly time is mainly dependent on the speed of the UAV. In other words, the slower the UAV fly, the higher energy consumption of the battery energy. From Figure 16, we notice that, the data collection has very little difference when UAV fly at $3\,\text{ms}^{-1}$ and $5\,\text{ms}^{-1}$. Therefore, under given constrictions, the higher the fly speed of the UAV, the more saved battery energy.

The fly height has very little impact on data collection in simulations because of the same transmission rate is adopted. However, the fly height has a huge impact on data collection in experiments because a real and complex antenna system are conducted among the transmissions between the node and the UAV. The higher the flying height, the less interference from external factors (e.g., buildings, etc.). Thus, the better the transmission, the higher the network performance.

6. Conclusions

In this paper, we developed two mechanisms: PCdFS and PMCdFS. PCdFS mechanism is used to build the scheduling communications when the nodes are only covered by one of the UAVs. PMCdFS is used to balance the communication between the nodes and multi-UAVs when these nodes within the range of multi-UAVs at the same time. Based on the two mechanisms, we proposed the Balance algorithm which highly enhances the network fairness in the applications where both the nodes and the collectors are mobile. Two key mechanisms for designing Balance algorithm are: (i) divide the interesting areas into different priority areas and (ii) provide an independent priority value for each node in the same priority group according to their contact duration with the UAVs. We examined the performance of proposed algorithms through extensive simulations, and real experiments. In the experiments, we used 15 mobile nodes at a path with several intersections and one island at the Tongji campus in Shanghai, China. We also confirm the applicability of the proposed algorithm in a challenging and realistic scenario through numerous experiments. Both simulation results and experiment results present that the proposed PCdFS algorithm enhanced the network performance efficiently. The backhaul dimensioning is an interesting problem that we will address in our future work. It depends on the used backhaul type (either satellite or terrestrial) and on the allocation that is reserved to the network slice dedicated for Machine Type Communication (MTC) traffic.

Author Contributions: Conceptualization, X.M., T.L., R.K., and R.D.; methodology, X.M. and T.L.; software, X.M. and T.L.; validation, X.M. and T.L.; formal analysis, X.M. and T.L.; resources, T.L.; data curation, X.M. and T.L.; writing–original draft preparation, X.M. and T.L.; writing–review and editing, X.M., T.L., R.K. and R.D.; funding acquisition, S.L. All authors have read and agreed to the published version of the manuscript.

Funding: This research received no external funding.

Conflicts of Interest: Authors declare no conflict of interest.

References

1. Ali, F. UAV-enabled healthcare architecture: Issues and challenges. *Future Gener. Comput. Syst.* **2019**, *97*, 425–432.

2. Liu, X.; Ansari, N. Resource Allocation in UAV-Assisted M2M Communications for Disaster Rescue. *IEEE Wirel. Commun. Lett.* **2019**, *8*, 580–583. [CrossRef]

3. Samir Labib, N.; Danoy, G.; Musial, J.; Brust, M.R.; Bouvry, P. Internet of Unmanned Aerial Vehicles—A Multilayer Low-Altitude Airspace Model for Distributed UAV Traffic Management. *Sensors* **2019**, *19*, 4779. [CrossRef] [PubMed]

4. Mozaffari, M.; Saad, W.; Bennis, M.; Nam, Y.; Debbah, M. A Tutorial on UAVs for Wireless Networks: Applications, Challenges, and Open Problems. *IEEE Commun. Surv. Tutor.* **2019**, *21*, 2334–2360. [CrossRef]

5. Poudel, S.; Moh, S. Medium Access Control Protocols for Unmanned Aerial Vehicle-Aided Wireless Sensor Networks: A Survey. *IEEE Access* **2019**, *7*, 65728–65744. [CrossRef]

6. Samir, M.; Sharafeddine, S.; Assi, C.M.; Nguyen, T.M.; Ghrayeb, A. UAV Trajectory Planning for Data Collection from Time-Constrained IoT Devices. *IEEE Trans. Wirel. Commun.* **2020**, *19*, 34–46. [CrossRef]

7. Liu, J.; Wang, X.; Bai, B.; Dai, H. Age-optimal trajectory planning for UAV-assisted data collection. In Proceedings of the 2018 IEEE INFOCOM-IEEE Conference on Computer Communications Workshops (INFOCOM WKSHPS), Honolulu, HI, USA, 15–19 April 2018; pp. 553–558.

8. Tong, P.; Liu, J.; Wang, X.; Bai, B.; Dai, H. UAV-Enabled Age-Optimal Data Collection in Wireless Sensor Networks. In Proceedings of the 2019 IEEE International Conference on Communications Workshops (ICC Workshops), Shanghai, China, 20–24 May 2019; pp. 1–6.

9. Di Francesco, M.; Das, S.K.; Anastasi, G. Data Collection in Wireless Sensor Networks with Mobile Elements: A Survey. *ACM Trans. Sensor Netw.* **2011**, *8*, 7.1–7.31. [CrossRef]

10. Zhao, M.; Ma, M.; Yang, Y. Efficient Data Gathering with Mobile Collectors and Space-Division Multiple Access Technique in Wireless Sensor Networks. *IEEE Trans. Comput.* **2011**, *60*, 400–417. [CrossRef]
11. Pang, Y.; Zhang, Y.; Gu, Y.; Pan, M.; Han, Z.; Li, P. Efficient data collection for wireless rechargeable sensor clusters in Harsh terrains using UAVs. In Proceedings of the 2014 IEEE Global Communications Conference, Austin, TX, USA, 8–12 December 2014; pp. 234–239.
12. Ma, X.; Kacimi, R.; Dhaou, R. Fairness-aware UAV-assisted data collection in mobile wireless sensor networks. In Proceedings of the 2016 International Wireless Communications and Mobile Computing Conference (IWCMC), Paphos, Cyprus, 5–9 September 2016; pp. 995–1001.
13. MA, X. Data Collection of Mobile Sensor Networks by Drones. 2017. Available online: http://docplayer.fr/18339015-These-en-vue-de-l-obtention-du-doctorat-de-l-universite-detoulouse.html (accessed on 31 March 2020).
14. Ho, D.; Park, J.; Shimamoto, S. Performance evaluation of the PFSC based MAC protocol for WSN employing UAV in rician fading. In Proceedings of the 2011 IEEE Wireless Communications and Networking Conference, Cancun, Mexico, 28–31 March 2011; pp. 55–60.
15. Ho, D.; Shimamoto, S. Highly reliable communication protocol for WSN-UAV system employing TDMA and PFS scheme. In Proceedings of the 2011 IEEE GLOBECOM Workshops (GC Wkshps), Houston, TX, USA, 5–9 December 2011; pp. 1320–1324.
16. Ma, X.; Chisiu, S.; Kacimi, R.; Dhaou, R. Opportunistic communications in WSN using UAV. In Proceedings of the 2017 IEEE Annual Consumer Communications and Networking Conference (CCNC), Las Vegas, NV, USA, 8–11 January 2017; pp. 510–515.
17. Alterio, F.D.; Ferranti, L.; Bonati, L.; Cuomo, F.; Melodia, T. Quality Aware Aerial-to-Ground 5G Cells through Open-Source Software. In Proceedings of the 2019 IEEE Global Communications Conference (GLOBECOM), Waikoloa, HI, USA, 9–13 December 2019; pp. 1–6.
18. Raveneau, P.; Chaput, E.; Dhaou, R.; Dubois, E.; Gelard, P.; Beylot, A.L. Carreau: CARrier REsource access for mUle, DTN applied to hybrid WSN/satellite system. In Proceedings of the 2013 IEEE Vehicular Technology Conference (VTC Fall), Las Vegas, NV, USA, 2–5 September 2013; pp. 1–5.
19. Raveneau, P.; Chaput, E.; Dhaou, R.; Dubois, E.; Gelard, P.; Beylot, A. Martinet: A disciplinarian protocol for resource access in DTN. In Proceedings of the 2013 IFIP Wireless Days (WD), Valencia, Spain, 13–15 November 2013; pp. 1–3.
20. Zhan, C.; Zeng, Y. Completion Time Minimization for Multi-UAV-Enabled Data Collection. *IEEE Trans. Wirel. Commun.* **2019**, *18*, 4859–4872. [CrossRef]
21. Shen, C.; Chang, T.; Gong, J.; Zeng, Y.; Zhang, R. Multi-UAV Interference Coordination via Joint Trajectory and Power Control. *IEEE Trans. Signal Process.* **2020**, *68*, 843–858. [CrossRef]
22. Ebrahimi, D.; Sharafeddine, S.; Ho, P.; Assi, C. UAV-Aided Projection-Based Compressive Data Gathering in Wireless Sensor Networks. *IEEE Internet Things J.* **2019**, *6*, 1893–1905. [CrossRef]
23. Zhan, C.; Huang, R. Energy Minimization for Data Collection in Wireless Sensor Networks with UAV. In Proceedings of the 2019 IEEE Global Communications Conference, Waikoloa, HI, USA, 9–13 December 2019; pp. 1–6.
24. Zhan, C.; Lai, H. Energy Minimization in Internet-of-Things System Based on Rotary-Wing UAV. *IEEE Wirel. Commun. Lett.* **2019**, *8*, 1341–1344. [CrossRef]
25. Gong, J.; Chang, T.; Shen, C.; Chen, X. Flight Time Minimization of UAV for Data Collection Over Wireless Sensor Networks. *IEEE J. Sel. Areas Commun.* **2018**, *36*, 1942–1954. [CrossRef]
26. Zong, J.; Shen, C.; Cheng, J.; Gong, J.; Chang, T.; Chen, L.; Ai, B. Flight Time Minimization via UAV's Trajectory Design for Ground Sensor Data Collection. In Proceedings of the 2019 International Symposium on Wireless Communication Systems (ISWCS), Oulu, Finland, 27–30 August 2019; pp. 255–259.
27. Popescu, D.; Stoican, F.; Stamatescu, G.; Chenaru, O.; Ichim, L. A Survey of Collaborative UAV-WSN Systems for Efficient Monitoring. *Sensors* **2019**, *19*, 4690. [CrossRef]
28. Lin, H.; Yan, Z.; Chen, Y.; Zhang, L. A Survey on Network Security-Related Data Collection Technologies. *IEEE Access* **2018**, *6*, 18345–18365. [CrossRef]
29. Jing, X.; Yan, Z.; Pedrycz, W. Security Data Collection and Data Analytics in the Internet: A Survey. *IEEE Commun. Surv. Tutor.* **2019**, *21*, 586–618. [CrossRef]

30. Li, J.; Zhao, H.; Wang, H.; Gu, F.; Wei, J.; Yin, H.; Ren, B. Joint Optimization on Trajectory, Altitude, Velocity, and Link Scheduling for Minimum Mission Time in UAV-Aided Data Collection. *IEEE Internet Things J.* **2020**, *7*, 1464–1475. [CrossRef]

31. Wang, Z.; Liu, R.; Liu, Q.; Thompson, J.S.; Kadoch, M. Energy-Efficient Data Collection and Device Positioning in UAV-Assisted IoT. *IEEE Internet Things J.* **2020**, *7*, 1122–1139. [CrossRef]

32. You, C.; Zhang, R. 3D Trajectory Optimization in Rician Fading for UAV-Enabled Data Harvesting. *IEEE Trans. Wirel. Commun.* **2019**, *18*, 3192–3207. [CrossRef]

33. Hu, Q.; Cai, Y.; Yu, G.; Qin, Z.; Zhao, M.; Li, G.Y. Joint Offloading and Trajectory Design for UAV-Enabled Mobile Edge Computing Systems. *IEEE Internet Things J.* **2019**, *6*, 1879–1892. [CrossRef]

34. Hu, Y.; Yuan, X.; Xu, J.; Schmeink, A. Optimal 1D Trajectory Design for UAV-Enabled Multiuser Wireless Power Transfer. *IEEE Trans. Commun.* **2019**, *67*, 5674–5688. [CrossRef]

35. Qin, Z.; Dong, C.; Wang, H.; Li, A.; Dai, H.; Sun, W.; Xu, Z. Trajectory Planning for Data Collection of Energy-Constrained Heterogeneous UAVs. *Sensors* **2019**, *19*, 4884. [CrossRef] [PubMed]

36. Alessandro, R. A Risk-Aware Path Planning Strategy for UAVs in Urban Environments. *J. Intell. Robot. Syst.* **2019**, *9*, 629–643

37. Zeng, Y.; Xu, X.; Zhang, R. Trajectory Design for Completion Time Minimization in UAV-Enabled Multicasting. *IEEE Trans. Wirel. Commun.* **2018**, *17*, 2233–2246. [CrossRef]

38. Wu, Q.; Zeng, Y.; Zhang, R. Joint Trajectory and Communication Design for Multi-UAV Enabled Wireless Networks. *IEEE Trans. Wirel. Commun.* **2018**, *17*, 2109–2121. [CrossRef]

39. Luitpold, B. Coordinated Target Assignment and UAV Path Planning with Timing Constraints. *J. Intell. Robot. Syst.* **2019**, *94*, 857–869.

40. Hu, J.; Zhang, H.; Song, L. Reinforcement Learning for Decentralized Trajectory Design in Cellular UAV Networks With Sense-and-Send Protocol. *IEEE Internet Things J.* **2019**, *6*, 6177–6189. [CrossRef]

41. Liu, Y.; Gao, C.; Zhang, Z.; Lu, Y.; Chen, S.; Liang, M.; Tao, L. Solving NP-Hard Problems with Physarum-Based Ant Colony System. *IEEE/ACM Trans. Comput. Biol. Bioinform.* **2017**, *14*, 108–120. [CrossRef]

42. Lesko, J.; Schreiner, M.; Megyesi, D.; Kovacs, L. Pixhawk PX-4 Autopilot in Control of a Small Unmanned Airplane. In Proceedings of the 2019 Modern Safety Technologies in Transportation (MOSATT), Kosice, Slovakia, 3–7 June 2019; pp. 90–93.

43. PixhawkWeb. Available online: https://pixhawk.org/ (accessed on 31 March 2020).

44. MAVLINK. Available online: https://mavlink.io/en/messages/common.html (accessed on 31 March 2020).

Article

Trajectory Planning for Data Collection of Energy-Constrained Heterogeneous UAVs

Zhen Qin [1], Chao Dong [2,*], Hai Wang [1], Aijing Li [1], Haipeng Dai [3] and Weihao Sun [1] and Zhengqin Xu [1]

[1] College of Communications Engineering, Army Engineering University of PLA, Nanjing 210042, China; qzqzla912@163.com (Z.Q.); haiw.wang@gmail.com (H.W.); lishan_wh@163.com (A.L.); imx_101@163.com (W.S.); zhengqinxu66@163.com (Z.X.)

[2] College of Electronic and Information Engineering, Nanjing University of Aeronautics and Astronautics, Nanjing 211106, China

[3] Department of Computer Science and Technology, Nanjing University, Nanjing 210023, China; haipengdai@nju.edu.cn

* Correspondence: dch@nuaa.edu.cn

Received: 29 September 2019; Accepted: 7 November 2019; Published: 8 November 2019

Abstract: Nowadays, Unmanned Aerial Vehicles (UAVs) have received growing popularity in the Internet-of-Things (IoT) which often deploys many sensors in a relatively wide region. Since the battery capacity is limited, sensors cannot transmit over a long distance. It is necessary for designing efficient sensor data collection mechanisms to prolong the lifetime of the IoT and enhance data collection efficiency. In this paper, we consider a UAV-enabled data collection scenario, where multiple heterogeneous UAVs with different energy constraints are employed to collect data from sensors. The value of data depends on the importance of the monitoring area of the sensor and the freshness of collected data. Our objective is to maximize the data collection utility by jointly optimizing the communication scheduling and trajectory of each UAV. The data collection utility is determined by the amount and value of the collected data. This problem is a variant of multiple knapsack problem, which is a classical NP-hard problem. First, we transform the initial problem into a submodular function maximization problem under energy constraints, and then we design a novel trajectory planning algorithm to maximize the data collection utility, while accounting for different values of data and different energy constraints of heterogeneous UAVs. Finally, under different network settings, the performance of the proposed trajectory planning algorithm is evaluated via extensive simulations. The results show that the proposed algorithm can obtain maximum data collection utility.

Keywords: unmanned aerial vehicles; trajectory planning; sensors; data collection utility

1. Introduction

Thanks to its tremendous application potentials in civilian, commercial and military related fields, the Internet of things (IoT) has attracted increased attention in many applications, e.g., natural disaster prediction, smart city, environmental monitoring, and reconnaissance [1–5]. The IoT often deploys many energy-constrained sensors in a relatively wide region. The task of the sensor is to collect data from the monitoring area, and then it uses multi-hop transmission mode to transmit data to the base station or sink node. Since the battery capacity is limited, sensors cannot transmit over a long distance. It is necessary for designing efficient sensor data collection mechanisms to prolong the lifetime of the IoT and enhance data collection efficiency [6].

In order to achieve efficient data collection, more and more people exploit Unmanned Aerial Vehicles (UAVs) to collect data from sensors, which will probably be a prospective technology for realizing the future IoT [6,7]. The heterogeneity and multi-domain nature of UAVs are indispensable in the IoT environment [8–10]. Thus, it is necessary for the IoT environment to use multiple heterogeneous UAVs with different capabilities. Different from traditional Wireless Sensor Networks (WSNs), the UAV-enabled data collection system uses mobile data collection devices installed on UAVs to communicate with sensors directly through the UAV-sensor channels, which are dominated by line-of-sight (LoS). UAVs can move towards the sensors and establish reliable connections with them due to their flexible deployment and high mobility [11]. UAV-enabled data collection system can reduce the energy consumption of sensors and improve the throughput and coverage.

There are many studies on studying how to use UAVs to collect reliable data from sensors. They mainly focus on two aspects of optimization. On the one hand, some works focus on solving the energy limitation problem of sensors in WSN [6,12–14]. They aim to optimize the wake-up schedule of sensor nodes, reduce the transmission power of sensors or improve the energy efficiency of data collection. However, these works rarely consider the value of data and distinguish the data collected from each sensor. For example, in reconnaissance application, the monitoring data of enemy command center is more important than that of living area. The value of data depends on the importance of monitoring area of the sensor and elapsed time after the previous collecting time (i.e., the freshness of collected data). On the other hand, some works focus on improving energy efficiency of UAV system [15–17]. They mainly aim to optimize the deployment of UAV, the trajectory of UAV, and the velocity of UAV. In these works, they use a single UAV or homogeneous UAVs. There are few studies that consider multiple heterogeneous UAVs with different energy constraints and power efficiency. Multiple heterogeneous UAVs can not only solve the energy limitation problem of a single UAV, but also fully utilize the capability characteristics of heterogeneous UAVs to implement complementary performance.

In this paper, we study a UAV-enabled data collection scenario, where multiple heterogeneous UAVs with different energy constraints are employed to collect data from sensors. The UAVs are responsible for transmitting data from sensors to the base station or sink node. We aim to maximize the data collection utility by planning trajectories of UAVs. The data collection utility is calculated by the value of data and amount of data. The value of data collected from the sensor depends on the importance of the monitoring area of the sensor and the freshness of collected data. Our problem contains three main technical challenges:

- Each UAV has different energy constraints and power efficiency. Thus, it is difficult to plan trajectory and assign tasks under their respective energy constraints.
- The value of data collected from each sensor is different, which depends on the importance of the monitoring area of the sensor and elapsed time after the previous collecting time (i.e., the freshness of collected data).
- Joint consideration of communication scheduling and trajectory optimization is a variant of the multiple knapsack problem, which is a classical NP-hard problem.

To solve this problem, we transform the initial problem into a submodular function maximization problem under energy constraints, and then, to maximize the data collection utility, we propose a novel trajectory planning algorithm. The main contributions of this paper are summarized as follows:

- Considering the different values of data and different energy constraints of heterogeneous UAVs, we focus on using multiple heterogeneous UAVs to collect data from sensors. Our objective is to maximize the data collection utility by jointly optimizing the communication scheduling and trajectories of UAVs. This problem is a variant of the multiple knapsack problem, which is a classical NP-hard problem [18,19].
- We prove that the data collection utility function is a submodular function, and transform the initial problem into the problem of maximizing a submodular function under energy constraints, and then we propose a novel trajectory planning algorithm to maximize the data collection utility, while accounting for different energy constraints of heterogeneous UAVs.

- Sufficient simulations are performed to demonstrate the validity and applicability of the proposed algorithm. The data collection utility of our algorithm can be increased by 134% at most, and the proposed algorithm is the closest to the optimal scheme compared with other algorithms.

The rest of this paper is organized as follows. In Section 2, we introduce the related work about the UAV-enabled data collection system and trajectory planning. In Section 3, we present the system model and problem formulation. Then we propose a solution for the formulated problem in Section 4. Simulation results are provided and analyzed in Section 5. Discussion is provided in Section 6. Finally, we conclude the paper in Section 7.

2. Related Work

2.1. UAV-Enabled Data Collection

There are many works on studying how to use the UAV to collect data from sensors. In [20], the authors considered that UAVs were used for collecting imagery information from nodes, and then, the UAVs transmitted information to the ground station. They proposed a predictive compression policy to maximize the end-to-end image quality. Gong et al. utilized a UAV to collect data from sensors which are deployed on a straight line [16]. The authors minimized the flight time of the UAV from a starting location to an ending location, and they jointly optimized the transmit power of sensors, the speed of the UAV and the data collection intervals. Ebrahimi et al. considered a scenario where UAVs collected the data in dense WSNs [6]. The authors used a novel solution methodology which is called projection-based Compressive Data Gathering (CDG). CDG aggregated data from sensors to the selected projection nodes which acted as cluster heads. Next, the UAV transferred the aggregated data from selected nodes to the sink node. In [21], the authors proposed a novel UAV-assisted backscatter communication. The UAV collected data from terrestrial backscattering tags, and then uploaded the collected data when it flied to the coverage area of the base station. Liu et al. proposed a UAV trajectory design for data collection to reduce redundant data and improving energy efficiency [22]. In [23], the authors deployed multi-UAV to serve vehicles on a highway. They utilized UAVs to deliver critical data to the vehicles crossing the given highway segment. By planning the trajectory of each UAV and optimizing the radio resource allocation, they aimed to minimize the number of UAVs to serve all vehicles. Sanaa et al. deployed UAVs as base stations to provide instant recovery via temporary wireless coverage [24]. They minimized the number of UAVs and optimized the positions of them in selected locations to enhance performance. Yang et al. studied a UAV-enabled data collection system, in which the UAV was employed to gather data from ground users. The sensors have limited battery and lower power. To prolong the lifetime of sensors, UAVs can move close to sensors to collect their information with minimum transmit power [11]. However, these works rarely consider distinguishing the data collected from each sensor. The value of data collected from each sensor is different, which depends on the importance of the monitoring area of the sensor and the freshness of collected data.

2.2. Trajectory Planning

Although people have strong interest in UAVs, studies on the location optimizing and trajectory planning of UAVs are still in progress. These studies are different in the optimization method and objective function because they assume different environments. These works are mainly divided into two types: single UAV trajectory planning and multi-UAV trajectory planning. Hu et al. considered a UAV used for the mobile edge computing system, where the mobile UAV equipped with computing resources provided service for many ground users [13]. By jointly optimizing the ratio of offloading tasks, the trajectories of UAVs, and the user scheduling variables, the authors minimized the maximum delay of all users. In the IoT system, Zhan et al. used a rotary-wing UAV for collecting the data from the IoT devices [14]. Under the energy constraint of the UAV, the authors minimized the maximum energy consumption of all IoT devices. Moataz et al. utilized a UAV to collect data from time-constrained IoT devices [25]. These devices with limited buffer sizes had their own target data upload deadline, and thus data needed to be collected before it lost its value. Their goal was to maximize the number of served IoT devices by jointly optimizing the radio resource allocation and the trajectory of the UAV. This paper took into account the change in the value of data. It provided a basis for us to consider the value of a sensor's data. Hu et al. studied a UAV-enabled wireless power system, where the UAV provided wireless energy supply for ground users with a linear topology. The authors maximized the minimum received energy of ground users by optimizing the trajectory of the UAV [26]. They first presented the globally optimal one-dimensional (1D) trajectory solution to the minimum received energy maximization problem. Zeng et al. studied a multicasting system which utilized the UAV to transmit the file to all ground users [27]. By designing the UAV's trajectory, the authors minimized the mission completion time of the UAV. Meanwhile, they guaranteed that each ground user can successfully recover the file. However, in some applications, a single UAV has been unable to meet the demands of missions. There are many works on studying how to design the trajectories of multi-UAV. Under urban environments, in order to minimize the risk to the population, the authors proposed a risk-aware trajectory planning algorithm for multi-UAV [28]. Islam et al. proposed a task-oriented trajectory planning scheme for multi-UAV [29]. The UAVs taken autonomous decisions to find their trajectories for flying to the mission area while avoiding collision to barriers. In [30], the authors aimed to minimizing the mission time by planning the trajectory of each UAV, while satisfying the time requirements. Under the same test scenarios, Christian et al. presented advancements over the A* and the smoothing algorithms [31]. Hu et al. exploited the nested Markov chains to analyze the probability for successful data transmission, and then, for real-time sensing missions, the authors proposed a sense-and-send protocol [32]. To solve the decentralized UAV trajectory planning problem, they proposed a multi-UAV Q-learning algorithm. Wu et al. used multi-UAV as mobile base stations which provided the service to the ground users [33]. The authors optimized the trajectory of each UAV to maximize the minimum throughput of ground users. Zhan et al. employed multi-UAVs to collect data from sensors in WSN [17]. By jointly optimizing the trajectories of UAVs, wake-up association and scheduling for sensors, the author minimized the maximum mission completion time of all UAVs. However, there are few studies that consider multiple heterogeneous UAVs with different energy constraints and power efficiency. Multi-heterogeneous UAVs not only can solve the energy limitation problem of a traditional single UAV, but also make use of the capability characteristics of heterogeneous UAVs to achieve complementary performance.

3. System Model and Problem Formulation

3.1. System Model

3.1.1. Network Model

We consider a UAV-enabled data collection scenario, where k heterogeneous UAVs with different energy constraints are used for collecting the data from sensors to a remote base station or sink node as shown in Figure 1. In the UAV-enabled data collection system, since sensors are employed in a large area, it is inconvenient for the UAVs to fly over each sensor to collect data. In order to achieve efficient and scalable performance, more and more people adopt a clustering approach in WSN. In this paper, an overlapping clustering method is used for dividing the sensors on the ground [34,35]. Sensors transmit data to cluster heads, and then UAVs move towards cluster heads to collect data. The characteristic of overlapping clustering is that a sensor may belong to multiple clusters at the same time, which is different from traditional clustering algorithms. The cluster head can receive data from all sensors in its coverage. In other words, the sensor will transmit its data to each cluster head which it belongs to. For example, if a sensor fits in two overlapping clusters, it will transmit its data to two cluster heads. Establishing overlapping clusters can improve the success rate and robustness of data collection. For convenience, Table 1 provides major notations used in this paper.

Table 1. Major notations.

Notation	Definition
U	UAV set
S	Sensor set
C	Cluster head set
L	UAV trajectory set
u_i	UAV i
s_j	Sensor j
c_a	Cluster head a
L_i	Trajectory of UAV i
k	Number of UAVs, or number of trajectories
n	Number of sensors
m	Number of cluster heads
$E_{\max,i}$	Energy budget of UAV i
$p_{\min,m}$	Minimum motion power
$E_{m,i}$	Motion energy consumption of UAV i
$p_{m,i}$	Actual motion power of UAV i
b_i	Length of trajectory L_i
v_i	Velocity of UAV i
η_i	Power efficiency of UAV i
$E_{h,i}$	Hovering energy consumption of UAV i
$N_{a,i}$	Amount of data collected by UAV i in cluster head a
$p_{\min,h}$	Minimum hovering power
$p_{h,i}$	Actual hovering power of UAV i
$R_{a,i}$	Data transmission rate between UAV i and cluster head a
$E_{c,i}$	Communication energy consumption of UAV i
α	Path loss exponent
e_x	Communication energy consumption parameter
$V_j(t)$	Value of data collected from sensor j
T_j	Recovery interval of sensor j
t'_j	Time of previous data collection from sensor j
$q_{a,i}$	Data utility of cluster head a collected by UAV i
Q_i	Data collection utility of UAV i
Q	Overall data collection utility
$v_{\max}$	Maximum velocity
$l_i(t)$	u_i trajectory projected on the ground
$\dot{l}_i(t)$	Time derivative of $l_i(t)$
$d_{\min}$	Minimum inter-UAV distance
H	Altitude of UAVs

The UAV, sensor and cluster head sets are denoted as $U = \{u_1, ..u_i.., u_k\}$, $S = \{s_1, ..s_j.., s_n\}$ and $C = \{c_1, ..c_a.., c_m\}$, respectively. In addition, ground sensors can be partitioned into m sets, $S_1, S_2, ...S_m$. Each UAV u_i is constrained with an energy budget $E_{\max,i}$. In this paper, the UAV mainly consumes communication-related energy and propulsion energy [36–38]. The communication-related energy is used for transmitting the collected data. The propulsion energy includes motion energy and hovering energy. The UAV consumes motion energy for flying between clusters, and hovering energy for hovering at cluster heads to collect data.

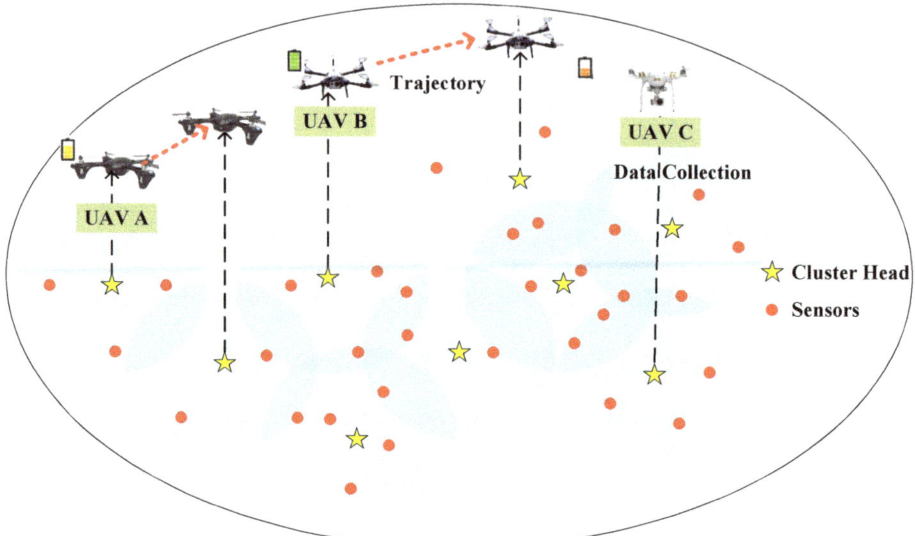

Figure 1. Monitoring scenario.

3.1.2. Propulsion Energy Consumption Model

The motion energy is spent to overcome the gravity and drag forces caused by forward motions and wind. The motion energy consumption is calculated by minimum motion power $p_{\min,m}$ and the length of a UAV's trajectory [38,39]. It can be expressed as

$$E_{m,i} = \frac{p_{m,i} \cdot b_i}{v_i} = \frac{p_{\min,m} \cdot b_i}{\eta_i \cdot v_i}, \tag{1}$$

where $p_{m,i}$ is the actual motion power consumption of UAV u_i, η_i is the UAV's power efficiency, v_i is the velocity of the UAV u_i and b_i is the length of trajectory L_i.

The hovering energy consumption depends on the hovering time and actual hovering power $p_{h,i}$. The actual hovering power relates to the power efficiency and minimum hovering power $p_{\min,h}$. The minimum hovering power relates to the density of air, diameter, thrust and the number of rotors [39,40]. The hovering time is calculated by amount of data $N_{a,i}$ which is collected from cluster head c_a by UAV u_i and data transmission rate $R_{a,i}$ between cluster head c_a and UAV u_i. Therefore, the hovering energy consumption of UAV u_i for collecting data from cluster head c_a can be calculated by

$$E_{h,i}^a = \frac{p_{h,i} \cdot N_{a,i}}{R_{a,i}} = \frac{p_{\min,h} \cdot N_{a,i}}{\eta_i \cdot R_{a,i}}. \tag{2}$$

In this paper, we mainly consider that UAVs are used for data collection application. This kind of applications commonly used small rotary-wing UAVs. For example, the mass of UAV is 2.07 kg, the number of rotors is 4, and the rotor diameter is 0.254 m [38]. According to references [39,40], the minimum motion power is set to 388.32 J/s, and the minimum hovering power is set to 308 J/s.

3.1.3. Communication-Related Energy Consumption Model

The communication-related energy consumption for transmitting data cannot be ignored when the transmission distance or the amount of data is large. The energy consumed for successful transmitting wireless data is affected by the channel between source and destination nodes, the transmission distance and other factors like interference, fading and noises. The communication energy consumption for transmitting $N_{a,i}$ bits over distance d can be calculated by [41]

$$E_{c,i}^a = N_{a,i} \cdot d^\alpha \cdot e_x, \tag{3}$$

where e_x and α are constants which depends on the characteristics of the communication channel. e_x is unit energy consumption which represents the energy consumption for transmitting one bit, measured in $J/(m^\alpha \cdot bit)$, and α is the path loss exponent which depends on the data transmission environment.

3.1.4. Utility Model

The data collection utility is calculated by the value of data and the amount of data. The value of data depends on the importance of the monitoring area of the sensor and the freshness of collected data. In fact, the importance of the monitoring area has different performance metrics in different applications and scenarios. For example, the importance of the monitoring area can be defined by traffic [42].

To calculate the data collection utility, we first define the value of data. On the one hand, the value of data collected from the sensor depends on the importance of monitoring area of the sensor. In this paper, the initial value of data from s_j is defined as $V_j^0 = V_j^{\max}$. Once the data of sensor s_j is collected, the value of data collected from s_j is set to $V_j^{\min}$. For any sensor s_j and $s_{j'}$, if the monitoring area of sensor s_j is more important than the monitoring area of sensor $s_{j'}$, the relations can be expressed by

$$V_j^{\max} > V_{j'}^{\max}, V_j^{\min} > V_{j'}^{\min}. \tag{4}$$

On the other hand, the value of data collected from sensor s_j depends on elapsed time after the previous collecting time (i.e., the freshness of collected data). For each sensor s_j, recovery interval T_j is different which depends on the importance of the monitoring area and the required monitoring interval of the sensor. At time t, the value of data collected from sensor s_j can be denoted as [43]

$$V_j(t) = \begin{cases} A \times \exp(t - t'_j) + B, & if \quad t - t'_j \leq T_j \\ V_j^{\max}, & otherwise \end{cases}, \tag{5}$$

$$\begin{cases} A = \dfrac{V_j^{\max} - V_j^{\min}}{e^{T_j} - 1}, \\ B = V_j^{\min} - A, \end{cases} \tag{6}$$

where t'_j is the time of previous data collection from sensor s_j. As we can see from Equation (5) and Equation (6), when the sensor's data is collected by one UAV, the value of data will decrease to the minimum value. As time elapses, the value of data increases exponentially until it reaches its maximum value. After the value of data reaches the maximum value, it remains until the sensor's data is collected by UAVs again.

In this paper, sensors transmit data to cluster heads, and then UAVs move towards cluster heads to collect data. The data collection utility mainly depends on the amount of data and its value. The data collection utility of the selected cluster head c_a which is served by a UAV u_i can be given by

$$q_{a,i} = \sum_{s_j \in S_a} N_{a,i}(s_j) \cdot V_j(t_{a,i}), \tag{7}$$

where S_a is the set of all sensors in cluster c_a, $N_{a,i}(s_j)$ represents the amount of data of sensor s_j included in cluster head c_a which is served by UAV u_i. Meanwhile, since the time of data collection is relatively short, we do not consider the changes of the data's amount and value in the process of data collection. $t_{a,i}$ represents the time when a UAV u_i starts to collect data from cluster head c_a. The data collection utility of UAV u_i can be calculated by

$$Q_i = \sum_{c_a \in P_i} q_{a,i}, \tag{8}$$

where P_i is the set of cluster head that is served by UAV u_i. Therefore, the overall utility of data collection mission can be calculated by

$$Q(P) = \sum_{i=1}^{k} Q_i = \sum_{i=1}^{k} \sum_{c_a \in P_i} q_{a,i}. \tag{9}$$

3.2. Problem Formulation

In this paper, the flying altitude of the UAVs is assumed to be a constant altitude H. We assume that $r = (x_r, y_r, H)$ is the initial location of all UAVs. The total energy consumption E_i includes the hovering energy consumption, motion energy consumption and communication energy consumption, which can be expressed by

$$E_i = E_{m,i} + E_{h,i} + E_{c,i}. \tag{10}$$

Denote the trajectory of UAV u_i projected on the ground as $l_i(t) = [x_i(t), y_i(t)]^T \in \mathbb{R}^{2 \times 1}$, where $0 \leq t \leq T$. The trajectory of each UAV is subject to the velocity constraints, which can be given by

$$\left\| \dot{l}_i(t) \right\| \leq v_{\max}, \forall i, t \in [0, T], \tag{11}$$

where $\dot{l}_i(t)$ is the time derivative of $l_i(t)$ and $v_{\max}$ is the maximum velocity of UAVs.

Our goal is to plan the trajectories of heterogeneous UAVs with different energy constraints to maximize the overall data collection utility. Therefore, the optimization problem can be formulated as

$$P1: \max_{P \in C, L_i, 1 \leq i \leq k} Q(P) \tag{12}$$

$$s.t. \quad E_i \leq E_{\max,i}, \forall i, \tag{13}$$

$$\left\| \dot{l}_i(t) \right\| \leq v_{\max}, \forall i, t \in [0, T], \tag{14}$$

$$L_i(0) = L_i(T) = r, \forall i, \tag{15}$$

$$\| L_i(t) - L_{i'}(t) \| \geq d_{\min}, \forall i \neq i', t \in [0, T]. \tag{16}$$

where P represents the selected cluster heads, L_i represents the trajectory of UAV u_i, r is the initial location of all UAVs and $d_{\min}$ denotes the minimum distance between UAVs to ensure collision avoidance. Constraint (13) implies the energy consumption of UAV u_i cannot be greater than its maximum energy constraint $E_{\max,i}$. In (15), it ensures that each UAV needs to return to initial location r by the end of data collection mission. When all UAVs fly at the same altitude H, the trajectories of UAVs are also constrained by collision avoidance (16).

4. Solution

4.1. Hardness Analysis

The formulated problem combines two-level optimizations. The objective of upper level optimization is to select cluster heads and the objective of lower level optimization is to design trajectories for energy-constraint heterogeneous UAVs. The results of each level optimization problem would directly affect another level optimization. If we select cluster heads without considering trajectory planning, it will consume much motion energy. If we do not consider to select appropriate cluster heads in trajectory planning, the data collection utility will not be maximized. Therefore, the two-level optimizations are coupled with each other and cannot be solved separately.

Without considering the motion energy consumption, the upper cluster head selection problem can be regarded as a simplified form of the formulated problem

$$P2: \max_{P \in C, L_i, 1 \leq i \leq k} Q(P) \tag{17}$$

$$s.t. \quad E_{h,i} + E_{c,i} \leq E_{\max,i}, \forall i, \tag{18}$$

$$\left\| \dot{l_i}(t) \right\| \leq v_{\max}, \forall i, t \in [0, T], \tag{19}$$

$$L_i(0) = L_i(T) = r, \forall i, \tag{20}$$

$$\|L_i(t) - L_{i'}(t)\| \geq d_{\min}, \forall i \neq i', t \in [0, T]. \tag{21}$$

This problem can be modeled as a multiple capacity-constraint knapsack problem. When $k = 1$, this problem is a knapsack problem, which is a classical NP-hard problem. Therefore, when the value of k is greater than 1, our problem is also NP-hard. The knapsack problem is a combinatorial optimization problem: under the given weight limit, its objective is to select items which have unique weight and value to maximize the total value [18,19].

In addition, if we consider the motion energy consumption, we would calculate the k closed trajectories including all cluster heads in the selected set P. This problem can be formulated as multiple Travelling Salesman Problem, which is also a NP-hard problem. Therefore, the original optimization problem is difficult to solve, which combines two coupling NP-hard problems.

4.2. Submodular Analysis

To solve this problem, we transform the initial problem into the problem of maximizing a submodular function with energy constraints. We prove the data collection utility function has three tractable properties: submodularity, nonnegativity and monotonicity. We first give some definitions to facilitate further analysis.

Definition 1. *(monotonicity, nonnegativity, and submodularity) given a finite set Ψ, a submodular function is a set function $f : 2^{\Psi} \to R$. f is called monotonicity (nondecreasing), nonnegativity, and submodularity if and only if it can satisfy the following requirements, respectively.*

- $f(X) \leq f(Y)$ *for all $X \subseteq Y \subseteq \Psi$ or equivalently: $f(X \cup \{x\}) - f(X) > 0$ for all $X \subseteq \Psi$ and $x \in \Psi \backslash X$ (monotonicity);*
- $f(\varnothing) = 0$ *and $f(X) \geq 0$ for all $X \subseteq \Psi$ (nonnegativity);*
- $f(X) + f(Y) \geq f(X \cup Y) + f(X \cap Y)$, *for any $X, Y \subseteq \Psi$ or equivalently: $f(X \cup \{x\}) - f(X) \geq f(Y \cup \{x\}) - f(Y)$, $X \subseteq Y \subseteq \Psi$, $x \in \Psi \backslash Y$ (submodularity).*

Theorem 1. *The constructed objective function is submodular, monotone and nonnegative.*

Proof. According to the definition of the data collection utility function, $Q(P) \geq 0$, then it is nonnegative.

The data collection utility $Q(P)$ increases as the number of cluster heads collected by UAVs increases. According to the utility model, for the set $X \subseteq Y \subseteq C$, we can obtain the following inequation

$$Q(X) \leq Q(Y), \tag{22}$$

where implies $Q(P)$ is monotone. Next, we prove that $Q(P)$ is a submodular function by proving the following inequation

$$Q(X \cup \{c\}) - Q(X) \geq Q(Y \cup \{c\}) - Q(Y), X \subseteq Y \subseteq C, x \in C \backslash Y, \tag{23}$$

where X and Y represent the set of cluster heads in WSN. We denote S_X as the sensors covered by the set of cluster heads X. We prove the inequation under two cases.

Case 1 $(S_c \cap S_Y = \varnothing)$: In this case, the data of sensors included in the newly added cluster has never been collected. Therefore, the value of data of sensors covered by cluster head c can reach their maximum value $V^{\max}$. We can obtain

$$Q(X \cup \{c\}) - Q(X) = Q(Y \cup \{c\}) - Q(Y). \tag{24}$$

Case 2 $(S_c \cap S_Y \neq \varnothing)$: In this case, the data of some sensors included in the newly added cluster has been collected. Once the data of sensor is collected, the value of data will be set to $V^{\min}$. Therefore, we can obtain

$$Q(X \cup \{c\}) - Q(X) \geq Q(Y \cup \{c\}) - Q(Y). \tag{25}$$

Therefore, we prove that $Q(P)$ is a submodular function. To solve this problem, the initial problem is transformed into a submodular function maximization problem with energy constraints. We propose a novel trajectory planning algorithm which refers to the idea of [44,45]. It aims to maximize the overall data collection utility, while accounting for cluster head selection and diffiernt energy constraints of heterogeneous UAVs. $\square$

4.3. Algorithm

Based on the submodular function, we jointly consider the upper level optimization and lower level optimization, and then we design a simple but efficient algorithm referring to the idea of [44,45]. The algorithm attempts to select appropriate cluster heads to collect data and design the collecting sequence. The core idea of our algorithm is to iteratively select a new cluster head $c_{j'}$ by greedy method, which has the maximum utility-cost ratio. For example, in iteration j, the selected cluster head can be expressed as follows

$$c_{j'} = \underset{c \in I \backslash P_{j-1}}{\arg \max} \frac{Q(P_{j-1} \cup \{c\}) - Q(P_{j-1})}{E(P_{j-1} \cup \{c\}) - E(P_{j-1})}. \tag{26}$$

Algorithm 1 consists of a parent loop and a child loop. After inputting and initializing relative parameters, we use the parent loop to select cluster heads and plan trajectories (Line 2–Line 12). When all cluster heads have been traversed, the parent loop is no longer executed. In the parent loop, there is a child loop for selecting the cluster head which has the maximum utility-cost ratio (Line 3–Line 7). In each iteration, we use Algorithm 2 to calculate the energy cost and the trajectory, which considers the energy constraint of each UAV. The utility and energy cost are calculated according to the previously selected cluster heads plus possible cluster head c_j, and then, we pick up the cluster head with the highest utility ratio (Line 7). Afterwards, we check whether the UAVs satisfy the respective energy constraints (Line 8–Line 10). Next, it deletes this cluster head and starts next parent loop. Each parent loop returns a solution which is better than previous solution and the nature of the result depends on the quality of Algorithm 2. Finally, we can obtain the selected cluster heads P and the trajectories of UAVs L which satisfy the respective energy constraints.

Algorithm 1 Cluster Head Selection and Trajectory Planning Algorithm

---

Input: Cluster head set C, energy constraints $E_{max,i}, 1 \le i \le k$.

Output: Selected cluster heads $P \subseteq C$ and k trajectories of UAVs.

1: Initialize $I \leftarrow C, P_0 \leftarrow \varnothing, E(P_0) \leftarrow 0, Q(P_0) \leftarrow 0, L \leftarrow \varnothing, j \leftarrow 1;$

2: **while** $I \ne \varnothing$ **do**

3:     **for** $i = 1$ to $|I|$ **do**

4:         Calculate $Q(P_{j-1} \cup \{c_i\})$ and $Q(P_{j-1})$;

5:         Using Algorithm 2 to get the trajectories and the energy cost $E(P_{j-1} \cup \{c_i\})$;

6:     **end for**

7:     $c_{j'} = \underset{c \in I}{\arg\max} \ \dfrac{Q(P_{j-1} \cup \{c\}) - Q(P_{j-1})}{E(P_{j-1} \cup \{c\}) - E(P_{j-1})};$

8:     **if** $E(P_{j-1} \cup \{c_{j'}\}) \ne Inf$ **then**

9:         $P_j \leftarrow P_{j-1} \cup \{c_{j'}\}, j \leftarrow j+1, L \leftarrow L_{c_{j'}};$

10:    **end if**

11:    $I \leftarrow I \backslash c_{j'};$

12: **end while**

13: Output $P \leftarrow P_{j-1}, L.$

Algorithm 2 Multiple Energy-constrained Heterogeneous UAV Trajectory Planning Algorithm

---

Input: $P, E_{max,i}, 1 \le i \le k$, starting location r.

Output: k trajectories of UAVs and the energy cost E.

1: Initialize $Y \leftarrow P, E \leftarrow \varnothing, MAXE = 0, L \leftarrow \varnothing;$

2: Sort the UAVs, based on $E_{max,i}$, in descending order in a list ζ;

3: **while** $Y \ne \varnothing$ && $MAXE \ne Inf$ **do**

4:     **for** $i = 1$ to $|\zeta|$ **do**

5:         Initialize $X_i \leftarrow \{r\}, L_i \leftarrow \varnothing, E_i \leftarrow 0;$

6:         **for** $j = 1$ to $|Y|$ **do**

7:             Use the TSP algorithm to calculate the energy cost $E_{i,j}$ and the trajectory $L_{i,j}$ which covers

                all cluster heads in $X_i \cup c_j;$

8:             **if** $E_{i,j} \le E_{max,i}$ **then**

9:                 $X_i \leftarrow X_i \cup c_j, E_i \leftarrow E_{i,j}, L_i \leftarrow L_{i,j};$

10:            **end if**

11:         **end for**

12:         $Y \leftarrow Y \backslash X_i, E \leftarrow E + E_i, L \leftarrow L \cup L_i;$

13:     **end for**

14:     **if** $Y \ne \varnothing$ **then**

15:         $E = Inf;$

16:     **end if**

17: **end while**

---

5. Simulation Results

In this paper, our algorithm aims to maximize the data collection utility by optimizing the trajectory of each UAV. The data collection utility is calculated by the value and amount of data. The value of data depends on the importance of the monitoring area of the sensor and the freshness of collected data.

5.1. Simulation Setup

We consider a mission area of size 1 km × 1 km. The simulations are performed according to parameters specified in Table 2 [38–41]. The time requirement for data uploading is not a constant. It can be changed depending on the amount of data and data transmission rate. In fact, sensors continuously monitor the area and generate new data. However, since the time of data collection is relatively short, we do not consider the changes of data's amount and value in the process of data collection. We assume that communication links between UAVs and sensors are dominated by the LoS links where the channel quality mainly depends on the UAV-sensor distance [16,33,46]. Meanwhile, since the UAVs fly at a fixed altitude, we can set the data transmission rate to be 2 Mbps. Furthermore, the simulation results are averaged over extensive simulation runs.

Table 2. Major notations.

Definition	Parameters	Values
Area size	D	1 km × 1 km
Minimum inter-UAV distance	d_{min}	100 m
Altitude of UAVs	H	50 m
Maximum velocity	v_{max}	20 m/s
Minimum motion power	$p_{min,m}$	388.32 J/s
Minimum hovering power	$p_{min,h}$	308 J/s
Power efficiency of UAV	η	70%–80%
Data transmission rate	R	2 Mbps
Path loss exponent	α	2
Communication energy consumption parameter	e_x	$10\ pJ/(m^\alpha \cdot bit)$

5.2. Baseline Setup

To demonstrate the performance of the proposed UAV trajectory planning algorithm (UE), we compare and implement the following four benchmark schemes:

- Optimal scheme (OPT): To evaluate how the proposed algorithm approaches the optimal performance, we use brute-force searching method to get the optimal scheme which maximizes the data collection utility.
- RAN algorithm: Multi-UAV randomly select cluster heads for data collection. Based on the selected data collection points, we consider the energy constraint of each UAV to plan the trajectories.
- EC algorithm: The main purpose of this algorithm is to collect as much data as possible from sensors. However, it does not take into account the value of data.
- GU algorithm: This algorithm selects a cluster head which has maximum data collection utility in each iteration. However, the algorithm does not consider the energy consumption of the UAV when it selects cluster heads.

5.3. Different Number of Sensors

In this simulation, we set the number of UAVs to 3. Figure 2 shows the trend of data collection utility as the number of sensors changes. We can observe that the data collection utility gradually increases as the number of sensors increases. Our algorithm achieves almost the same performance with the optimal scheme when the number of sensors is small. However, the gap between our algorithm and the optimal scheme increases as the number of sensors increases, from 5.2% to 66.6%. Meanwhile, the proposed algorithm shows better performance when the number of sensors is large. Compared with RAN algorithm, the data collection utility of our algorithm is improved by 103%–134%. This is reasonable since it chooses the data collection points with the highest utility-cost ratio each time, which saves the energy of UAVs and improves the data collection utility of UAVs. Compared with EC algorithm, the data collection utility of our algorithm is improved by 49%–62%. Because our algorithm considers the value of data when selecting cluster heads, not only the amount of data collection. Furthermore, GU algorithm chooses a cluster which has the most data collection utility. However, it does not consider the energy consumption for collecting data from this cluster. Compared with the GU algorithm, the data collection utility of our algorithm is improved by 72%–102%. Our proposed algorithm makes reasonable and effective use of UAV's energy to collect more valuable data of sensors.

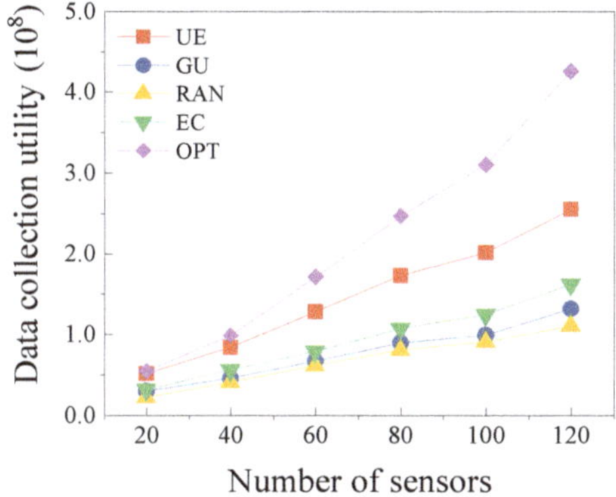

Figure 2. Different Number of Sensors.

In Figure 3, we illustrate the convergence of our algorithm under different number of sensors. From the figure, we note that our algorithm achieves fast convergence in three cases. Meanwhile, we can obtain that the number of iterations of the proposed algorithm is related to the number of sensors.

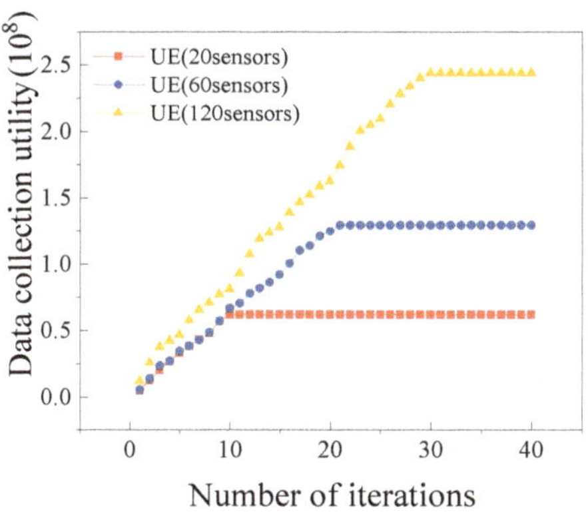

Figure 3. Convergence of UE Algorithm ($n = 20, 60, 120.$)

5.4. Different Number of UAVs

In this simulation, the number of sensors is set to 100. As shown in Figure 4, as the number of UAVs increases, the advantages of our algorithm are more obvious. This is because we fully consider the energy constraint and power efficiency of each UAV in trajectory planning. Therefore, as the number of heterogeneous UAVs increases, the data collection utility gradually increases, and our algorithm is closer to the optimal scheme than other three algorithms. The data collection utility of the optimal scheme is 17.6%–38.9% higher than that of our algorithm. In practical applications, we should consider the mission requirements and existing equipment to dispatch an appropriate number of UAVs to perform mission. Using too many UAVs may bring economic pressure and reduce the energy efficiency. Figure 5 shows the convergence of the proposed algorithm under different number of UAVs. We can observe that our algorithm achieves fast convergence. Meanwhile, we can also find that the number of UAVs has little effect on the convergence of the proposed algorithm.

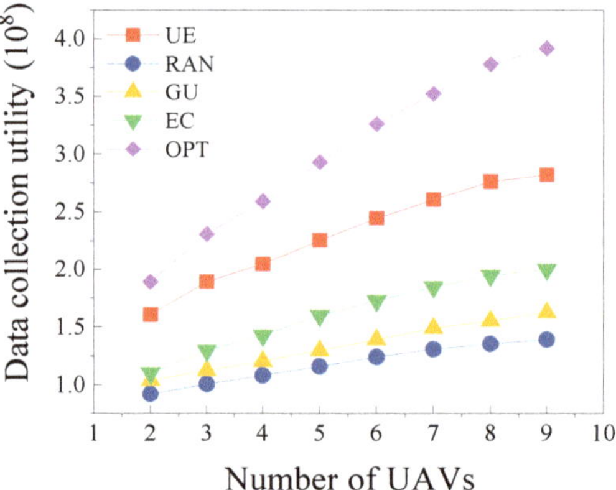

Figure 4. Different Numberof UAVs.

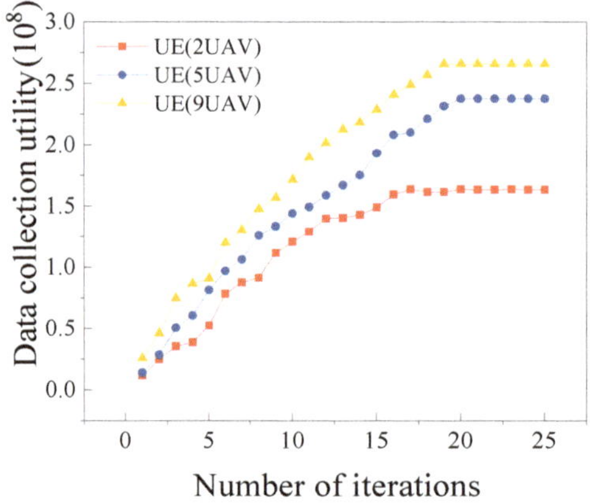

Figure 5. Convergence of UEAlgorithm ($k = 2, 5, 9$).

5.5. Different Mission Area Sizes

Figure 6 shows the trend of data collection utility as the size of mission area changes. We set the number of sensors to 100, and the number of UAVs to 3. We assume the mission area is a square area, and the variable in this simulation is the side length of the mission area. As shown in Figure 6, with the expansion of the mission area, the data collection utility gradually decreases. This is reasonable since the UAVs need to consume more energy to fly between data collection points when the sensors are distributed in a large mission area. Under this scenario, the energy used for data collection will be reduced, leading to the decrease of data collection utility. However, when the mission area is large, the data collection utility of our algorithm is also higher than other algorithms. For example, when the mission area is 1500 m × 1500 m, the data collection utility of our algorithm is 52%–134% higher than compared algorithms. Meanwhile, the data collection utility of the optimal scheme is 8.7%–35.1% higher than that of our algorithm as the mission area expands. In Figure 7, we illustrate the convergence of our algorithm under different mission area sizes. We can see that the algorithm can converge quickly in different data collection areas.

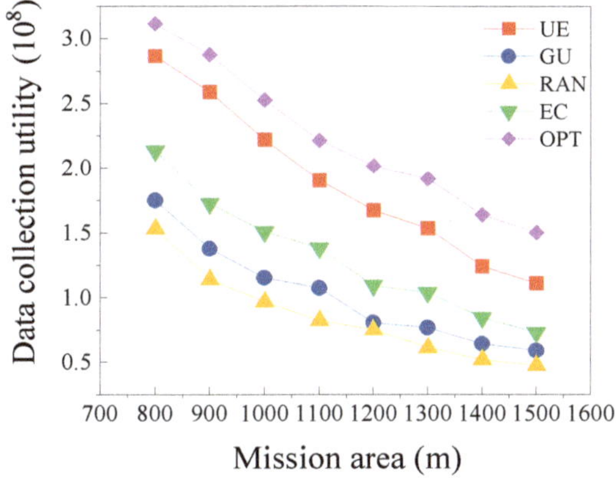

Figure 6. Different Mission Area Sizes.

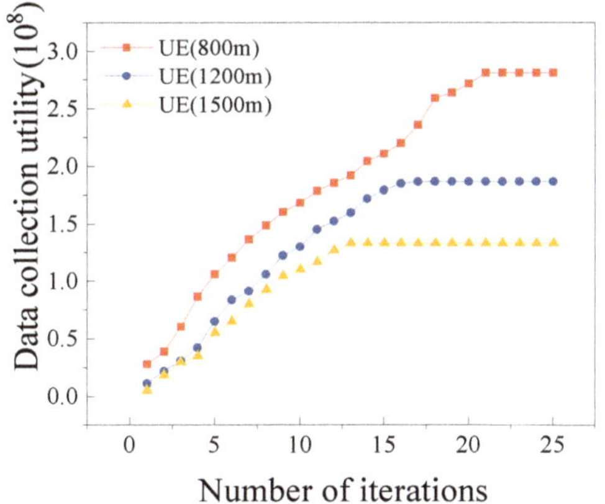

Figure 7. Convergence of UE Algorithm (800 m, 1200 m, 1500 m).

5.6. Trajectories of UAVs

In this subsection, we use Figure 8 to show the resulting trajectories by each of the algorithms. The serial number of the cluster represents the importance of its coverage area. UE algorithm chooses the data collection points with the highest utility-cost ratio each time, which saves the energy of UAVs and improves the data collection utility of UAVs. Compared with the other three algorithms, our algorithm takes into account the data collection utility and the energy consumption.

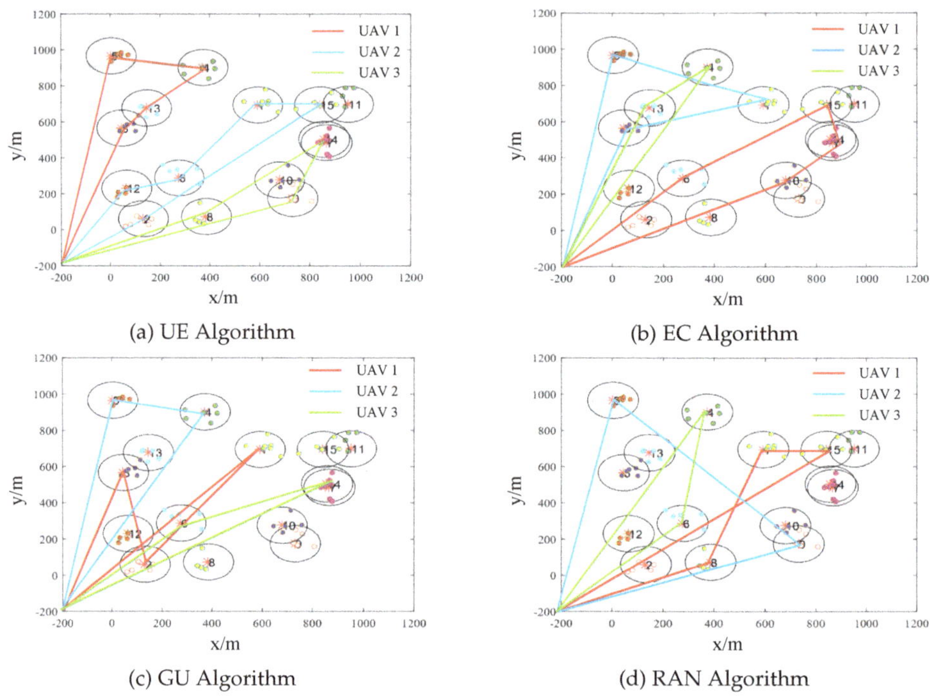

(a) UE Algorithm

(b) EC Algorithm

(c) GU Algorithm

(d) RAN Algorithm

Figure 8. Trajectories of Three UAVs.

6. Discussion

In this paper, we mainly focus on two-dimensional trajectory planning of UAVs. In fact, it is worthwhile to optimize UAV's altitude. However, the optimization of flight altitude will bring some challenges. First, the ascend and descend of UAVs will bring extra energy consumption. Second, the flight altitude of UAVs can influence the quality of communication channel. Third, it will bring new optimization variables and increase the search space. We need to further study to solve these problems. In the future, to further improve the performance of multi-UAV data collection system, we will present a new design framework of three-dimensional UAV trajectory.

7. Conclusions

In this paper, we consider exploiting UAVs to collect data from sensors. The value of data collected from each sensor is different. It depends on the importance of the monitoring area of the sensor and the freshness of collected data. To improve the data collection utility, we optimize the trajectory planning, communication scheduling and sensor node association. The data collection utility is determined by the amount and value of data. First, we formulate this problem as a variant of multiple knapsack problem, which is a classical NP-hard problem. We transform the initial problem into the problem of maximizing a submodular function under energy constraints. To maximize the data collection utility, we propose a novel trajectory planning algorithm, while accounting for different value of data and different energy constraints of heterogeneous UAVs. Sufficient simulations are performed to demonstrate the validity and applicability of the proposed algorithm. The results show that the data collection utility of our algorithm can be increased by 134% at most.

Author Contributions: Conceptualization, Z.Q., C.D. and H.W.; methodology, H.D. and Z.Q.; software, A.L. and W.S.; validation, Z.Q. and C.D.; formal analysis, H.W. and A.L.; investigation, A.L., W.S. and Z.X.; writing–original draft preparation, Z.Q.; writing–review and editing, C.D., H.D. and A.L.

Funding: This work was supported in part by the National Natural Science Foundation of China under Grants (No. 61931011, No. 61872178, No. 61827801, No. 61702545, No. 61702525, and No. 61631020), in part by the Fundamental Research Funds for the Central Universities No. 021014380079, and in part by the Natural Science Foundation of Jiangsu Province under Grant No. BK20181251.

References

1. Liu, F.; Yang, J.; Gui, G. DSF-NOMA: UAV-assisted emergency communication technology in a heterogeneous Internet of things. *IEEE Internet Things J.* **2019**, *6*, 5508–5519. [CrossRef]
2. Al-Fuqaha, A.; Guizani, M.; Mohammadi, M.; Aledhari, M.; Ayyash, M. Internet of things: A survey on enabling technologies, protocols, and applications. *IEEE Commun. Surv. Tutorials* **2015**, *17*, 2347–2376. [CrossRef]
3. Zhang, R.; Wang, M.; Shen, X.; Xie, L.L. Probabilistic analysis on QoS provisioning for Internet of things in LTE-A heterogeneous networks with partial spectrum usage. *IEEE Internet Things J.* **2016**, *3*, 354–365. [CrossRef]
4. Liu, J.; Nishiyama, H.; Kato, N.; Guo, J. On the outage probability of device-to-device communication enabled multi-channel cellular networks: A RSS threshold-based perspective. *IEEE J. Sel. Areas Commun.* **2016**, *34*, 163–175. [CrossRef]
5. Jiang, B.; Yang, J.; Xu, H.; Song, H.; Zheng, G. Multimedia data throughput maximization in Internet-of-things system based on optimization of cache-enabled UAV. *IEEE Internet Things J.* **2018**, *6*, 3525–3532. [CrossRef]
6. Ebrahimi, D.; Sharafeddine, S.; Ho, P. ; Assi, C. UAV-aided projection-based compressive data gathering in Wireless Sensor Networks. *IEEE Internet Things J.* **2018**, *6*, 1893–1905. [CrossRef]
7. Li, G.; He, J.; Peng, S.; Jia, W.; Wang, C.; Niu, J.; Yu, S. Energy efficient data collection in large-scale Internet of things via computation offloading. *IEEE Internet Things J.* **2018**, *6*, 4176–4187. [CrossRef]
8. Kim, H.; Ben-othman, J. On collision-free reinforced barriers for multi domain IoT with heterogeneous UAVs. *IEEE Commun. Lett.* **2018**, *22*, 2587–2590. [CrossRef]
9. Eiaz, W.; Naeem, M.; Shahid, A.; Anpalagan, A.; Jo, M. Efficient energy management for the Internet of things in smart cities. *IEEE Commun. Mag.* **2017**, *55*, 84–91.
10. Motlagh, N.H.; Taleb, T.; Arouk, O. Low-altitude Unmanned Aerial Vehicles-based Internet of things services: Comprehensive survey and future perspectives. *IEEE Internet Things J.* **2016**, *3*, 899–922. [CrossRef]
11. Yang, D.; Wu, Q.; Zeng, Y.; Zhang, R. Energy trade-off in Ground-to-UAV communication via trajectory design. *IEEE Trans. Veh. Techn.* **2018**, *67*, 6721–6726. [CrossRef]
12. Zhan, C.; Zeng, Y.; Zhang, R. Energy-efficient data collection in UAV enabled Wireless Sensor Network. *IEEE Commun. Lett.* **2017**, *7*, 328–331. [CrossRef]
13. Hu, Q.; Cai, Y.; Yu, G.; Qin, Z.; Zhao, M.; Li, G. Joint offloading and trajectory design for UAV-enabled mobile edge computing systems. *IEEE Internet Things J.* **2018**, *6*, 1879–1892. [CrossRef]
14. Zhan, C.; Lai, H. Energy minimization in Internet-of-things system based on rotary-wing UAV. *IEEE Wirel. Commun. Lett.* **2019**. [CrossRef]
15. You, C.; Zhang, R. 3D trajectory optimization in rician fading for UAV-enabled data harvesting. *IEEE Trans. Wirel. Commun.* **2019**, *18*, 3192–3207. [CrossRef]
16. Gong, J.; Chang, T.-H.; Shen, C.; Chen, X. Flight time minimization of UAV for data collection over wireless sensor networks. *IEEE J. Sel. Areas Commun.* **2018**, *36*, 1942–1954. [CrossRef]
17. Zhan, C.; Zeng, Y. Completion time minimization for multi-UAV enabled data collection. *IEEE Trans. Wirel. Commun.* **2019**, *18*, 4859–4872. [CrossRef]

18. Liu, Y.; Gao, C.; Zhang, Z.; Lu, Y.; Chen, S.; Liang, M.; Tao, L. Solving NP-hard problems with physarum-based ant colony system. *IEEE/ACM Trans. Comput. Biol. Bioinf.* **2015**, *14*, 108–120. [CrossRef] [PubMed]

19. Chen, Y.; Hao, J. Memetic search for the generalized quadratic multiple knapsack problem. *IEEE Trans. Evol. Comput.* **2016**, *20*, 908–923. [CrossRef]

20. Rovira-Sugranes, A.; Afghah, F.; Razi, A. Optimized compression policy for flying ad hoc networks. In Proceedings of the 2019 16th IEEE Annual Consumer Communications & Networking Conference (CCNC), Las Vegas, NV, USA, 11–14 January 2019; pp. 1–2.

21. Yang, S.; Deng, Y.; Tang, X.; Ding, Y.; Zhou, J. Energy efficiency optimization for UAV-assisted backscatter communications. *IEEE Commun. Lett.* **2019**. [CrossRef]

22. Liu, X.; Liu, Y.; Zhang, N.; Wu, W.; Liu, A. Optimizing trajectory of Unmanned Aerial Vehicles for efficient data acquisition: A matrix completion approach. *IEEE Internet Things J.* **2019**, *6*, 1829–1840. [CrossRef]

23. Samir, M.; Sharafeddine, S.; Assi, C.; Nguyen, T.M.; Ghrayeb, A. Trajectory planning and resource allocation of multiple UAVs for data delivery in vehicular networks. *IEEE Networking Lett.* **2019**, *1*, 107–110. [CrossRef]

24. Sharafeddine, S.; Islambouli, R. On-demand deployment of multiple aerial base stations for traffic offloading and network recovery. *Comput. Networks* **2019**, *156*, 52–61. [CrossRef]

25. Samir, M.; Sharafeddine, S.; Assi, C.; Nguyen, T.M.; Ghrayeb, A. UAV trajectory planning for data collection from time-constrained IoT devices. *IEEE Trans. Wirel. Commun.* **2019**. . [CrossRef]

26. Hu, Y.; Yuan, X.; Xu, J.; Schmeink, A. Optimal 1D trajectory design for UAV-enabled multiuser wireless power transfer. *IEEE Trans. Commun.* **2017**, *67*, 5674–5688. [CrossRef]

27. Zeng, Y.; Xu, X.; Zhang, R. Trajectory design for completion time minimization in UAV-enabled multicasting. *IEEE Trans. Wirel. Commun.* **2018**, *17*, 2233–2246. [CrossRef]

28. Stefano, P.; Giorgio, G.; Alessandro, R. A risk-aware path planning strategy for UAVs in urban environments. *J. Intell. Rob. Syst.* **2018**, *95*, 629–643.

29. Islam, S.; Razi, A. A path planning algorithm for collective monitoring using autonomous drones. In Proceedings of the 2019 53rd Annual Conference on Information Sciences and Systems (CISS), Baltimore, MD, USA, 20–22 March 2019; pp. 1–6.

30. Babel, L. Coordinated target assignment and UAV path planning with timing constraints. *J. Intell. Rob. Syst.* **2019**, *94*, 857–869. [CrossRef]

31. Christian, Z.; Kampen, E.V. Advancements for A* and RRT in 3D path planning of UAVs. Available online: https://arc.aiaa.org/doi/abs/10.2514/6.2019-0920 (accessed on 8 September 2019).

32. Hu, J.; Zhang, H.; Song, L. Reinforcement learning for decentralized trajectory design in cellular UAV networks with sense-and-send protocol. *IEEE Internet Things J.* **2018**, *6*, 6177–6189. [CrossRef]

33. Wu, Q.; Zeng, Y.; Zhang, R. Joint trajectory and communication design for multi-UAV enabled wireless networks. *IEEE Trans. Wirel. Commun.* **2018**, *17*, 2109–2121. [CrossRef]

34. Youssef, M.; Youssef, A.; Younis, M. Overlapping multihop clustering for Wireless Sensor Networks. *IEEE Trans. Parallel Distrib. Syst.* **2009**, *20*, 1844–1856. [CrossRef]

35. Youssef, A.; Younis, M.; Youssef, M.; Agrawala, A. Distributed formation of overlapping multi-hop clusters in Wireless Sensor Networks. In Proceedings of the 2006 IEEE Global Communications Conference, San Francisco, CA, USA, 27 November–1 December 2006.

36. Zeng, Y.; Xu, J.; Zhang, R. Energy minimization for wireless communication with rotary-wing UAV. *IEEE Trans. Wirel. Commun.* **2019**, *18*, 2329–2345. [CrossRef]

37. Thammawichai, M.; Baliyarasimhuni, S.P.; Kerrigan, E.C. Optimizing communication and computation for multi-UAV information gathering applications. *IEEE Trans. Aerosp. Electron. Syst.* **2018**, *54*, 601–615. [CrossRef]

38. Monwar, M.; Semiari, O.; Saad, W. Optimized path planning for inspection by Unmanned Aerial Vehicles swarm with energy constraints. In Proceedings of the 2018 IEEE Global Communications Conference, Abu Dhabi, UAE, 9–13 December 2018; pp. 1–6.

39. Stolaroff, J.K.; Samaras, C.; Oneill, E.R.; Lubers,A.; Mitchell, A.S.; Ceperley, D. Energy use and life cycle greenhouse gas emissions of drones for commercial package delivery. *Nature Commun.* **2018**, *9*, 1–13.

40. Driessens, S.; Pounds, P.E.I. Towards a more efficient quadrotor configuration. In Proceedings of the 2013 IEEE/RSJ International Conference on Intelligent Robots and Systems, Tokyo, Japan, 3–7 November 2013; pp. 1–7.

41. Tang, C.; Mckinley, P.K. Energy optimization under informed mobility. *IEEE Trans. Parallel Distrib. Syst.* **2016**, *17*, 947–962. [CrossRef]
42. Ruan, L.; Wang, J.; Chen, J.; Xu, Y.; Yang, Y.; Jiang, H.; Zhang, Y.; Xu, Y. Energy-efficient multi-UAV coverage deployment in UAV networks: A game-theoretic framework. *China Commun.* **2018**, *15*, 194–209. [CrossRef]
43. Na, H.J.; Yoo, S. PSO-based dynamic UAV positioning algorithm for sensing information acquisition in Wireless Sensor Networks. *IEEE Access* **2019**, *7*, 77499–77513. [CrossRef]
44. Wu, T.; Yang, P.; Dai, H.; Xu, W.; Xu, M. Collaborated tasks-driven mobile charging and scheduling: a near optimal result. In Proceedings of the IEEE Conference on Computer Communications, Paris, France, 29 April–2 May 2019; pp. 1–9.
45. Zhang, H.; Vorobeychik, Y. Submodular optimization with routing constraints. In Proceedings of the AAAI, Phoenix, AZ USA, 12–17 February 2016; pp. 1–9.
46. Zeng, Y.; Zhang, R.; Teng, J.L. Wireless communications with unmanned aerial vehicles: Opportunities and challenges. *IEEE Commun. Mag.* **2016**, *54*, 36–42. [CrossRef]

Article

Completion Time Minimization for Multi-UAV Information Collection via Trajectory Planning

Zhen Qin [1], Aijing Li [1], Chao Dong [2,*], Haipeng Dai [3] and Zhengqin Xu [1]

[1]   College of Communications Engineering, Army Engineering University of PLA, Nanjing 210042, China;
       qzqzla912@163.com (Z.Q.); lishan_wh@163.com (A.L.); zhengqinxu66@163.com (Z.X.)
[2]   College of Electronic and Information Engineering, Nanjing University of Aeronautics and Astronautics,
       Nanjing 211106, China
[3]   Department of Computer Science and Technology, Nanjing University, Nanjing 210023, China;
       haipengdai@nju.edu.cn
*   Correspondence: dch@nuaa.edu.cn

Received: 5 August 2019; Accepted: 16 September 2019; Published: 18 September 2019

Abstract: Unmanned Aerial Vehicles (UAVs) are widely used as mobile information collectors for sensors to prolong the network time in Wireless Sensor Networks (WSNs) due to their flexible deployment, high mobility, and low cost. This paper focuses on the scenario where rotary-wing UAVs complete information collection mission cooperatively. For the first time, we study the problem of minimizing the mission completion time for a multi-UAV system in a monitoring scenario when considering the information collection quality. The mission completion time includes flying time and hovering time. By optimizing the trajectories of all UAVs, we minimize the mission completion time while ensuring that the information of each sensor is collected. This problem can be formulated as a mixed-integer non-convex one which has been proved to be NP-hard. To solve the formulated problem, we first propose a hovering point selection algorithm to select appropriate hovering points where the UAVs can sequentially collect the information from multiple sensors. We model this problem as a BS coverage problem with the information collection quality in consideration. Then, we use a min-max cycle cover algorithm to assign these hovering points and get the trajectory of each UAV. Finally, with the obtained UAVs trajectories, we further consider the UAVs can also collect information when flying and optimize the time allocations. The performance of our algorithm is verified by simulations, which show that the mission completion time is minimum compared with state-of-the-art algorithms.

Keywords: wireless sensor networks; unmanned aerial vehicle; mission completion time; trajectory planning

---

1. Introduction

Due to their tremendous application potentials in military and civilian related applications, Wireless Sensor Networks (WSNs) have attracted increasing research interest in many fields, such as industrial process control, environmental monitoring, and battlefield surveillance [1–4]. Generally, WSNs are composed of small, battery-limited sensors which cannot transmit over a long distance [5]. There is a great interest recently in utilizing rotary-wing Unmanned Aerial Vehicles (UAVs) as mobile information collectors in WSNs [6–8]. They can move towards the sensors and establish reliable connections with them due to their high mobility and flexible deployment [9]. Rotary-wing UAVs are equipped with propellers that enable them to hover over fixed locations. This advantage makes them suitable candidates for mobile information collectors as they can hover over sensors to collect information [10,11]. Nowadays, with the popularity of WSNs, a single UAV cannot meet the demands of monitoring missions over larger monitoring areas. Information collection of WSNs with multi-UAV cooperation is more desired. Currently, there are two types of methods to collect the information from WSNs with multiple UAVs. The first is to use static UAVs to form a network to cover all sensors, which is suitable for small area monitoring [12–15]. The second is to use mobile UAVs to cover a larger area [16–18]. As shown in Figure 1, we mainly focus on the second scenario in this paper.

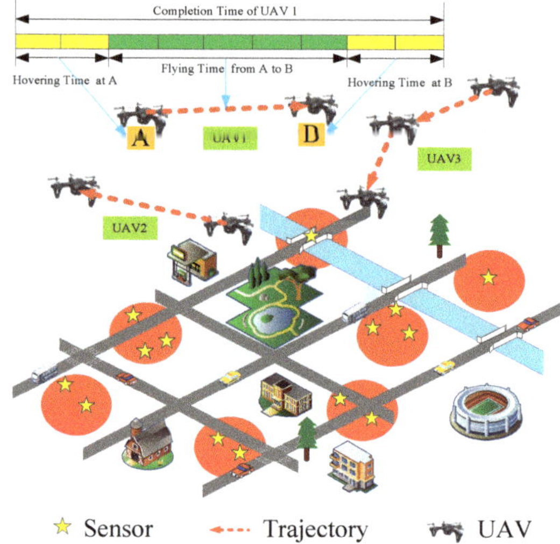

Figure 1. Monitoring scenario.

For many WSNs applications, the timeliness of monitoring information is critical. Therefore, the mission completion time of multi-UAV cooperative information collection is very important. For example, in a city security monitoring system, if the control center gets fire information as soon as possible, the city may avoid serious economic loss and casualties. In traffic monitoring applications, real-time road information helps cars effectively avoid traffic congestion. In this paper, the mission completion time includes flying time and hovering time, both of which are related to the trajectory planning of the UAVs. Therefore, the trajectories need to be carefully planned such that UAVs can complete theirs missions quickly and satisfy the communication requirements along their entire trajectory. By optimizing the trajectory of each UAV, we pursue minimizing the mission completion time of multi-UAV system.

Some works have studied how to minimize a single UAV's the mission completion time [16,17,19–21]. However, these works cannot be used for information collection with multi-UAV directly. In fact, a multi-UAV system requires solving two sub-problems, task assignment and trajectory optimization. Generally, we cannot address one of problems individually because the two sub-problems are tightly coupled. If task assignment is not considered in trajectory planning, it may result in unreasonable task assignment. For example, a UAV may be responsible for too many tasks, which can lead to longer mission completion time. If trajectory optimization is not considered in task assignment, it may result in long flying distance, which may also lead to longer mission completion time. Therefore, we cannot solve the two sub-problems separately. In other words, we cannot assign tasks first, and then apply the algorithm of a single UAV to solve this problem. Since we use a multi-UAV system to collect information, the completion time of the entire mission is the time of all tasks completed by UAVs. In other words, it is the maximum completion time among the UAVs. The completion time of a UAV is mainly divided into two parts, flying time and hovering time. As shown in Figure 1, we give a toy example illustrating flying time and hovering time. The flying time is the time taken for the UAV to fly from point A to point B. The hovering time is the time for the UAV staying at the hovering point (A or B) to collect information from sensors. There are two main differences from previous works. First, some works only consider flying time or only consider hovering time in UAV-enabled network [22–24]. Second, some works assume that UAVs only collect information from sensors when hovering [25–30]. If flying time is fully utilized, the mission completion time can be reduced. Therefore, we consider that the UAVs can collect information when they are hovering and flying.

In this paper, our objective is to plan the trajectories of all UAVs to minimize the mission completion time, while ensuring that the information of each sensor is collected. Rotary-wing UAVs can collect information when they are hovering and flying. This problem can be formulated as a mixed-integer non-convex problem. Since the problem involves infinitely many variables over time, it is difficult to optimize trajectories, and hovering and flying communication scheduling at the same time. To tackle this problem, we propose an improved fly-hover-fly trajectory planning algorithm. We first propose a hovering point selection algorithm to get hovering points. Then, based on these hovering points, we plan the trajectory of each UAV. Intuitively, it can use the algorithm of Multiple Traveling Salesman Problem (MTSP) to plan the trajectories of multi-UAV. However, MTSP's goal is to minimize the total completion time of multi-UAV. Different from MTSP, our problem aims to collect all sensors' information with multi-UAV such that the maximum completion time among UAVs is minimized. By using the Base Station (BS) coverage algorithm and min-max cycle cover algorithm, we select appropriate hovering points where the UAVs can sequentially collect the information from multiple sensors and get the trajectory of each UAV. Finally, based on the obtained trajectories, we propose an effective method to optimize UAVs hovering and fly-collection time allocations. The main contributions of this paper are summarized as follows:

- To the best of our knowledge, this is the first paper to study how to minimize the mission completion time considering the information collection quality when multiple rotary-wing UAVs collect information cooperatively in WSNs. We model it as a mixed-integer non-convex problem, which has been proved to be NP-hard.
- We first propose the FHF algorithm to optimize the hovering points and obtain the trajectories. We select appropriate hovering points where the UAVs can sequentially collect the information from multiple sensors. We model this problem as a BS coverage problem with the information collection quality in consideration. Then, we use a min-max cycle cover algorithm to assign these hovering points and get the trajectory of each UAV.
- With the obtained UAVs trajectories, we further consider that the UAVs can also collect information when flying. We propose an effective method to optimize UAVs hovering- and fly-collection time allocations.
- Extensive simulation results show the mission completion time of the proposed algorithm is minimum and closer to the optimal scheme compared with other algorithms. Furthermore, we also verify the performance of the algorithms under different settings.

The rest of this paper is organized as follows. In Section 2, we introduce the related work. The system model and problem formulation are presented in Section 3. Then, we propose the FHF algorithm to optimize the hovering points and obtain the trajectories in Section 4. In Section 5, we optimize the time allocations. Simulation results are provided and analyzed in Section 6. Discussion is provided in Section 7. Finally, we conclude the paper in Section 8.

2. Related Work

2.1. Single UAV's Mission Completion Time Problem

There are many studies on minimizing a single UAV's mission completion time. In [16], the authors studied a UAV-enabled multicasting system and planned the trajectory of the UAV for minimizing mission completion time, while ensuring that ground terminals successfully recovered files. To reduce the single UAV's mission completion time and periodic flight duration, Zhang et al. [17]

optimized the trajectory of UAV and communication resource allocation, while satisfying the data requirement of ground users. Zhang et al. [19] focused on a scenario where a single UAV communicated with multiple ground base stations. They aimed to minimize the mission completion time by planning the trajectory of UAV, while the UAV was also subject to practical communication connectivity constraint with ground base stations. One of the most important reconnaissance missions of the UAV is to take pictures of interesting targets in a wide area. Under this scenario, Lim et al. [20] proposed a trajectory generation algorithm for reconnaissance UAV with time constraints. Zhang et al. [21] proposed sparse A^* search approach for a single UAV route planning with time constraints. It can be seen that the above works only discuss a single UAV's mission completion time in different scenarios. However, these works cannot be used for information collection with multi-UAV directly. Therefore, the mission completion time problem of multi-UAV is worth being further studied.

2.2. Multi-UAV's Mission Completion Time Problem

There are some studies on completing missions by multi-UAV within the specified time. Pohl et al. [31] studied the routing of multi-UAV, while the trajectories of UAVs met the constraints of total path cost optimization, total mission time, enemy radar avoidance, and time on target. This work aimed to minimize the total distance by planning trajectories. Shi et al. [32] proposed a statistical physics method for UAVs cooperative reconnaissance mission planning, while the trajectories satisfied the reconnaissance time requirements. However, in many applications, it is more important to complete the missions as soon as possible. In [23], the authors introduced a geometric path planning method and proposed a resource allocation algorithm for multi-UAV. However, they only considered the flying time. Hayat et al. [33] optimized the trajectory of each UAV and synchronized multi-UAV so that the UAVs can fly to the next task only after all UAVs had completed their previous task. Mousavi et al. [34] aimed to form UAV coalition formation to reduce the mission completion time. Wang et al. [35] found the optimal schedule to perform diverse missions of different time windows at various locations using fixed-wing heterogeneous UAVs. The solutions of minimizing the mission completion time for multi-UAV system are not directly applicable to our problem.

3. System Model and Problem Formulation

3.1. System Model

As shown in Figure 1, we consider a monitoring scenario where multiple rotary-wing UAVs are employed to collect information from sensors whose distributions follow the homogeneous Poisson Point Process. Traditionally, sensors transmit their monitored information to a base station via multi-hop transmissions. Therefore, a sensor is required to not only transmit its own information, but also relay the information of other sensors. As a consequence, the sensors battery may drain quickly and the multi-hop network connection may be lost [22]. By utilizing the UAVs as mobile information collectors, sensors can directly send monitored information to UAVs. However, as UAVs have limited on-board battery, it is important to reduce the completion time needed for the information collection mission. In this paper, we aim to minimize the maximum completion time among all UAVs by optimizing the trajectories of UAVs to complete the information collection mission as soon as possible. We assume the UAVs fly at a fixed altitude. The UAV collects information from sensors via time-division multiple access (TDMA) that means the UAV only communicates with at most one sensor at each time. Meanwhile, we consider a scenario where ground users are mainly composed of sensors. The extension to the three-dimensional (3D) trajectory planning and other multiple access schemes will be left for our future work. For convenience, Table 1 provides major notations used in this paper.

Table 1. Major notations .

Notation	Definition
U	UAV set
P	Sensor set
u_i	UAV i
p_j	Sensor j
k	Number of UAVs, or number of trajectories
n	Number of sensors
$l_i(t)$	u_i trajectory projected on the ground
$\dot{l}_i(t)$	Time derivative of $l_i(t)$
c_j	Position of p_j
$d_{i,j}(t)$	Distance between UAV i and sensor j
$v_{\max}$	Maximum velocity
$d_{\min}$	Minimum inter-UAV distance
$h(t)$	Instantaneous channel power gain
$R_{i,j}(t)$	Instantaneous channel capacity
$\bar{R}_j$	Total amount of data of p_j
B	Bandwidth
p	Transmission power
σ^2	White Gaussian noise power
H	Altitude of UAV
$f_{i,j}(t)$	Binary variable
r	Communication radius
T_F^i	Flying time
T_H^i	Hovering time
T_i	Completion time of u_i
T	Mission Completion time

The UAV and sensor sets are denoted as $U \triangleq \{u_1, ..., u_k\}$ and $P \triangleq \{p_1, ..., p_n\}$, respectively. Considering a 3D Cartesian coordinate system, sensor p_j is fixed at $c_j = (x_j, y_j, 0)$. Denote the trajectory of UAV u_i projected on the ground as $l_i(t) = [x_i(t), y_i(t)]^T \in \mathbb{R}^{2 \times 1}$, where $0 \leq t \leq T$. The trajectory of each UAV should satisfy the following constraint

$$l_i(0) = l_i(T), \forall i, \tag{1}$$

which means that the UAVs need to fly back to their starting location by the end of monitoring mission. Furthermore, the trajectory of each UAV is subject to the velocity constraints, which can be given by

$$\left\| \dot{l}_i(t) \right\| \leq v_{\max}, \forall i, t \in [0, T], \tag{2}$$

where $v_{\max}$ denotes the maximum velocity of UAVs and $\dot{l}_i(t)$ is the time derivative of $l_i(t)$. As multi-UAV fly at the constant altitude, the trajectory of each UAV is expected to satisfy the collision avoidance constraint [36]

$$\|l_i(t) - l_{i'}(t)\| \geq d_{\min}, \forall i \neq i', t \in [0, T], \tag{3}$$

where $d_{\min}$ is the minimum distance between the UAVs for avoiding collision. In addition, the time-varying distance between UAV u_i and sensor p_j can be expressed as

$$d_{i,j}(t) = \sqrt{H^2 + (x_i(t) - x_j)^2 + (y_i(t) - y_j)^2}. \tag{4}$$

Due to scattering and ground reflection, the communication channels between UAVs and sensors include both Line-of-Sight (LoS) and Non-Line-of Sight (NLoS) links [16]. The UAV–ground link channel is characterized by the presence of strong LoS path. The Rician channel model is an appropriate choice, as it can effectively reflect the combination of scattering and LoS that exists in the links between the UAVs and sensors [13]. In this paper, we consider the more practically Rician fading channels between the UAVs and sensors [13,16,37–39]. The instantaneous channel gains between the UAV u_i and sensor p_j can be expressed as [16,37]

$$h_{i,j}(t) = \sqrt{\beta_{i,j}(t)}g_{i,j},\tag{5}$$

where $g_{i,j}$ is the small-scale fading coefficient. $\beta_{i,j}(t)$ is the average channel power gain, which can be expressed by [16,37]

$$\beta_{i,j}(t) = \beta_0 d_{i,j}^{-\alpha}(t) = \frac{\beta_0}{\left[H^2 + (x_i(t) - x_j)^2 + (y_i(t) - y_j)\right]^{\alpha/2}},\tag{6}$$

where α is the path loss exponent and β_0 denotes the average channel power gain at the reference distance $d_0 = 1m$. $g_{i,j}$ is a random variable that corresponds to the effects of the small-scale fading such that $E\left[|g_{i,j}|^2\right] = 1$, which can be given by [16,37]

$$g_{i,j} = \sqrt{\frac{K_R}{K_R + 1}}g + \sqrt{\frac{1}{K_R + 1}}\tilde{g},\tag{7}$$

where K_R is the Rician factors of the channel and g denotes the deterministic LoS channel component with $|g| = 1$. $\tilde{g}$ represents the random scattered component which is a zero-mean unit-variance circularly symmetric complex Gaussian (CSCG) random variable. For Rician fading, the cdf of $|g_{i,j}|^2$ is explicitly represented as [37,38]

$$F_{|g_{i,j}|^2}(x) = 1 - Q_1(\sqrt{2K_R}, \sqrt{2(K_R + 1)x}),\tag{8}$$

where $Q_1(a,b)$ is the standard Marcum-Q function. We assume that p is the transmission power of sensors. The instantaneous channel capacity is expressed as [36]

$$R_{i,j}(t) = B\log_2(1 + \frac{p|h_{i,j}(t)|^2}{\sum\limits_{m=1,m\neq j}^{n} p|h_{i,j}(t)|^2 x_m(t) + \sigma^2}),\tag{9}$$

where B is the channel bandwidth and σ^2 denotes the white Gaussian noise power at the receiver. The term $\sum\limits_{m=1,m\neq j}^{n} p|h_{i,m}(t)|^2 x_m(t)$ in Equation (9) represents the co-channel interference caused by the transmissions of other sensors in time t. We denote a binary variable $x_m(t)$ to indicate whether sensor p_m transmits information to UAVs at time t, which can be given by

$$x_m(t) = \begin{cases} 1 & p_j \ transmits \ information \ to \ UAVs \\ 0 & otherwise \end{cases}.\tag{10}$$

The maximum communication radius projected on the ground is expressed by r, which depends on the maximum communication distance $d_{U-S}^{\max}$ and UAV altitude H, which is determined as [15,40]

$$r = \sqrt{d_{U-S}^{\max\,2} - H^2},\tag{11}$$

We assume that the UAVs fly at a determined altitude H when they collect information from the sensors. Therefore, r mainly depends on the maximum communication distance. To successfully collect information from sensors, the received power of UAV must be higher than or equal to the minimum decodable power $p_{U,\min}$ [40]. The maximum communication distance between the sensor and the UAV can be calculated as

$$d_{U-S}^{\max} = \sqrt{\frac{G_S^t \cdot G_U^r \cdot \lambda^2 \cdot p}{(4\pi)^2 \cdot p_{U,\min}}},$$

(12)

where G_S^t is the transmit antenna gain of the sensor, G_U^r is the receive antenna gain of the UAV. λ represents the wavelength of the signal transmitted by a sensor which is calculated by the frequency of signal. Based on the maximum communication radius r, the communication condition can be expressed as

$$\left\| l_i(t) - c_j' \right\| \leq r,$$

(13)

where c_j' is the coordinate representation of sensor p_j on a two-dimensional plane. It is worth noting that, if UAV u_i needs to collect information from sensor p_j at time t, in order to ensure reliable transmission, the distance between them must satisfy Equation (13).

We define a binary variable $f_{i,j}(t)$, and it indicates whether sensor p_j is served by UAV u_i at time t

$$f_{i,j}(t) = \begin{cases} 1 & u_i \; collect \; information \; from \; p_j \\ 0 & otherwise \end{cases}.$$

(14)

The binary variables specify not only the communication scheduling across the different time, but also the association between UAVs and sensors. In this paper, the UAV collects information from sensors via TDMA that means the UAV only communicates with at most one sensor at each time. Meanwhile, each sensor is served by only one UAV at a time instance, but can be served by different UAVs over different time slots. These constraints can be expressed as

$$\sum_{j=1}^{n} f_{i,j}(t) \leq 1, \forall i, j,$$

(15)

$$\sum_{i=1}^{k} f_{i,j}(t) \leq 1, \forall i, j,$$

(16)

Furthermore, the total amount of data $\bar{R}_j$ transmitted from sensor p_j to UAV u_i is a function of UAV trajectory $l_i(t)$, which is expressed as

$$\bar{R}_j(l_i(t)) = \int_0^T \mathrm{B}\log_2\left(1 + \frac{p\left|h_{i,j}(t)\right|^2}{\sum\limits_{m=1,m\neq j}^{n} p\left|h_{i,j}(t)\right|^2 x_m(t) + \sigma^2}\right) f_{i,j}(t) dt.$$

(17)

3.2. Problem Formulation

The throughput requirement corresponding to sensor p_j is assumed to be C_j bits. By optimizing the trajectories of UAVs, our objective is to minimize mission completion time, while ensuring that the information of sensors is collected. The mission completion time refers to the time taken by all UAVs to complete respective tasks. The completion time of a UAV is mainly divided into two parts, flying time and hovering time. Therefore, the completion time of a single UAV can be represented as

$$T_i = T_F^i + T_H^i,$$

(18)

where T_F^i represents the flying time and T_H^i represents the hovering time.

We denote $\mathbf{F} = \{f_{i,j}(t), \forall i, j, t\}$ and $\mathbf{L} = \{l_i(t), \forall i, t\}$. Assuming the sensors' locations are given, we aim to minimize the mission completion time by jointly optimizing the UAVs' trajectories (i.e., $\mathbf{L}$), communication scheduling and association (i.e., $\mathbf{F}$). Define $T(\mathbf{F}, \mathbf{L}) = \max_{1 \le i \le k} T_i$ as the mission completion time which is a function of $\mathbf{F}$ and $\mathbf{L}$. The optimization problem can be formulated as

$$\min_{\mathbf{F}, \mathbf{L}, \{T_H\}, \{T_F\}} T \tag{19}$$

$$s.t. \quad l_i(0) = l_i(T), \forall i, \tag{20}$$

$$\bar{R}_j \ge C_j, \forall j, \tag{21}$$

$$\sum_{i=1}^{k} f_{i,j}(t) \le 1, \forall i, j, t \in [0, T], \tag{22}$$

$$\sum_{j=1}^{n} f_{i,j}(t) \le 1, \forall i, j, t \in [0, T], \tag{23}$$

$$\left\| \dot{l}_i(t) \right\| \le v_{\max}, \forall i, t \in [0, T], \tag{24}$$

$$\| l_i(t) - l_{i'}(t) \| \ge d_{\min}, \forall i \ne i', t \in [0, T], \tag{25}$$

$$0 \le \| l_i(t) - c_j' \| f_{i,j}(t) \le r, \forall i, j, t \in [0, T]. \tag{26}$$

This problem is challenging to be solved due to three main reasons. First, the problem needs to optimize continuous functions $\mathbf{F}$ and $\mathbf{L}$, which essentially involve an infinite number of optimization variables that are closely coupled with each other. Second, since the optimization variables $\mathbf{F}$ for UAV–sensor association and communication scheduling are binary, the formulated problem is a non-convex problem. Third, it is difficult to optimize trajectories, hovering and flying communication scheduling at the same time. To solve this problem, we propose an improved fly-hover-fly trajectory planning algorithm (FLY), in which the UAVs successively hover at a finite number of hovering points each for an optimized hovering duration. Different from other traditional fly-hover-fly trajectory designs [25–30], UAVs can collect information from sensors not only in hovering, but also in flying. First, we propose the FHF algorithm to optimize the hovering points and obtain the trajectories in Section 4. Then, under the obtained trajectories, we further consider the UAVs can also collect information when flying and optimize the time allocations in Section 5.

4. FHF Trajectory Planning Algorithm

As shown in Figure 2, we propose FHF algorithm which includes two steps. First, we propose a hovering point selection algorithm to select appropriate hovering points with considering the information collection quality. Red dots represent sensors and blue dots represent optimized hovering points. The number of hovering points is less than that of ground sensors. Not stopping above each sensor node, we select appropriate hovering points where the UAVs can sequentially collect the information from multiple sensors. By hovering at different points, which are close to different subsets of sensors, the UAVs can obtain better wireless communication links and decrease the flying time compared with the policy of hovering above each sensor. Moreover, the policy of hovering above each sensor was used as a comparison algorithm in our simulations.

Second, we use min-max cycle cover algorithm to assign these hovering points and get the trajectory of each UAV. Based on the hovering points optimized in the first step, we use Traveling Salesman Problem (TSP) algorithm to calculate a cycle covering all hovering points. Then, we use a novel k-cycles algorithm for getting the trajectory of each UAV, which can minimize the maximum time of the cycles including flying time and hovering time.

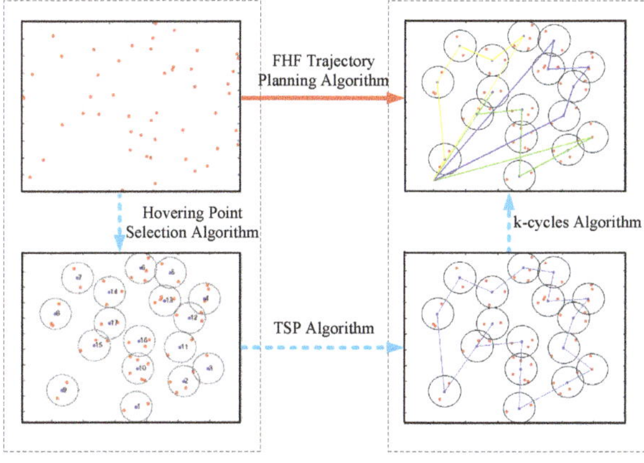

Figure 2. FHF trajectory planning algorithm.

4.1. Hovering Point Selection

For a given SNR threshold, it is unnecessary for UAVs to fly over all sensors in general. We model hovering point selection problem as a BS coverage problem [15]. Specifically, given sensors' locations and UAV's communication radius, the hovering point selection problem's objective is to obtain a minimum number of hovering points and respective locations, while ensuring that all sensors are covered by at least one hovering point. As shown in Algorithm 1, we use the min-max location algorithm as sub-algorithm to solve this problem. Meanwhile, major notations used in Algorithm 1 are provided in Table 2.

Algorithm 1 Hovering point selection algorithm.

Input: r, sensor set P with known locations $\{c_j\}_{p_j \in P}$.
Output: Hovering point set V, with optimized locations $\{h_m\}_{v_m \in V}$.
 1: Initialize $X \leftarrow \varnothing, Y \leftarrow P, V \leftarrow \varnothing, m \leftarrow 1$;
 2: **while** $|X| < n$ **do**

 3:     Calculate boundary sensor set $Y_{bo} \subseteq Y$, update inner sensor set $Y_{in} \leftarrow Y \backslash Y_{bo}$;
 4:     Randomly select a sensor $b_0 \in Y_{bo}$. Denote $\rho \leftarrow \{b \,|\, \|b_0 - b\| \leq 2r, b \in Y_{bo}\}$;
 5:     Use the min-max location algorithm to find hovering point's location h_m to cover ρ. Denote the maximum distance is d;
 6:     **while** $d > r$ **do**

 7:         $\rho \leftarrow \rho \backslash \arg\max_{b \in \rho} \|b - h_m\|$. Repeat step 5;
 8:     **end while**
 9:     $\rho \leftarrow \rho \cup \{b \,|\, \|b - h_m\| \leq r, b \in Y_{in}\}$;
10:     $\theta \leftarrow \{b \,|\, r < \|b - h_m\| \leq 2r, b \in Y_{in}\}$;
11:     **while** $\theta \neq \varnothing$ **do**

12:         Find a sensor $b' \in \theta$ which has the shortest distance to h_m;
13:         Remove (add) b' from (to) θ (ρ) if b' is covered by changing h_m via using min-max location algorithm. Stop otherwise;
14:     **end while**
15:     $X \leftarrow X \cup \rho$;
16:     $Y \leftarrow Y \backslash \rho$;
17:     $V \leftarrow V \cup \{v_m\}, m \leftarrow m + 1$.
18: **end while**

Table 2. Major notations in Algorithm 1.

Notation	Definition
V	Hovering point set
X	Covered sensor set
Y	Uncovered sensor set
h_m	Hovering point's location
Y_{bo}	Boundary sensor set
Y_{in}	Inner sensor set
b_o	A boundary sensor
d	The farthest distance to h_m

Now, we describe the hovering point selection algorithm in detail. First, we define Y as the set of uncovered sensors, which is initialized to be equal to P; X as the set of covered sensor, which is initialized to an empty set; and V as the set of hovering points, which is initialized to empty set (Line 1). We utilize the convex hull to define boundary sensors, whereas other boundary sensors definitions can also be used to produce similar results. In this algorithm, to reduce the occurrence of outlier sensors, which require one hovering point to cover it, we give higher priority to the boundary sensors so that boundary sensors are guaranteed to be covered firstly. The boundary sensors exist in Y_{bo} and the inner sensors exist in Y_{in} (Line 3). We randomly choose a boundary sensor b_0 that exists in Y_{bo}. The sensors that are less than $2r$ away from b_0 are put in ρ (Line 4).

At this time, it can be converted into the min-max location problem to find the position of the hovering point. In operations research of facilities location type, the min-max location problem is a classical combinatorial optimization problem. The problem can be described as follows: given a function to calculate the cost between demand points and a facility, a space of feasible locations of a facility, and a set of n demand points, find the facility's location which minimizes the maximum facility-demand point cost [41,42]. The simple special case when the demand points and feasible locations are in the plane with Euclidean distance as cost, it is also known as the smallest circle problem [43]. Our goal is to minimize the maximum sensor-hovering point distance. Therefore, we use the min-max location algorithm to find hovering point's location h_m to cover all sensors in ρ [44] (Line 5). This algorithm minimizes the maximum distance from sensors to h_m. The farthest distance is defined as d.

If the value of d exceeds r, ρ needs to be rebuilt until the farthest distance is less than r. In each iteration, the sensor which is the farthest away from h_m is deleted (Lines 6–8). Update ρ by adding sensors which satisfy the distance conditions (Line 9). This check reduces the execution times of the min-max location algorithm in the next step. The sensors in Y_{in} whose distance to h_m are in the range of $(r, 2r)$ are put in θ (Line 10). We find a sensor $b' \in \theta$ which is the closest to h_m. Remove (add) b' from (to) θ (ρ) if b' is covered by changing h_m via using min-max location algorithm. When θ is empty or new h_m is not found, the mth hovering point's location h_m is determined (Line 11–14). X, Y and V are updated based on the results we got (Line 15–17). When all sensors are covered, the algorithm will stop. In Figure 3, we show an example of Algorithm 1. There are seventy sensors in a square area of 5 km $\times$ 5 km and r equals 500 m. By utilizing Algorithm 1, we obtain twenty hovering points and respective locations.

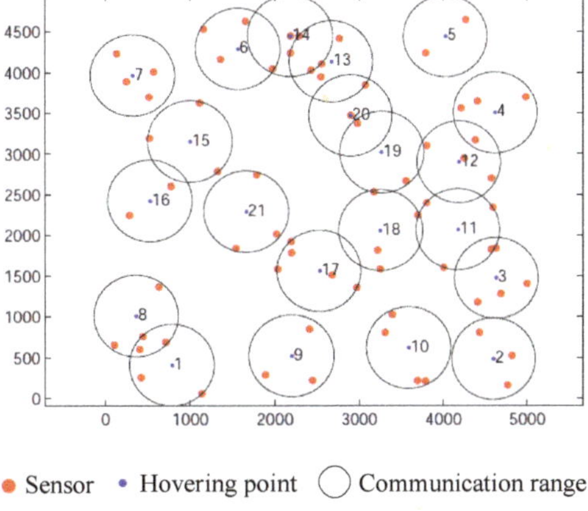

Figure 3. Hovering points.

For hovering point selection algorithm, the convex hull algorithm of n points to find boundary sensors has complexity $O(n \log n)$ [15,45]. In addition, each hovering point executes the min-max location algorithm for up to $O(n)$ times. Meanwhile, the complexity of the min-max location algorithm is $O(n^2)$ [44]. Since the number of hovering points is at most $O(n)$, the overall complexity of the hovering point selection algorithm is upper-bounded by $O(n[n \log n + n^3])$. To sum up, the complexity of the hovering point selection algorithm is upper-bounded by $O(n^4)$.

4.2. Min-Max Cycle Cover

The multi-UAV system is modeled as a complete undirected graph $G = (V, E)$, where vertices in V represent hovering points for UAV and edges in E represent flying paths of the UAV. For each edge $e(v_m, v_{m+1}) \in E$, $v_m, v_{m+1} \in V$, an edge weight $w(v_m, v_{m+1})$ is given to represent the flying time. For each vertex v_m, a vertex weight $h(v_m)$ is given to represent the hovering time to collect information. Since we use a multi-UAV system to collect information, the completion time of the entire mission is the time of tasks completed by all UAVs. In other words, it is the maximum completion time of UAVs. The minimum mission completion time problem aims to cover hovering points with multi-UAV such that the maximum weight of the cycles including flying time and hovering time is minimized. To tackle this problem, we formulate it as a min-max cycle cover problem. The min-max cycle cover problem is an extension of the classical TSP. Since the TSP is NP-hard when $k = 1$, the min-max cycle cover problem is also NP-hard for any $k \geq 1$ [46].

The heuristic algorithm we propose includes two steps to plan the trajectory of each UAV. Firstly, the ant colony algorithm is used for obtaining a cycle C covering all hovering points. As a global searching algorithm, the ant colony algorithm can solve TSP problems [47]. Then, we use a novel k-cycles algorithm for getting the trajectory of each UAV [48]. This algorithm mainly decomposes cycle C into k segments.

The k-cycles algorithm for obtaining k trajectories of UAVs is shown as Figure 4. This algorithm takes the number of UAVs, cycle C and the complete undirected graph $G = (V, E)$ as input. In k-cycles algorithm, to decompose cycle C into k trajectories of UAVs, we first compute the bound vector $Q = (Q_1, ..., Q_i, ..., Q_k)$. Q_i can be calculated on the basis of the total weight of cycle C, $Q_i = \frac{i}{k}W(C)$, for all $1 \leq i \leq k$, where $W(C) = h(C) + w(C)$. $h(C)$ and $w(C)$ represent the weight of all vertices and the weight of all edges in cycle C, respectively. Second, along the cycle, we build a mixed set of edges and vertices containing $2|V| + 1$ elements, denoting it as $VE = (ve_0 = v_0, ve_1 = e(v_0, v_1), ve_2 = v_1, ve_3 = e(v_1, v_2), ..., ve_{2|V|+1} = v_0)$. $wh(ve_0, ..., ve_m)$ represents the sum weight of vertices and edges. Compared with Author1 [49], the weight of vertices, which represents the hovering time, is calculated one time in k-cycles algorithm. By constructing a mixed set of vertices and edges, the weight of vertices can be taken into account in the cycle splitting. Next, from ve_0, we aim to find an element $ve_{m(i)}$ along cycle C such that the weight of segment $wh(ve_0, ve_1, ..., ve_{m(i)})$ satisfies $wh(ve_0, ve_1, ..., ve_{m(i)}) \leq Q_i$, for each i, $1 \leq i \leq k$. There are two possibilities for the demarcation point $ve_{m(i)}$.

- If $ve_{m(i)}$ is a path, we would delete this path when we decompose the cycle. Then, we define $ve_{m(i)-1}$ as the ending point of this segment, $ve_{m(i)+1}$ as the starting point of the next segment.
- If $ve_{m(i)}$ is a hovering point, we would define it as the starting point of the next segment (i.e., $ve'_{m(i)} = ve_{m(i)}$). Then, we take $ve_{m(i)-2}$ as the end of this segment.

Then, we obtain k segments of cycle C (i.e., $\{C_1, ..., C_i, ..., C_k\}$). Finally, the segments add the starting location to build the closed trajectory for each UAV.

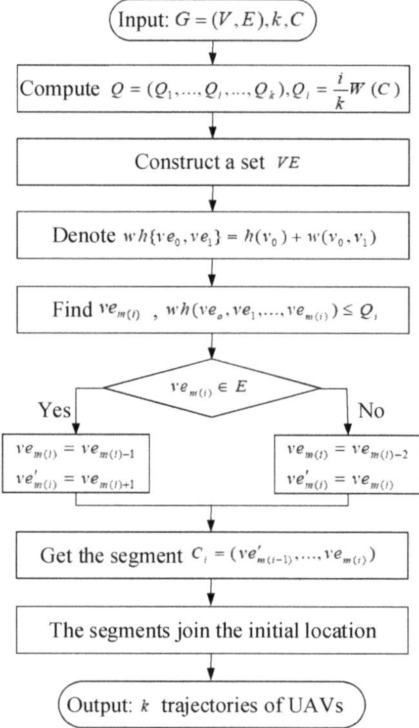

Figure 4. k-cycles algorithm.

The complexity of the ant colony algorithm is related to the number of ants, the number of hovering points and the number of iterations. The number of hovering points cannot exceed n. Thus, the complexity of the ant colony algorithm is upper-bounded by $O(iter \cdot A \cdot n^2)$, where $iter$ is the number of iterations and A is the number of ants. Furthermore, the complexity of k-cycles algorithm depends on the number of sensors. Since k-cycles algorithm is a polynomial-time algorithm, its complexity is $O(n)$. In addition, the complexity of the hovering point selection algorithm is upper-bounded by $O(n^4)$. To sum up, the complexity of the hover-collection UAV trajectory planning algorithm is upper-bounded by $O(n^2[n^2 + iter \cdot A])$.

5. Time Allocation

With the obtained UAVs trajectories, we further consider the UAVs can also collect information when flying. We propose an effective method to optimize UAVs fly-collection and hovering time allocations.

We define $\tilde{T}_{H,m} \geq 0$ as the new allocated time for the UAVs to collect information when hovering at location v_m. Since UAVs can collect information when flying, we can obtain a constraint $\tilde{T}_{H,m} \leq T_{H,m}, \forall m$. The time allocation problem is challenging to be solved due to three main reasons. First, the time allocation problem needs to optimize continuous functions $\mathbf{F}$, which essentially involves an infinite number of optimization variables. Second, the mission completion time is unknown. Therefore, the time allocation problem cannot be solved by widely time discretization method. Third, the UAV–sensor association and communication scheduling are binary, thus the formulated problem is a non-convex problem. To tackle this problem, we use a new discretization method, called path discretization [50,51].

With the UAVs' trajectories UT and hovering points V given, UAV's trajectory UT_i can be discretized into Z^i line segments, which are represented by the Z^i sampling points $\{q_{z^i}\}$. The length of the z^ith line segment can be expressed as $\mu_{z^i} = \left\| q_{z^i+1} - q_{z^i} \right\|$. Note that Z^i needs to be chosen to be sufficiently large so that $\mu_{z^i} \leq \mu_{\max}^i$. $\mu_{\max}^i$ is an appropriately chosen value so that the distance between each sensor and UAV u_i is approximately unchanged under each line segment. Furthermore, we assume the time for flying in the z^ith line segment is π_{z^i}. Meanwhile, λ_{z^i} is defined as the allocated information collection time when u_i flies on the z^ith line segment, where $\lambda_{z^i} \leq \pi_{z^i}$. When information collection occurs in the mobile state, the UAV sends an activation signal to a sensor which satisfies SNR conditions. Then, the UAV collects information from the activated sensor.

The formulated problem reduces to optimizing the communication scheduling $\mathbf{F}$, the hovering time $\{\tilde{T}_H\}$ and fly-collection time $\{\lambda_{z^i}\}$. Therefore, we can obtain

$$\min_{\mathbf{F}, \{\tilde{T}_H\}, \{\lambda_{z^i}\}} T \tag{27}$$

$$s.t. \quad \bar{R}_j \geq C_j, \forall j, \tag{28}$$

$$\sum_{i=1}^{k} f_{i,j}(t) \leq 1, \forall i, j, t \in [0, T], \tag{29}$$

$$\sum_{j=1}^{n} f_{i,j}(t) \leq 1, \forall i, j, t \in [0, T], \tag{30}$$

$$0 \leq \left\| l_i(t) - c_j' \right\| f_{i,j}(t) \leq r, \forall i, j, t \in [0, T], \tag{31}$$

$$\tilde{T}_{H,m} \leq T_{H,m}, \forall m, \tag{32}$$

$$\lambda_{z^i} \leq \pi_{z^i}, \forall z, i. \tag{33}$$

With the given trajectories, since the velocity of UAVs remains unchanged, the new hovering time cannot exceed the initial value. The new formulated problem is an integer programming problem, which can use the existing software toolbox such as CVX to be efficiently solved. Therefore, under the given trajectories, we can get fly-collection and hovering time allocations. Furthermore, joint optimization of multi-UAV trajectories, communication scheduling and time allocation will be left for our future works.

6. Simulation Results

In this section, we present the simulation results to validate the proposed FLY algorithm, which mainly includes the FHF algorithm and time allocations.

6.1. Simulation Setup

We assumed that the size of the monitoring area was 5 km × 5 km. The total bandwidth was 1 MHz, and the average channel power gain at 1 m was −50 dBm [52]. The maximum velocity of a UAV was set to 50 m/s. Furthermore, the minimum distance between UAVs was set to 100 m [36] and UAVs flew at a constant altitude of 50 m. We assumed that the transmission power equaled 10 dBm [16]. The white Gaussian noise power was equal to −110 dBm [16]. Following the authors of [15,40], the communication radius was 500 m, which was calculated by the altitude and maximum communication radius. As stated in [36], the Rician channel model can be utilized for estimating the link performance for UAV-to-ground links. Simulations were conducted using the parameters specified in Table 3.

Table 3. Simulation parameters.

Definition	Parameters	Values
Monitoring area size	D	5 km × 5 km
Altitude of UAV	H	50 m
Maximum velocity	$v_{\max}$	50 m/s
Minimum inter-UAV distance	$d_{\min}$	100 m
Path loss	α	2.6
Bandwidth	B	1 MHz
Transmission power	p	10 dBm
Unit channel power gain	β_0	−50 dBm
White Gaussian noise power	σ^2	−110 dBm
Rician factor	K_R	2
Minimum decodable power	$p_{U,\min}$	−84 dBm
Frequency of signal	f_λ	2.4 GHz
Transmit antenna gain	G_S^t	1
Receive antenna gain	G_U^r	1

6.2. Baseline Setup

To demonstrate the performance of the proposed information-collection UAV trajectory planning algorithm (FLY), we compared and implemented the following six benchmark schemes:

- FHF algorithm: It is the subalgorithm of the FLY algorithm. This algorithm only considers the UAVs collecting information when they are hovering.
- Hovering above each sensor (SHP): The feasible trajectory of the UAV needs to ensure the throughput requirement of each sensor is satisfied. One straightforward approach is to let the UAVs sequentially visit all sensors. First, SHP algorithm calculates a cycle which covers all sensors using TSP algorithm. Then, it uses k-cycles algorithm to obtain the trajectories of UAVs.
- Sensor balanced algorithm (PB): This algorithm first calculates a cycle by using TSP algorithm. Then, it decomposes the cycle into k trajectories that contain the same number of sensors.
- KTSP-based algorithm (KTSP): This algorithm designs the trajectories of multi-UAV by using K-Traveling Salesman Problem (KTSP) algorithm [53].
- Kmeans-based algorithm (KMEAN): This algorithm uses kmeans algorithm to obtain hovering points. Then, it uses the min-max cycle cover algorithm to obtain the trajectories of UAVs [54].
- Optimal scheme (OPT): To evaluate how the proposed algorithms approach the optimal performance, we use brute-force searching method to obtain the optimal scheme that minimizes the mission completion time.

6.3. Different Number of Sensors

In this simulation, with different number of sensors, the number of UAVs was fixed at three and the seven algorithms were compared. As the number of sensors changes, the trend of mission completion time is shown in Figure 5. It was observed that the algorithms we proposed significantly outperformed four other algorithms. FHF and FLY algorithms were the closest to the optimal solution compared with four other algorithms. FLY algorithm achieved almost the same performance with the optimal scheme when $n < 20$, and the mission completion time difference was less than 6%. When the number of sensors was greater than 20, FLY algorithm was also closer to the optimal scheme than five other algorithms. For example, when $n = 120$, FLY algorithm was 119% of the optimal solution, which was the biggest gap between FLY algorithm and the optimal scheme in this simulation. It was also observed that the completion time of mission conducted only in static case cOULD not be smaller than in both static and mobile cases. FLY algorithm was reduced by 7.0–14.1% compared with FHF algorithm. That was expected; since UAVs utilized the flying phase to collect information, the hovering time was reduced. Meanwhile, we found that our subalgorithm (FHF) performed better than the other compared algorithms. FHF algorithm was reduced by 3.5–46.9% compared with four other algorithms. When the number of sensors was 10, FHF algorithm was reduced by 3.5% compared with KMEAN algorithm. When the number of sensors was 120, FHF algorithm was reduced by 46.9% compared with KTSP algorithm. By hovering at a point where it could communicate with multiple sensors, the UAV could decrease the flying time compared with SHP algorithm, PB algorithm and KTSP algorithm. In particular, when the number of sensors increased, the UAV could communicate with more sensors at one hovering point. Therefore, our algorithms performed better when the number of sensors was large. Moreover, SHP algorithm and PB algorithm use the same TSP algorithm, thus the performance was relatively close. Meanwhile, KMEAN algorithm selected hovering points without considering the information collection quality of UAVs. There was retransmission in the actual transmission, which increased the information collection time and KMEAN algorithm did not reach the effect of selecting the hovering points. Therefore, KMEAN algorithm was close to SHP algorithm and PB algorithm.

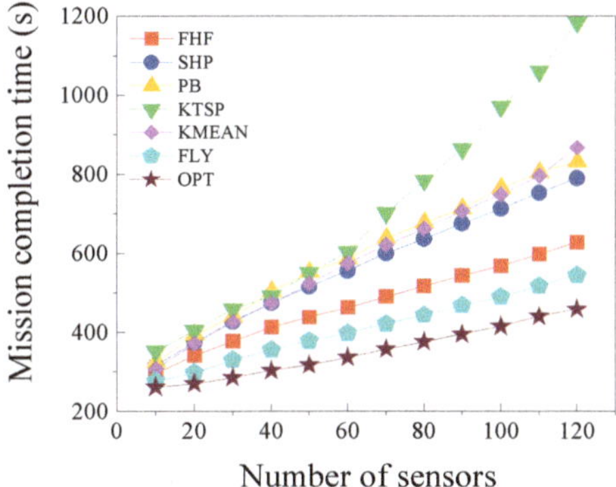

Figure 5. Different number of sensors.

6.4. Different Number of UAVs

Under different number of UAVs, the number of sensors was set to be 100 and we compared the seven algorithms. As the number of UAVs changes, the trend of mission completion time is shown in Figure 6. Obviously, increasing the number of UAVs could improve the speed of monitoring, and the mission could be completed in less time. However, increasing the number of UAVs also brought some overhead and economic pressure. As shown in Figure 6, from one UAV to four UAVs, the mission completion time dropped significantly. When the number of UAVs exceeded five, the changes in mission completion time tended to be flat. It was obvious that more UAVs could share the monitoring mission to reduce mission completion time. However, some sensors were far from the initial location, and the UAV might take too much time going back and forth. Therefore, the maximum completion time of UAVs had a lower bound, and it was impossible to keep falling as the number of UAVs increased. Compared with four other algorithms, our algorithms were closer to optimal scheme. As shown in Figure 6, the gap between our algorithms and the optimal solution decreased as the number of UAVs increased. When $k = 1$, FHF algorithm was 86.7% higher than the optimal scheme and FLY algorithm was 52.4% higher than it. When $k = 10$, FHF algorithm was 12% higher than the optimal scheme and FLY algorithm was 3.1% higher than it. As the number of UAVs changed, FLY algorithm was reduced by 8.0 18.1% compared with FHF algorithm. Such results demonstrate the flying and hovering information collection solution is better than hovering information collection solution.

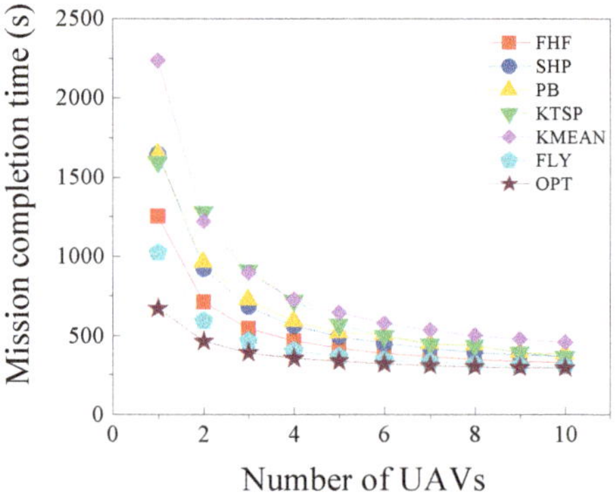

Figure 6. Different number of UAVs.

6.5. Different Communication Radius

Under different communication radius, we set the number of UAVs at 3 and the number of sensors at 120, and compared the seven algorithms. Figure 7 shows the trend of mission completion time as the communication radius changes. Under different communication radius, our algorithms were closer to the optimal scheme than four other algorithms. As the communication radius increased, our algorithms had obvious advantages. When the communication radius was 100 m, FLY algorithm was 145% of the optimal algorithm, and FHF algorithm was 169% of the optimal algorithm. However, when the communication radius was 800 m, FLY algorithm was 116% of the optimal algorithm, and FHF algorithm was 130% of the optimal algorithm. As shown in Figure 7, the mission completion time of our algorithms and KMEAN algorithm gradually decreased as the communication radius increased, and other algorithms' change trends were relatively small. The reason is that the change in communication radius mainly affected the hovering point selection algorithm, which is an important difference between three algorithms and three other algorithms. Since we considered the information collection quality in hovering point selection algorithm, our algorithms had a higher performance improvement compared with KMEAN algorithm. Of course, in practical applications, the communication range of the UAV is related to the altitude and the communication equipment of the UAV.

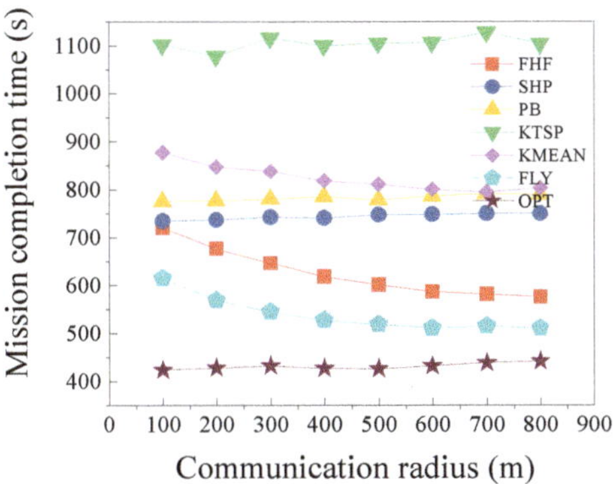

Figure 7. Different communication radius.

6.6. Different Monitoring Area Size

Under different monitoring area size, we set the number of UAVs at 3 and the number of sensors at 100 and compared the seven algorithms. Figure 8 shows the trend of mission completion time as the monitoring area size changes. It was observed that FHF and FLY algorithms were closer to the optimal solution than four other algorithms. The gap between FHF algorithm and the optimal solution increased as the monitoring area expanded, from 21.6% to 67.7%. The gap between FLY algorithm and the optimal solution increased as the monitoring area expanded, from 10.3% to 41.3%. As the monitoring area became larger, the advantage of FLY algorithm was more obvious. FLY algorithm was reduced by 6–15.6% compared with FHF algorithm. That was expected; since the larger area caused a long flying time, the UAVs could obtain more time to collect information when flying. Furthermore, FHF algorithm was reduced by 9.7–40.2% compared with four other algorithms. As the monitoring area size increased, the flying scope of the UAV became larger, and the flying time became longer, resulting in longer mission completion time. As shown in Figure 8, FLY algorithm performed the best under different monitoring area size. Since it fully considers the information collection quality and the positions of sensors, the algorithm can be applied to different monitoring area size. Based on the simulation results in the previous subsection, if the communication range of UAVs is appropriately changed, the mission completion time will be shorter.

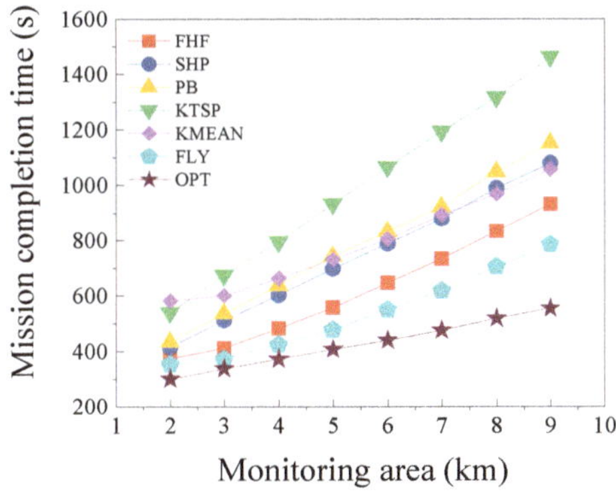

Figure 8. Different monitoring area size.

7. Discussion

7.1. Trade-Offs in Communication and Trajectory Design

There are some new and interesting trade-offs among the energy, delay and throughput in UAV communication and trajectory design which are different from traditional terrestrial communication [55]. In this section, we mainly discuss the trade-offs of UAVs when they are used as mobile information collectors. First, there is a trade-off between delay and throughput in UAV-enabled wireless network. To maximize throughput, UAVs always fly sufficiently close to the sensors to improve the link capacity. Although this method improves the throughput, it also brings delay due to the movement of the UAV. Second, there is also an interesting trade-off between energy consumption and throughput in UAVs. To gain higher throughput, UAVs generally fly close to sensors which consume more energy to move. Third, the above two trade-offs naturally imply an energy-delay trade-off. To reduce the delay in UAV–sensor communication, UAVs need to move faster to sensors ,which brings more energy consumption. In fact, by planning the trajectory of UAV, the energy, delay and throughput can be traded off among each other. Finally, there is a trade-off between sensors and UAVs [9]. The sensors have limited battery and lower power. To prolong the lifetime of sensors, UAVs can move close to sensors to collect their information with minimum transmit power. However, UAVs will consume more energy to move. In this paper, we mainly focus on minimizing mission completion time on the premise of achieving throughput requirements. In the future, we will consider these trade-offs in trajectory design.

7.2. Multiple Access Schemes

In this paper, we consider that the UAVs collect information from sensors via time-division multiple access (TDMA) that means only one sensor is scheduled for information collection at each time [36,37,39,55]. There are some works using other multiple access schemes [25,52]. For example, the authors considered a scenario that a UAV is dispatched as the mobile BS to provide service for ground users via orthogonal frequency division multiple access (OFDMA) [52]. Besides orthogonal multiple access schemes such as TDMA considered above, non-orthogonal multiple access schemes also are considered in UAV-enabled wireless networks. To further improve the capacity limits, Wu et al. found that non-orthogonal multiple access schemes based on superposition coding (SC) can be jointly designed with the UAV trajectory [25]. The authors considered a two-user broadcast channel. To achieve the capacity region, they proposed a practical fly-hover-fly trajectory with SC. However, they only considered two users. In the future, for the next-generation wireless networks with massive connectivities, we will study designing the efficient trajectory algorithm joint with other multiple access schemes.

7.3. The Case of a Large Number of Sensors

In practical application, there may be many sensors deployed in the monitoring area. The complexity of hovering point selection problem becomes extremely high when the number of points n is very large. Therefore, the application scope of the proposed algorithm will be limited. To expand the application scope of the algorithm, we propose an efficient method. When the number of sensors is large, the monitoring area can be divided into some subareas. In other words, ground sensors can be partitioned into M disjoint sets, $P_1, P_2, \ldots, P_M$ with each set corresponding to the sensors in its subarea. Then, the hovering point selection algorithm can be implemented for sensors in each subarea. By dividing the monitoring area, the complexity of trajectory planning problem can be reduced, and the application scope of the proposed algorithm can be improved.

8. Conclusions

In this paper, we aim to minimize the mission completion time of multi-UAV by optimizing the trajectories of UAVs to complete the mission as soon as possible. We formulate this problem as a mixed-integer non-convex problem. It is difficult to optimize trajectories, hovering and flying communication scheduling at the same time. To tackle the formulated problem, we propose an improved fly-hover-fly trajectory planning algorithm, which includes two steps. First, we propose the FHF algorithm to optimize the hovering points and obtain the trajectories. Second, with the obtained UAVs trajectories, we further consider that the UAVs can also collect information when flying. We propose an effective method to optimize UAVs fly-collection and hovering time allocations. The simulation results show that the mission completion time of our algorithm is minimum compared with other algorithms.

Since we add some constraints in the model, there are some limitations in practical application. For example, we assume that the UAVs fly at a fixed altitude. In fact, UAV's altitude will directly affect Rician factor and information transmission rate. It is also worthwhile to optimize UAV's altitude. In the future, we will exploit the vertical trajectory of the UAV and present a new design framework of 3D UAV trajectory to further improve the performance of multi-UAV information collection system. In addition, for the next-generation wireless networks with massive connectivities, we will study designing the efficient trajectory algorithm joint with other multiple access schemes. In our future works, other practical constraints on UAV's trajectory will be considered, such as the maximum acceleration, the maximum instantaneous output power of the engine and the maximum turning angle.

Author Contributions: Conceptualization, Z.Q. and C.D.; methodology, H.D. and Z.Q.; software, A.L. and Z.X.; validation, Z.Q. and C.D.; formal analysis, A.L.; investigation, A.L. and Z.X.; writing—original draft preparation, Z.Q.; and writing—review and editing, C.D., H.D. and A.L.

Funding: This work was supported in part by the National Natural Science Foundation of China under Grants (No. 61931011, No. 61872178, No. 61827801, No. 61702545, No. 61702525 and No. 61631020), in part by the Fundamental Research Funds for the Central Universities No. 021014380079 and in part by the Natural Science Foundation of Jiangsu Province under Grant No. BK20181251.

Conflicts of Interest: The authors declare no conflict of interest.

References

1. Ma, C.; Liang, W.; Zheng, M.; Sharif, H. A connectivity-aware approximation algorithm for relay node placement in Wireless Sensor Networks. *IEEE Sens. J.* **2015**, *16*, 515–528. [CrossRef]
2. Zhao, M.; Li, J.; Yang, Y. A framework of joint mobile energy replenishment and data gathering in Wireless Rechargeable Sensor Networks. *IEEE Trans. Mob. Comput.* **2014**, *13*, 2689–2705. [CrossRef]
3. Kim, H.; Han, S. An efficient sensor deployment scheme for large-scale Wireless Sensor Networks. *IEEE Commun. Lett.* **2015**, *19*, 98–101. [CrossRef]
4. Bukhari, S.H.R.; Rehmani, M.H.; Siraj, S. A survey of channel bonding for wireless networks and guidelines of channel bonding for futuristic cognitive radio sensor networks. *IEEE Commun. Surv. Tutor.* **2015**, *18*, 924–948. [CrossRef]
5. Mozaffari, M.; Saad, W.; Bennis, M.; Debbah, M. Mobile Unmanned Aerial Vehicles (UAVs) for energy-efficient Internet of Things communications. *IEEE Trans. Wirel. Commun.* **2017**, *16*, 7574–7589. [CrossRef]
6. Abdulla, A.E.A.A.; Fadlullah, Z.M.; Nishiyama, H.; Kato, N.; Ono, F.; Miura, R. An optimal data collection technique for improved utility in UAS-aided networks. In Proceedings of the 2014 IEEE International Conference on Computer Communications (INFOCOM), Toronto, ON, Canada, 27 April–2 May 2014.
7. Zhan, C.; Zeng, Y.; Zhang, R. Energy-efficient data collection in UAV enabled wireless sensor network. *IEEE Commun. Lett.* **2017**, *7*, 328–331. [CrossRef]
8. Zeng, Y.; Zhang, R.; Teng, J.L. Wireless communications with Unmanned Aerial Vehicles: Opportunities and challenges. *IEEE Commun. Mag.* **2016**, *54*, 36–42. [CrossRef]
9. Yang, D.; Wu, Q.; Zeng, Y.; Zhang, R. Energy trade-off in ground-to-UAV communication via trajectory design. *IEEE Trans. Veh. Technol.* **2018**, *67*, 6721–6726. [CrossRef]
10. Zhan, C.; Lai, H. Energy minimization in Internet-of-Things system based on rotary-wing UAV. *IEEE Commun. Lett.* **2019**. [CrossRef]

11. Hou, H.; Wang, L. Analysis on time-variant air-to-ground radio communication channel for rotary-wing UAVs. In Proceedings of the 2019 IEEE 89th Vehicular Technology Conference (VTC2019-Spring), Kuala Lumpur, Malaysia, 28 April–1 May 2019.

12. Al-Hourani, A.; Kandeepan, S.; Lardner, S. Optimal LAP altitude for maximum coverage. *IEEE Wirel. Commun. Lett.* **2014**, *3*, 569–572. [CrossRef]

13. Azari, M.M.; Rosas, F.; Chen, K.C.; Pollin, S. Optimal UAV positioning for terrestrial-aerial communication in presence of fading. In Proceedings of the 2016 IEEE Global Communication Conference (GLOBECOM), Washington, DC, USA, 4–8 Decmber 2016.

14. Alzenad, M.; El-Keyi, A.; Lagum, F.; Yanikomeroglu, H. 3-D placement of an Unmanned Aerial Vehicle Base Station (UAV-BS) for energy-efficient maximal coverage. *IEEE Wirel. Commun. Lett.* **2017**, *6*, 434–437. [CrossRef]

15. Lyu, J.; Zeng, Y.; Zhang, R.; Lim, T.J. Placement optimization of UAV-mounted mobile base stations. *IEEE Wirel. Commun. Lett.* **2017**, *21*, 604–607. [CrossRef]

16. Zeng, Y.; Xu, X.; Zhang, R. Trajectory design for completion time minimization in UAV-enabled multicasting. *IEEE Trans. Wirel. Commun.* **2018**, *17*, 2233–2246. [CrossRef]

17. Zhang, J.; Zeng, Y.; Zhang, R. UAV-enabled radio access network: Multi-mode communication and trajectory design. *IEEE Trans. Signal Proces.* **2018**, *66*, 5269–5284. [CrossRef]

18. Zeng, Y.; Zhang, R. Energy-efficient UAV communication with trajectory optimization. *IEEE Trans. Wirel. Commun.* **2017**, *16*, 3747–3760. [CrossRef]

19. Zhang, S.; Zeng, Y.; Zhang, R. Cellular-enabled UAV communication: A connectivity-constrained trajectory optimization perspective. *IEEE Trans. Commun.* **2018**, *67*, 2580–2604. [CrossRef]

20. Lim, C.; Park, S.; Ryoo, C.; Choi, K. A path planning algorithm for surveillance UAVs with timing mission constrains. In Proceedings of the 2010 IEEE International Conference on Control, Automation and Systems (ICCAS), Gyeonggi-do, Korea, 27–30 October 2010.

21. Zhang, C.; Meng, X. Spare A^* search approach for UAV route planning. In Proceedings of the 2017 IEEE International Conference on Unmanned Systems (ICUS), Beijing, China, 27–29 October 2017.

22. Gong, J.; Chang, T.; Shen, C.; Chen, X. Flight time minimization of UAV for data collection over wireless sensor networks. *IEEE J. Sel. Areas Commun.* **2018**, *36*, 1942–1954. [CrossRef]

23. Shin, H.; Leboucher, C.; Tsourdos, A. Resource allocation with cooperative path planning for multiple UAVs. In Proceedings of the 2012 IEEE UKACC International Conference on Control, Cardiff, UK, 3–5 September 2012.

24. Mozaffari, M.; Saad, W.; Bennis, M.; Debbah, M. Wireless communication using Unmanned Aerial Vehicles (UAVs): Optimal transport theory for hover time optimization. *IEEE Trans. Wirel. Commun.* **2017**, *16*, 8052–8066. [CrossRef]

25. Wu, Q.; Xu, J.; Zhang, R. Capacity characterization of UAV-enabled two-user broadcast channel. *IEEE J. Sel. Areas Commun.* **2018**, *36*, 1955–1971. [CrossRef]

26. Hu, Y.; Yuan, X.; Xu, J.; Schmeink, A. Optimal 1D trajectory design for UAV-enabled multiuser wireless power transfer. *IEEE Trans. Commun.* **2019**, *67*, 5674–5688. [CrossRef]

27. Xie, L.; Xu, J.; Zhang, R. Throughput maximization for UAV-enabled wireless powered communication networks. *IEEE IoT J.* **2018**, *6*, 1690–1703. [CrossRef]

28. Wu, Y.; Xu, J.; Qiu, L.; Zhang, R. Capacity of UAV-enabled multicast channel: Joint trajectory design and power allocation. In Proceedings of the 2018 IEEE International Conference on Communications (ICC), Kansas City, MO, USA, 20–24 May 2018.

29. Yang, G.; Dai, R.; Liang, Y. Energy-efficient UAV backscatter communication with joint trajectory and resource optimization. In Proceedings of the 2019 IEEE International Conference on Communications (ICC), Shanghai, China, 20–24 May 2019.

30. Xu, J.; Zeng, Y.; Zhang, R. UAV-enabled wireless power transfer: Trajectory design and energy optimization. *IEEE Trans. Wirel. Commun.* **2018**, *17*, 5092–5106. [CrossRef]

31. Pohl, A.J.; Lamont, G.B. Multi-objective UAV mission planning using evolutionary computation. In Proceedings of the 2008 IEEE Winter Simulation Conference, Miami, FL, USA, 7–10 December 2008.

32. Shi, Z.; Huang, X.; Hua, Y.; Xu, D. Statistical physics method for multi-base multi-UAV cooperative reconnaissance mission planning. In Proceedings of the 2015 IEEE Advanced Information Technology, Electronic and Automation Control Conference, Chongqing, China, 19–20 December 2015.

33. Hayat, S.; Yanmaz, E.; Brown, T.X.; Bettstetter, C. Multi-objective UAV path planning for search and rescue. In Proceedings of the 2017 IEEE International Conference on Robotics and Automation, Singapore, 20 May–3 June 2017.

34. Mousavi, S.; Afghah, F.; Ashdown, J.D.; Turck, K. Leader-follower based coalition formation in large-scale UAV networks, a quantum evolutionary approach. In Proceedings of the 2018 IEEE International Conference on Computer Communications Workshops (INFOCOM WKSHPS), Honolulu, HI, USA, 15–19 April 2018.

35. Wang, J.J.; Zhang, Y.F.; Geng, L.; Fuh, J.Y.H.; Teo, S.H. Mission planning for heterogeneous tasks with teterogeneous UAVs. In Proceedings of the 2014 IEEE International Conference on Control Automation Robotics & Vision, Singapore, 10–12 December 2014.

36. Wu, Q.; Zeng, Y.; Zhang, R. Joint trajectory and communication design for multi-UAV enabled wireless networks. *IEEE Trans. Wirel. Commun.* **2018**, *17*, 2109–2121. [CrossRef]

37. You, C.; Zhang, R. 3D trajectory optimization in Rician fading for UAV-enabled data harvesting. *IEEE Trans. Wirel. Commun.* **2019**, *18*, 3192–3207. [CrossRef]

38. Yang, L.; Chen, J.; Hasna, M.O.; Yang, H. Outage performance of UAV-assisted relaying systems with RF energy harvesting. *IEEE Wirel. Commun. Lett.* **2018**, *22*, 2471–2474. [CrossRef]

39. Mishra, D.; Lohan, P.; Devi, L.N. Coverage-constrained utility maximization of UAV. In Proceedings of the 2019 IEEE International Conference on Communications (ICC), Shanghai, China, 20–24 May 2019.

40. NA, H.J.; Yoo, S. PSO-based dynamic UAV positioning algorithm for sensing information acquisition in Wireless Sensor Networks. *IEEE Access* **2019**, *7*, 77499–77513. [CrossRef]

41. Burkard, R.E.; Dollani, H. A note on the robust 1-center problem on trees. *Ann. Oper. Res.* **2002**, *110*, 69–82. [CrossRef]

42. Colebrook, M.; Gutiérrez, J.; Alonso, S.; Sicilia, J. A new algorithm for the undesirable 1-center problem on networks. *J. Oper. Res. Soc.* **2002**, *53*, 1357–1366. [CrossRef]

43. Megiddo, N. The weighted Euclidean 1-center problem. *Math. Oper. Res.* **1983**, *8*, 498–504. [CrossRef]

44. Elzinga, J.; Hearn, D.W. Geometrical solutions for some minimax location problems. *Transp. Sci.* **1972**, *6*, 379–394. [CrossRef]

45. Megiddo, N. Linear-time algorithms for linear programming in R^3 and related problems. *SIAM J. Comput.* **1983**, *12*, 759–776. [CrossRef]

46. Xu, W.; Liang, W.; Lin, X. Approximation algorithms for min-max cycle cover problems. *IEEE Trans. Comput.* **2015**, *64*, 600–613. [CrossRef]

47. Sun, Y.; Dong, W.; Chen, Y. An improved routing algorithm based on ant colony optimization in wireless sensor networks. *IEEE Wirel. Commun. Lett.* **2017**, *21*, 1317–1320. [CrossRef]

48. Qin, Z.; Dong, C.; Li, A.; Dai, H.; Wu, Q.; Xu, A. Trajectory planning for reconnaissance mission based on fair-energy UAVs cooperation. *IEEE Access* **2019**, *7*, 91120–91133. [CrossRef]

49. Xu, Z.; Xu, L. Approximation algorithms for min-max path cover problems with service handling time. In Proceedings of the 2009 International Symposium on Algorithms and Computation, Honolulu, HI, USA, 16–18 December 2009.

50. Zeng, Y.; Xu, J.; Zhang, R. Energy minimization for wireless communication with rotary-wing UAV. *IEEE Trans. Wirel. Commun.* **2019**, *18*, 2329–2345. [CrossRef]

51. Zeng, Y.; Xu, J.; Zhang, R. Rotary-wing UAV enabled wireless network: Trajectory design and resource allocation. In Proceedings of the 2018 IEEE Global Communication Conference (GLOBECOM), Abu Dhabi, UAE, 9–13 December 2018.

52. Wu, Q.; Zhang, R. Common throughput maximization in UAV-enabled OFDMA systems with delay consideration. *IEEE Trans. Commun.* **2018**, *66*, 6614–6627. [CrossRef]

53. Frieze, A.M. An extension of Christofides heuristic to the k-person travelling salesman problem. *Discret. Appl. Math.* **1983**, *6*, 79–83. [CrossRef]
54. Sampedro, C.; Bavle, H.; Sanchez-Lopez, J.L.; Fernandez, R.A.S.; Rodriguez-Ramos, A.; Molina, M.; Campoy, P. A flexible and dynamic mission planning architecture for UAV swarm coordination. In Proceedings of the 2016 IEEE International Conference on Unmanned Aircraft Systems, Arlington, VA, USA, 7–10 June 2016.
55. Wu, Q.; Liu, L.; Zhang, R. Fundamental trade-offs in communication and trajectory design for UAV-enabled wireless network. *IEEE Wirel. Commun.* **2019**, *26*, 36–44. [CrossRef]

Article

Aerial Cooperative Jamming for Cellular-Enabled UAV Secure Communication Network: Joint Trajectory and Power Control Design

Hanming Sun [1], Bin Duo [2,*], Zhengqiang Wang [3], Xiaochen Lin [4] and Changchun Gao [1]

[1] The Glorious Sun School of Business and Management, Donghua University, Shanghai 200051, China; 1149151@mail.dhu.end.cn (H.S.); gcc369@dhu.edu.cn (C.G.)

[2] College of Information Science & Technology, Chengdu University of Technology, Chengdu 610059, China

[3] School of Communication and Information Engineering, Chongqing University of Posts and Telecommunications, Chongqing 400065, China; wangzq@cqupt.edu.cn

[4] School of Electronic Information Engineering, Shanghai Dianji University, Shanghai 201306, China; linxc@sdju.edu.cn

* Correspondence: duobin@cdut.edu.cn

Received: 22 September 2019; Accepted: 11 October 2019; Published: 14 October 2019

Abstract: To improve the secrecy performance of cellular-enabled unmanned aerial vehicle (UAV) communication networks, this paper proposes an aerial cooperative jamming scheme and studies its optimal design to achieve the maximum average secrecy rate. Specifically, a base station (BS) transmits confidential messages to a UAV and meanwhile another UAV performs the role of an aerial jammer by cooperatively sending jamming signals to oppose multiple suspicious eavesdroppers on the ground. As the UAVs have the advantage of the controllable mobility, the objective is to maximize the worst-case average secrecy rate by the joint optimization of the two UAVs' trajectories and the BS's/UAV jammer's transmit/jamming power over a given mission period. The objective function of the formulated problem is highly non-linear regarding the optimization variables and the problem has non-convex constraints, which is, in general, difficult to achieve a globally optimal solution. Thus, we divide the original problem into four subproblems and then solve them by applying the successive convex approximation (SCA) and block coordinate descent (BCD) methods. Numerical results demonstrate that the significantly better secrecy performance can be obtained by using the proposed algorithm in comparison with benchmark schemes.

Keywords: UAV secure communication; secrecy rate maximization; jamming; trajectory design; power control

1. Introduction

Due to the many advantages of controllable mobility, such as on-demand fast deployment, wide coverage, low cost, and line-of-sight (LoS) transmission that offers good channel capacity, unmanned aerial vehicles (UAVs) have been extensively utilized in different scenarios, e.g., surveillance and monitoring [1–3], search and rescue [4,5], cargo transportation [6], data collection [7] and mobile relays [8].

Recently, UAVs have attracted increasing attention in wireless communications, and are anticipated to playing an important role in the next-generation wireless networks [9,10]. Generally, there are two promising solutions to UAV communication applications: cellular-enabled UAV communication (CEUC) and UAV-assisted terrestrial communication (UATC) networks [11]. In UATC, the UAVs are flexibly deployed as aerial base stations (BSs) or mobile relays to assist in providing reliable communication services for terrestrial networks [12–14]. By contrast, the UAVs are integrated into the wireless network scenarios as aerial users served by ground BSs in the CEUC system [15]. Owing much to the almost ubiquitous accessibility of the existing LTE (Long Term Evolution) and the forthcoming (beyond) fifth-generation ((B)5G) cellular networks, reliable communications can be supported between UAVs and their corresponding BSs [16,17]. The CEUC is anticipated to have a number of appealing advantages over the existing ground-to-UAV communications, including the ease of monitoring and management, ubiquitous accessibility, robust navigation and enhanced performance, etc. [11]. Despite its merits, the UAV communication based on the future (B)5G cellular networks is more susceptible to suspicious eavesdropping on the ground, which leads to a severe security challenge that is urged to be solved.

Currently, the UAV trajectory design combined with physical-layer security techniques, as a promising solution, has drawn significant attention to safeguard the UAV communication. Specifically, UAV secure communications are studied in [18,19], where the average secrecy rate is significantly improved via optimizing the trajectory of the UAV jointly with the power control for a finite mission duration. As cooperative jamming is one of the important physical-layer techniques that can enhance the secrecy performance, reference [20] proposed to employ a UAV as a friendly mobile jammer, to ensure the secrecy of the ground wiretap channel. In [21], a novel full-duplex operation was applied to the rotary-wing UAV to further improve the energy efficiency (EE) of UAV secrecy communications, and the EE was maximized by the joint optimization of the source transmit/UAV's jamming power and UAV trajectory. A four-node mobile relay and eavesdropper system is proposed in [22], where the UAV was employed as a mobile relay to assist in terrestrial communications. To cope with the non-convex secrecy rate maximization problem, an alternating optimization algorithm is designed by optimizing the power control and UAV trajectory alternatively. The authors in [23,24] proposed a dual-UAV UATC network to enhance the communication quality and improve the secrecy performance, where the downlink transmission from the UAVs is established by adaptively adjusting the UAVs' trajectories and transmit powers. Note that most of the above studies only focus on the security issues in UATC systems. However, how to design efficient anti-eavesdropping methods, to protect legitimate BS-to-UAV transmission in the CEUC networks, has not been investigated, and thus remains a challenging problem to address.

In light of the above, we propose an anti-eavesdropping scheme by employing an aerial UAV jammer in the CEUC network, where one UAV flies to receive confidential messages from a BS while the mobile UAV jammer confuses multiple suspicious eavesdroppers on the ground by sending jamming signals. Specifically, we take into account the joint optimization of both the UAVs' trajectories and the BS's/UAV jammer's power allocation, in order to maximize the average worst-case secrecy rate of the UAV receiver for a given finite period. In the proposed scheme, the UAVs are subject to the practical mobility as well as both the average and peak power constraints. In contrast with the above-mentioned existing works, the UAVs' trajectory design in our proposed CEUC network is particularly important, as the interference from other UAVs cannot be practically cancelled, which causes different objective function and constraints. Therefore, well-designed trajectories of the UAVs can not only avoid severe interference between UAVs, but also provide effective jamming signals to the eavesdroppers, which is expected to notably enhance the secrecy performance. As the formulated optimization problem is non-convex with the objective function as well as its constraints, it is very hard to obtain a globally optimal solution (Since the difficulty of the original problem is NP-hard, it is generally impossible to obtain the globally optimal solution by using the present optimization techniques.) To tackle this challenging problem, we first transform it into a lower bound expression with more tractability. Then, an efficient algorithm is designed by applying the block coordinate descent (BCD) method [25,26]. To be specific, we partition the total optimization variables into four blocks for the two UAVs' trajectories, BS's transmit power, and UAV jammer's jamming power control, respectively. Then, each block is alternatively optimized in each iteration with other blocks being fixed. Although we fix the other three blocks, the corresponding optimization problem remains intractable because of its non-convex. To obtain a high-quality approximately optimal solution, we thus introduce a series of slack variables and apply successive convex approximation (SCA) technique [27,28]. The proposed algorithm has the applicable complexity and guarantees to converge to a locally optimal solution to this problem. To best of knowledge, this is the first work that exploits the anti-eavesdropping UAV trajectory design to solve physical-layer security issues of the CEUC system. The numerical results illustrate that the designed algorithm achieves significantly better secrecy performance than all benchmarks without trajectory or power control design, especially the scheme without the UAV jammer, as in [18].

The rest of this paper is organized as follows. Section 2 gives the system model and problem formulation. In Section 3, a joint optimization algorithm is proposed and its complexity and convergence performance are also analyzed. The simulations are presented in Section 4 to verify the effectiveness of the proposed algorithm. Finally, Section 5 concludes the paper.

2. System Model and Problem Formulation

2.1. System Model

Consider a CEUC network, as shown in Figure 1, where a ground BS transmits confidential messages to a mobile UAV receiver (denoted by U) within a given UAV flight period T, while I malicious eavesdroppers on the ground, denoted by E_i for $i \in \mathcal{I} \triangleq \{1, \cdots, I\}$, intercept the messages from the valid UAV communication. To safeguard the legitimate transmission, the potential eavesdroppers are kept under surveillance by an aerial UAV jammer (denoted by J). The aim of the UAV J is to cooperatively send jamming signals to the eavesdroppers to resist their wiretapping. Notice that if there is no friendly UAV J and only one eavesdropper is considered, the proposed scenario reduces to the goround-to-UAV transmission in [18].

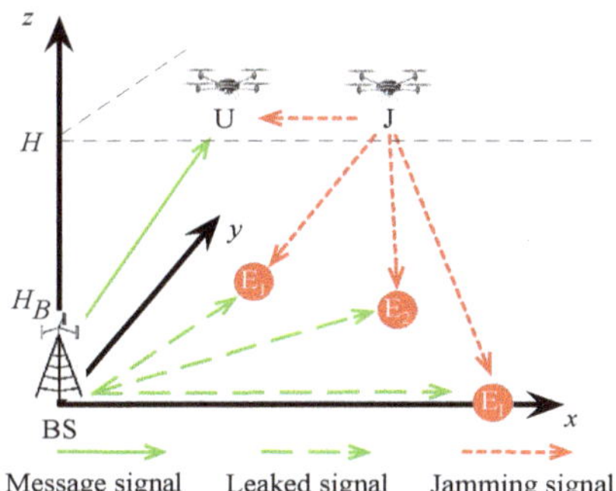

Figure 1. Cellular-enabled UAV secure communication network with aerial cooperative jamming.

Based on the three-dimensional Cartesian coordinate system, we denote $\mathbf{w}_B = [x_B, y_B]^T$ and $\mathbf{w}_{E_i} = [x_{E_i}, y_{E_i}]^T$ as the horizontal coordinates of the BS and E_i, respectively, which are assumed to be fixed and known beforehand to the UAVs. The assumption that $\mathbf{w}_{E_i}$ is known in the network is proper when E_i is an active ground node but untrusted by the UAV [29]. Therefore, E_i can be detected by the synthetic aperture radar or optical camera mounted on the UAV [18]. The initial and final locations of the UAVs are assumed to be pre-specified, which are denoted by $\mathbf{q}_{k,0} = [x_{k,0}, y_{k,0}]^T$ and $\mathbf{q}_{k,F} = [x_{k,F}, y_{k,F}]^T$ for $k \in \{U, J\}$, respectively. To make it more manageable, the period T is partitioned into N equal-length time slots, i.e., $T = \delta_t N$, where δ_t is the length of one time slot. As such, the UAV trajectory in time slot $n \in N$ can be represented approximately by $\mathbf{q}_k[n] \triangleq [x_k[n], y_k[n]]^T$ for $k \in \{U, J\}$, with a fixed altitude H. Let $\Omega = V_{\max}\delta_t$ be the maximum horizontal distance that the UAV can travel in a single time slot, where $V_{\max}$ is the maximum speed of the UAV. Practically, the UAVs should satisfy the following mobility constraints,

$$||\mathbf{q}_k[n+1] - \mathbf{q}_k[n]||^2 \leq \Omega^2, n = 0, \cdots, N-1, \tag{1}$$

$$\mathbf{q}_k[0] = \mathbf{q}_{k,0}, \mathbf{q}_k[N] = \mathbf{q}_{k,F}, \tag{2}$$

$$||\mathbf{q}_U[n] - \mathbf{q}_J[n]||^2 \geq d_{\min}^2, n = 0, \cdots, N, \tag{3}$$

where $d_{\min}$ is the minimum tolerable distance between the two UAVs that ensures the avoidance of a collision.

We assume that the ground-to-UAV and UAV-to-UAV transmissions are mainly governed by LoS channels [18,20,23,24]. Thus, the corresponding channel power gains in time slot n follow the free-space path loss model. (For the purpose of exposition, it is reasonable to assume that the ground-to-UAV follows the free-space LoS channel model when the UAV is deployed in the rural area with sufficiently high altitude. In this case, the probability of Non-LoS state is negligible compared to the dominant LoS state [30]. However, the proposed design is readily extendable to more general channel models in urban areas with Non-LoS effects, e.g., [30].), which are, respectively, given as below,

$$h_{BU}[n] = \rho_0 d_{BU}^{-2}[n] = \frac{\rho_0}{(H - H_B)^2 + ||\mathbf{q}_U[n] - \mathbf{w}_B||^2}, \tag{4}$$

$$h_{JE_i}[n] = \rho_0 d_{JE_i}^{-2}[n] = \frac{\rho_0}{H^2 + ||\mathbf{q}_J[n] - \mathbf{w}_{E_i}||^2}, \tag{5}$$

$$h_{JU}[n] = \rho_0 d_{JU}^{-2}[n] = \frac{\rho_0}{||\mathbf{q}_U[n] - \mathbf{q}_J[n]||^2}, \tag{6}$$

where $d_{BU}[n]$, $d_{JE_i}[n]$ and $d_{JU}[n]$ are the distances from the BS to the UAV U, from the UAV J to the eavesdropper E_i, and between the two UAVs in time slot n, respectively, ρ_0 is the channel power gain at the reference distance $d_0 = 1$ m and H_B is the altitude of the BS. The ground-to-ground transmission is assumed to follow the Rayleigh fading channel. As such, the channel power gain is denoted by

$$h_{BE_i} = \rho_0 \zeta_i d_{BE_i}^{-\kappa} = \frac{\rho_0 \zeta_i}{||\mathbf{w}_B - \mathbf{w}_{E_i}||^\kappa}, \tag{7}$$

where d_{BE_i} is the distance between the BS and the eavesdropper E_i, ζ_i is an exponentially distributed random variable with unit mean representing small-scale Rayleigh fading and $\kappa \geq 2$ is the distance-dependent path loss exponent.

Denote by $P[n]$ and $Q[n]$ the BS's transmit power and the UAV J's jamming power in time slot n, respectively. In practice, they should satisfy the respective average power constraint $\bar{P}$ or $\bar{Q}$, and peak power constraint $\hat{P}$ or $\hat{Q}$, i.e.,

$$\frac{1}{N} \sum_{n=1}^{N} P[n] \leq \bar{P}, 0 \leq P[n] \leq \hat{P}, \tag{8}$$

$$\frac{1}{N} \sum_{n=1}^{N} Q[n] \leq \bar{Q}, 0 \leq Q[n] \leq \hat{Q}, \tag{9}$$

where $\bar{P} \leq \hat{P}$ and $\bar{Q} \leq \hat{Q}$. Then, the achievable rate in bits/second/Hertz (bps/Hz) of the UAV U in time slot n is given by

$$R_U[n] = \log_2 \left(1 + \frac{P[n]h_{BU}[n]}{Q[n]h_{JU}[n] + \sigma^2} \right), \tag{10}$$

where $Q[n]h_{JU}[n]$ is the jamming interference from the UAV J, and σ^2 is the additive white Gaussian noise power at the receivers. Similarly, the achievable rate of the eavesdropper E_i in time slot n can be expressed as

$$
\begin{aligned}
R_{\mathrm{E}_i}[n] &= \mathbb{E}_{\zeta_i}\left[\log_2\left(1 + \frac{P[n]h_{\mathrm{BE}_i}}{Q[n]h_{\mathrm{JE}_i}[n]+\sigma^2}\right)\right] \\
&\leq \log_2\left(1 + \frac{P[n]\rho_0\|\mathbf{w_B}-\mathbf{w}_{\mathrm{E}_i}\|^{-\kappa}}{Q[n]h_{\mathrm{JE}_i}[n]+\sigma^2}\right) \\
&\triangleq \hat{R}_{\mathrm{E}_i}[n],
\end{aligned}
\tag{11}
$$

where $\mathbb{E}_{\zeta_i}[\cdot]$ is the expectation operator with respect to (w.r.t.) ζ_i. Note that $R_{\mathrm{E}_i}[n]$ is replaced by $\hat{R}_{\mathrm{E}_i}[n]$ based on Jensen' inequality and the concavity of $R_{\mathrm{E}_i}[n]$ w.r.t. ζ_i, and $\hat{R}_{\mathrm{E}_i}[n]$ is the largest rate that E_i can achieve. Therefore, in accordance with the theoretical results in [31], the worst-case secrecy rate for each time slot can be lower bounded by

$$
R_{\mathrm{wcs}}[n] = \max(R_{\mathrm{U}}[n] - \max_{i\in\mathcal{I}} \hat{R}_{\mathrm{E}_i}[n], 0).
\tag{12}
$$

Note that by adaptively setting $P[n] = 0$, the optimal solution to (12) is at least to be zero for any time slot n, without violating the power constraint (8). Therefore, the maximum operation can be dropped in the following optimization problems.

2.2. Problem Formulation

In this paper, we aim to maximize the average worst-case achievable secrecy rate from the BS to the UAV U over N time slots, by jointly optimizing the BS's transmit power $\mathbf{P} \triangleq \{P[n], n \in N\}$, the jamming power $\mathbf{Q} \triangleq \{Q[n], n \in N\}$ of the UAV J, and the UAV trajectory $\mathbf{q}_k = \{\mathbf{q}_k[n], n \in N\}$ for $k \in \{U, J\}$. Thus, this optimization problem can be formulated as

$$
\max_{\mathbf{P},\mathbf{Q},\mathbf{q}_U,\mathbf{q}_J} \frac{1}{N} \sum_{n=1}^{N} (R_{\mathrm{U}}[n] - R_{\mathrm{E}}[n])
\tag{13}
$$

$$
\text{s.t. } (1)\text{--}(3), (8)\text{--}(9).
$$

where we let $R_{\mathrm{E}}[n] = \max_{i\in\mathcal{I}} \hat{R}_{\mathrm{E}_i}[n]$, and thus $R_{\mathrm{E}}[n]$ corresponds to the maximum achievable rate among multiple eavesdroppers in time slot n. Optimally solving problem (13) is difficult, in general, due to the following two main reasons: (1) the objective function is not concave w.r.t the corresponding optimization variables even with fixed variables of other blocks and (2) the constraint in (3) is non-convex w.r.t. the UAVs' trajectory variables.

3. Joint Trajectory and Power Control Algorithm

In this section, an efficient algorithm is proposed to obtain the sub-optimal solution to problem (13). Specifically, we cope with problem (13) by solving four subproblems iteratively, i.e., the alternative optimization of the transmit power $\mathbf{P}$, jamming power $\mathbf{Q}$, UAV U's trajectory $\mathbf{q}_U$, and UAV J's trajectory $\mathbf{q}_J$, by fixing the other three optimization variables. Furthermore, the overall algorithm is presented, and its complexity and convergence are analyzed rigorously.

3.1. Transmit Power Optimization

For simplicity, let $a_n = \frac{\gamma_0}{d_{\mathrm{BU}}^2[n](1+Q[n]\gamma_0/\rho_0 d_{\mathrm{JU}}^2[n])}$, and $b_n = \frac{\gamma_0\|\mathbf{w_B}-\mathbf{w_E}\|^{-\kappa}}{1+Q[n]\gamma_0/d_{\mathrm{JE}}^2[n]}$, where $\gamma_0 = \rho_0/\sigma^2$ is the reference signal-to-noise ratio (SNR), and $\mathbf{w_E}$ is denoted as the horizontal location of the eavesdropper that achieves the largest rate and $d_{\mathrm{JE}}^2[n]$ is the distance from the UAV J to the eavesdropper $\mathbf{w_E}$. Thus, with given $\mathbf{Q}$, $\mathbf{q}_U$, and $\mathbf{q}_J$, problem (13) can be simplified as

$$
\max_{\mathbf{P}} \sum_{n=1}^{N} \left[\log_2\left(1 + a_n P[n]\right) - \log_2\left(1 + b_n P[n]\right)\right]
\tag{14}
$$

$$
\text{s.t. } (8).
$$

Based on the result in [18], the close-form solution to this problem is given by: $P^*[n] = \min([\Lambda_n]^+, \hat{P})$ if $a_n > b_n$; otherwise $P^*[n] = 0$, where $\Lambda_n = ((1/2b_n - 1/2a_n)^2 + (1/b_n - 1/a_n)/(\lambda \ln 2))^{\frac{1}{2}} - 1/2a_n - 1/2b_n$. The value of $\lambda \geq 0$ is a constant that ensures constraint (8) is met, which can be obtained cost-effectively via the bisection algorithm [32]. By obtaining the optimal transmit power variables $\mathbf{P}$, they can be seen as the given input for the jamming power optimization problem in the next subsections.

3.2. Jamming Power Optimization

Let $c_n = \frac{P[n]\gamma_0}{d_{\mathrm{BU}}^2[n]}$, $d_n = \frac{\gamma_0}{d_{\mathrm{JU}}^2[n]}$, $e_n = P[n]\gamma_0 ||\mathbf{w_B} - \mathbf{w_E}||^{-\kappa}$ and $f_n = \frac{\gamma_0}{d_{\mathrm{JE}}^2[n]}$. With given $\mathbf{P}$, $\mathbf{q_U}$, and $\mathbf{q_J}$, we can reformulate problem (13) as

$$\max_{\mathbf{Q}} \sum_{n=1}^{N} \left[\log_2 \left(1 + \frac{c_n}{1 + d_n Q[n]} \right) - \log_2 \left(1 + \frac{e_n}{1 + f_n Q[n]} \right) \right] \tag{15}$$

$$\text{s.t. (9).}$$

Problem (15) is a non-convex problem because of the non-convex objective function, which is actually difficult to solve for general N. However, the first term in (15) is convex w.r.t. $Q[n]$, and thus it can be approximated to a convex function within each iteration by applying the SCA method. It is known that the first-order Taylor expansion can be used to obtain the global under-estimator for any convex function at any point [32]. Thus, denoted by $\mathbf{Q}^l = \{Q^l[n], n \in N\}$, the given local point in the l-th iteration, we have

$$\log_2 \left(1 + \frac{c_n}{1 + d_n Q[n]} \right) \geq A_n^l + B_n^l (Q[n] - Q^l[n]) \tag{16}$$

where $A_n^l = \log_2 \left(1 + \frac{c_n}{1 + d_n Q^l[n]} \right)$ and

$$B_n^l = \frac{-c_n d_n}{\ln 2(1 + d_n Q^l[n])(1 + c_n + d_n Q^l[n])}.$$

With (16), problem (15) is lower bounded by the following problem for any given $\mathbf{Q}^l$,

$$\max_{\mathbf{Q}} \sum_{n=1}^{N} \left[B_n^l Q[n] - \log_2 \left(1 + \frac{e_n}{1 + f_n Q[n]} \right) \right] \tag{17}$$

$$\text{s.t. (9).}$$

Observe that this subproblem is concave w.r.t. $Q[n]$ and thus can be solved efficiently by the interior-point method [32]. After solving problem (17), the obtained jamming power $\mathbf{Q}$ serves as the given variables for the trajectory optimization problem of the UAVs.

3.3. Trajectory Optimization of the UAV U

Even with given $\mathbf{P}$, $\mathbf{Q}$, and $\mathbf{q_J}$, it is still hard to achieve the optimal solution to problem (13), due to the non-concavity of the objective function w.r.t. $\mathbf{q_U}$ and the non-convexity of the constraint (3). To tackle this subproblem, we first introduce the slack variables $\alpha = \{\alpha[n] = (H - H_\mathrm{B})^2 + ||\mathbf{q_U}[n] - \mathbf{w_B}||^2, n \in N\}$ and $\beta = \{\beta[n] = ||\mathbf{q_U}[n] - \mathbf{q_J}[n]||^2, n \in N\}$. After some simple transformations, solving problem (13) is equivalent to solve the following problem,

$$\max_{\mathbf{q_U},\alpha,\beta} \sum_{n=1}^{N} \left[\log_2 \left(1 + \frac{P[n]\gamma_0}{\alpha[n]} + \frac{Q[n]\gamma_0}{\beta[n]} \right) - \log_2 \left(1 + \frac{Q[n]\gamma_0}{\beta[n]} \right) \right] \tag{18}$$

$$\text{s.t. } \alpha[n] \geq (H - H_{\mathrm{B}})^2 + ||\mathbf{q}_{\mathrm{U}}[n] - \mathbf{w}_{\mathrm{B}}||^2, \tag{19}$$

$$\beta[n] \leq ||\mathbf{q}_{\mathrm{U}}[n] - \mathbf{q}_{\mathrm{J}}[n]||^2, \tag{20}$$

$$(1) - (3). $$

In fact, if $\alpha[n]$ ($\beta[n]$) is increased (decreased), the objective value of problem (13) will be decreased, and thus the constraints for α and β must satisfy the equalities. Problem (18) is still non-convex, because of the non-convex objective function in (18), and the constraints in (3) and (20). To tackle this difficulty, an important lemma is provided as below.

Lemma 1. *Given $K_1 > 0$ and $K_2 > 0$, the function $f(x,y) = \log_2\left(1 + \frac{K_1}{x} + \frac{K_2}{y}\right)$ is jointly convex w.r.t. $x > 0$ and $y > 0$.*

Proof. See Appendix A. $\square$

Based on Lemma 1, it is easy to prove the convexity of the first term in problem (18). By using the first-order Taylor expansions of a convex function $f(x,y)$ in a neighborhood of $(x,y) = (x_0, y_0)$, i.e., $f(x,y) = f(x_0, y_0) + f_x(x_0, y_0)(x - x_0) + f_y(x_0, y_0)(y - y_0)$, the first term in (18) at given local points denoted by $\boldsymbol{\alpha}^l = \{\alpha^l[n], n \in N\}$ and $\boldsymbol{\beta}^l = \{\beta^l[n], n \in N\}$ in the l-th iteration, can be given as follows,

$$\log_2\left(1 + \frac{P[n]\gamma_0}{\alpha[n]} + \frac{Q[n]\gamma_0}{\beta[n]}\right) \geq \log_2 C_n^l - \frac{D_n^l}{C_n^l \ln 2} \tag{21}$$

where $C_n^l = 1 + \frac{P[n]\gamma_0}{\alpha^l[n]} + \frac{Q[n]\gamma_0}{\beta^l[n]}$ and $D_n^l = P[n]\gamma_0(\alpha^l[n])^{-2}(\alpha[n] - \alpha^l[n]) + Q[n]\gamma_0(\beta^l[n])^{-2}(\beta[n] - \beta^l[n])$. Similarly, by using the first-order Taylor expansion at the given local point denoted by $\mathbf{q}_{\mathrm{U}}^l = \{\mathbf{q}_{\mathrm{U}}^l[n], n \in N\}$ in the l-th iteration, the convex function $||\mathbf{q}_{\mathrm{U}}[n] - \mathbf{q}_{\mathrm{J}}[n]||^2$, w.r.t. $\mathbf{q}_{\mathrm{U}}[n]$ in problem (3) and in problem (20) can be replaced by their convex lower bounds, i.e.,

$$||\mathbf{q}_{\mathrm{U}}[n] - \mathbf{q}_{\mathrm{J}}[n]||^2 \geq ||\mathbf{q}_{\mathrm{U}}^l[n] - \mathbf{q}_{\mathrm{J}}[n]||^2 + 2(\mathbf{q}_{\mathrm{U}}^l[n] - \mathbf{q}_{\mathrm{J}}[n])^T(\mathbf{q}_{\mathrm{U}}[n] - \mathbf{q}_{\mathrm{U}}^l[n]). \tag{22}$$

As a result, by applying SCA technique in each iteration, we approximate the original convex functions to more manageable functions at given local points. Therefore, with (21)–(22), we have the following optimization problem

$$\max_{\mathbf{q}_{\mathrm{U}}, \boldsymbol{\alpha}, \boldsymbol{\beta}} \sum_{n=1}^{N} \left[-\frac{D_n^l}{C_n^l \ln 2} - \log_2\left(1 + \frac{Q[n]\gamma_0}{\beta[n]}\right) \right] \tag{23}$$

$$\text{s.t. } \beta[n] \leq ||\mathbf{q}_{\mathrm{U}}^l[n] - \mathbf{q}_{\mathrm{J}}[n]||^2 + 2(\mathbf{q}_{\mathrm{U}}^l[n] - \mathbf{q}_{\mathrm{J}}[n])^T(\mathbf{q}_{\mathrm{U}}[n] - \mathbf{q}_{\mathrm{U}}^l[n]), \tag{24}$$

$$d_{\min}^2 \leq ||\mathbf{q}_{\mathrm{U}}^l[n] - \mathbf{q}_{\mathrm{J}}[n]||^2 + 2(\mathbf{q}_{\mathrm{U}}^l[n] - \mathbf{q}_{\mathrm{J}}[n])^T(\mathbf{q}_{\mathrm{U}}[n] - \mathbf{q}_{\mathrm{U}}^l[n]), \tag{25}$$

$$(1), (2), (19). $$

It is observed that problem (23) is now convex with all convex constraints. As such, the interior-point method can be used efficiently to solve this problem. Note that the lower bounds obtained by the Taylor expansions suggest that the optimal objective value by solving problem (23) is a lower bound of that of problem (18). In the next subsection, the solved $\mathbf{q}_{\mathrm{U}}$ is input to the trajectory optimization problem of the UAV J as the given variable.

3.4. Trajectory Optimization of the UAV J

With given $\mathbf{P}$, $\mathbf{Q}$, and $\mathbf{q}_U$, we let $\delta = \{\delta[n] = H^2 + ||\mathbf{q}_J[n] - \mathbf{w}_E||^2, n \in N\}$, to tackle the non-concavity of the objective function w.r.t. $\mathbf{q}_J$. Therefore, problem (13) can be rewritten as

$$\max_{\mathbf{q}_J, \beta, \delta} \sum_{n=1}^{N} \left[\log_2 \left(\beta[n] + c_n\beta[n] + Q[n]\gamma_0\right) - \log_2 \left(\beta[n] + Q[n]\gamma_0\right) \right.$$
$$\left. - \log_2 \left(1 + \frac{e_n\delta[n]}{\delta[n] + Q[n]\gamma_0}\right) \right] \tag{26}$$

$$\text{s.t. } \delta[n] \geq H^2 + ||\mathbf{q}[n] - \mathbf{w}_E||^2, \tag{27}$$
$$(1)\text{--}(3), (20).$$

Also, the constraint for δ holds with equalities, otherwise the objective value of problem (13) will be decreased by increasing $\delta[n]$. Similarly, by using the first-order Taylor expansion at given local points denoted by $\delta^l = \{\delta^l[n], n \in N\}$, $\beta^l = \{\beta^l[n], n \in N\}$ and $\mathbf{q}_J^l = \{\mathbf{q}_J^l[n], n \in N\}$ in the l-th iteration, the second and third terms in problem (26), and $||\mathbf{q}_U[n] - \mathbf{q}_J[n]||^2$ in (3) and in (20) can be substituted by their respective concave upper and convex lower bounds, i.e.,

$$\log_2 \left(\beta[n] + Q[n]\gamma_0\right) \leq \log_2 \left(\beta^l[n] + Q[n]\gamma_0\right) + \frac{\beta[n] - \beta^l[n]}{\ln 2(\beta^l[n] + Q[n]\gamma_0)}, \tag{28}$$

$$\log_2 \left(1 + \frac{e_n\delta[n]}{\delta[n] + Q[n]\gamma_0}\right) \leq E_n^l + F_n^l(\delta[n] - \delta^l[n]), \tag{29}$$

where $E_n^l = \log_2 \left(1 + \frac{e_n\delta^l[n]}{\delta^l[n] + Q[n]\gamma_0}\right)$,

$$F_n^l = \frac{e_n\gamma_0 Q[n]}{\ln 2(\gamma_0 Q[n] + (e_n + 1)\delta^l[n])(\gamma_0 Q[n] + \delta^l[n])},$$

and

$$||\mathbf{q}_U[n] - \mathbf{q}_J[n]||^2 \geq ||\mathbf{q}_U[n] - \mathbf{q}_J^l[n]||^2 - 2(\mathbf{q}_U[n] - \mathbf{q}_J^l[n])^T(\mathbf{q}_J[n] - \mathbf{q}_J^l[n]). \tag{30}$$

With problems (28)–(30), we approximate problem, (26) as the following optimization problem

$$\max_{\mathbf{q}_J, \beta, \delta} \sum_{n=1}^{N} \left[\log_2 \left(\beta[n] + c_n\beta[n] + Q[n]\gamma_0\right) - F_n^l\delta[n] - \frac{\beta[n]}{\ln 2(\beta^l[n] + Q[n]\gamma_0)} \right] \tag{31}$$

$$\text{s.t. } d_{\min}^2 \leq ||\mathbf{q}_U[n] - \mathbf{q}_J^l[n]||^2 - 2(\mathbf{q}_U[n] - \mathbf{q}_J^l[n])^T(\mathbf{q}_J[n] - \mathbf{q}_J^l[n]), \tag{32}$$
$$\beta[n] \leq ||\mathbf{q}_U[n] - \mathbf{q}_J^l[n]||^2 - 2(\mathbf{q}_U[n] - \mathbf{q}_J^l[n])^T(\mathbf{q}_J[n] - \mathbf{q}_J^l[n]), \tag{33}$$
$$(1), (2), (27).$$

Problem (31) is now a convex optimization problem that can be cost-effectively solved by the interior-point method. Furthermore, the Taylor expansions in problems (28)–(30) indicate that the objective value of problem (26) is at least the same as that by solving problem (31). Note that all the obtained variables $\mathbf{P}$, $\mathbf{Q}$, $\mathbf{q}_U$, and $\mathbf{q}_J$ are utilized as the given variables for the next iteration.

3.5. Overall Algorithm

In summary, the overall algorithm for obtaining the locally optimal solution to problem (13) is computed by the joint optimization of both the BS's transmit power $\mathbf{P}$ and the UAV J's jamming power

Q as well as the two UAVs' trajectories $\mathbf{q}_U$, and $\mathbf{q}_J$ variables, via alternatively solving subproblems (14), (17), (23) and (31) in an iterative way, respectively. The detailed procedure for solving problem (13) is summarized in Algorithm 1.

In the following, we analyze the computation complexity of Algorithm 1. In each iteration, the BS's transmit power, UAV J's jamming power, and the trajectories of UAVs U and J are optimized in sequence, based on the interior-point method by using existing solvers, such as CVX [33]. Therefore, the complexity for solving the four subproblems can be expressed by $O(\log N)$, $O(N^{3.5}\log(1/\epsilon))$, $O((3N)^{3.5}\log(1/\epsilon))$, and $O((3N)^{3.5}\log(1/\epsilon))$, respectively, for the given solution precision of $\epsilon > 0$ [34]. In addition, as the complexity for updating all variables in BCD iterations is in the order of $\log(1/\epsilon)$, the total computation complexity of the proposed algorithm is $O(N^{3.5}\log^2(1/\epsilon))$. Due to the polynomial time complexity, Algorithm 1 is applicable to the aerial cooperative jamming for cellular-enabled UAV networks.

Algorithm 1 Proposed algorithm for solving problem (13)

1: Initial $\mathbf{P}$, $\mathbf{Q}$, $\mathbf{q}_U$, $\mathbf{q}_J$, α, β and δ. Let $l = 0$.
2: **repeat**
3:　　Solve problem (14) with given $\mathbf{Q}^l$, $\mathbf{q}_U^l$, and $\mathbf{q}_J^l$,
　　　　and denote by $\mathbf{P}^{l+1}$ the optimal solution.
4:　　Solve problem (17) with given $\mathbf{P}^l$, $\mathbf{q}_U^l$, and $\mathbf{q}_J^l$,
　　　　and denote by $\mathbf{Q}^{l+1}$ the optimal solution.
5:　　Solve problem (23) with given $\mathbf{P}^l$, $\mathbf{Q}^l$, $\mathbf{q}_J^l$, α^l and β^l,
　　　　and denote by $\mathbf{q}_U^{l+1}$ the optimal solution.
6:　　Solve problem (31) with given $\mathbf{P}^l$, $\mathbf{Q}^l$, $\mathbf{q}_U^l$, β^l and δ^l,
　　　　and denote by $\mathbf{q}_J^{l+1}$ the optimal solution.
7:　　Update $l = l + 1$.
8: **until** Converge to a pre-specified precision $\epsilon > 0$.

Next, the convergence of Algorithm 1 is discussed as follows. Let $\psi(\mathbf{P}^l, \mathbf{Q}^l, \mathbf{q}_U^l, \mathbf{q}_J^l)$ denote the value of the objective function in problem (13) in the l-th iteration. Then, we have

$$\psi(\mathbf{P}^l, \mathbf{Q}^l, \mathbf{q}_U^l, \mathbf{q}_J^l) \leq \psi_{\mathbf{P}}(\mathbf{P}^{l+1}, \mathbf{Q}^l, \mathbf{q}_U^l, \mathbf{q}_J^l), \tag{34}$$

where $\psi_{\mathbf{P}}(\mathbf{P}^{l+1}, \mathbf{Q}^l, \mathbf{q}_U^l, \mathbf{q}_J^l)$ is defined as the obtained objective value of problem (14) and $\mathbf{P}^{l+1}$ is the optimal solution to problem (14). For the optimization of the jamming power $\mathbf{Q}$, the following equations hold,

$$\begin{aligned}
\psi(\mathbf{P}^{l+1}, \mathbf{Q}^l, \mathbf{q}_U^l, \mathbf{q}_J^l) &\overset{(j_1)}{=} \psi_{\mathbf{Q}}^{\mathrm{lb}}(\mathbf{P}^{l+1}, \mathbf{Q}^l, \mathbf{q}_U^l, \mathbf{q}_J^l) \\
&\overset{(j_2)}{\leq} \psi_{\mathbf{Q}}^{\mathrm{lb}}(\mathbf{P}^{l+1}, \mathbf{Q}^{l+1}, \mathbf{q}_U^l, \mathbf{q}_J^l) \\
&\overset{(j_3)}{\leq} \psi(\mathbf{P}^{l+1}, \mathbf{Q}^{l+1}, \mathbf{q}_U^l, \mathbf{q}_J^l),
\end{aligned} \tag{35}$$

where $\psi_{\mathbf{Q}}^{\mathrm{lb}}$ is denoted as the objective value of problem (17), (j_1) holds since the first-order Taylor expansion in (16) is tight at the local point $\mathbf{Q}^l$ in problem (17), (j_2) satisfies due to the optimal solution $\mathbf{Q}^{l+1}$ to problem (17), and (j_3) is because the computed objective value of problem (17) is lower bounded by that of problem (15). For the two UAVs' trajectories optimization, the similar derivation procedure as in (35) can be used, which are given as below,

$$\psi(\mathbf{P}^{l+1}, \mathbf{Q}^{l+1}, \mathbf{q}_U^l, \mathbf{q}_J^l) = \psi_{\mathbf{q}_U}^{\mathrm{lb}}(\mathbf{P}^{l+1}, \mathbf{Q}^{l+1}, \mathbf{q}_U^l, \mathbf{q}_J^l)$$
$$\leq \psi_{\mathbf{q}_U}^{\mathrm{lb}}(\mathbf{P}^{l+1}, \mathbf{Q}^{l+1}, \mathbf{q}_U^{l+1}, \mathbf{q}_J^l)$$
$$\leq \psi(\mathbf{P}^{l+1}, \mathbf{Q}^{l+1}, \mathbf{q}_U^{l+1}, \mathbf{q}_J^l), \tag{36}$$

$$\psi(\mathbf{P}^{l+1}, \mathbf{Q}^{l+1}, \mathbf{q}_U^{l+1}, \mathbf{q}_J^l) = \psi_{\mathbf{q}_J}^{\mathrm{lb}}(\mathbf{P}^{l+1}, \mathbf{Q}^{l+1}, \mathbf{q}_U^{l+1}, \mathbf{q}_J^l)$$
$$\leq \psi_{\mathbf{q}_J}^{\mathrm{lb}}(\mathbf{P}^{l+1}, \mathbf{Q}^{l+1}, \mathbf{q}_U^{l+1}, \mathbf{q}_J^{l+1})$$
$$\leq \psi(\mathbf{P}^{l+1}, \mathbf{Q}^{l+1}, \mathbf{q}_U^{l+1}, \mathbf{q}_J^{l+1}), \tag{37}$$

With (34)–(37), we finally obtain that

$$\psi(\mathbf{P}^l, \mathbf{Q}^l, \mathbf{q}_U^l, \mathbf{q}_J^l) \leq \psi(\mathbf{P}^{l+1}, \mathbf{Q}^{l+1}, \mathbf{q}_U^{l+1}, \mathbf{q}_J^{l+1}). \tag{38}$$

As a result, Algorithm 1 ensures that the obtained objective value of problem (13) is non-decreasing over the iterations, and thus it guarantees its convergence to the locally optimal solution to problem (13).

4. Numerical Results

In this section, we verify our joint trajectories and powers optimization (denoted as 2T&P) algorithm through simulations. Three benchmark schemes are taken into account as a comparison:

- UAVs' trajectories optimization without power control (denoted as 2T/NP);
- heuristic UAVs' trajectories with power control (2HT/P);
- joint optimization of the UAV U's trajectory and BS's power control without aerial cooperative jamming from the UAV J (denoted as 1T&P), which is identical with the algorithm proposed in [18].

Specifically, the 2T/NP scheme sets the powers of the BS and the UAV U as $P[n] = \bar{P}$ and $Q[n] = \bar{Q}, \forall n$, respectively, and the trajectories of the two UAVs are obtained by solving problems (23) and (31) iteratively until convergence. In the 2HT/P scheme, the UAV U flies directly to the top of the BS at its maximum speed, then stays hovering as long as possible, and finally travels directly to its destination at its maximum speed by the end of T. Different from UAV U, UAV J keeps hovering right above the eavesdropper with the largest achievable rate. Given heuristic trajectories in the 2HT/P, the powers $P[n]$ and $Q[n]$ can be obtained by solving problems (14) and (17), respectively. The initial UAV trajectory for the 2T&P and 2T/NP schemes are constructed by the heuristic UAV trajectories as in 2HT/P. The simulation parameters are specified in Table 1.

We first verify the convergence behaviour of the proposed Algorithm 1 versus the iteration numbers for different T in Figure 2. It is illustrated that the average secrecy rate increases quickly and converges within five iterations, and its performance increases significantly with T. This confirms that a locally optimal solution to problem (13) can be converged by using the proposed algorithm.

Figure 3 illustrates the optimized trajectories of the two UAVs by different schemes when T is sufficiently large, e.g., $T = 300$ s. It is observed that the hovering locations of all algorithms for the UAV U are directly above the BS. This occurs because the locations of the eavesdroppers are not related to the UAV U's trajectory due to the ground-to-air transmission, and thus the UAV U can obtain its maximum achievable rate hovering at the location on top of the BS. In addition, the the trajectories of the UAV U in 2T&P and 2T/NP show the curved paths in order to escape from the unintended interference caused by the UAV J. However, the trajectories of the UAV J present significant different. In particular, for our 2T&P scheme in Figure 3a, the UAV first flies along an arc-like path and reaches a certain point close to the eavesdropper E_1 to avoid a collision with the UAV U; then, it keeps static at this hovering location for a permission period, and finally reaches its destination by the end of T, also in an arc-like path to prevent it causing much interference for the UAV U. Notice that the hovering location of the UAV J is closer to E_1 compared to E_2, as the channel quality of BS-to-E_1 link is much better than that of BS-to-E_2 link. The BS-to-E_2 link can also be degraded if the UAV J can guarantee that the secrecy of the worst-case, i.e., BS-to-E_1 link transmission, by taking advantage of the dominant air-to-ground links. Moreover, at their hovering locations, the UAVs can achieve the better secrecy rate by effectively balancing between enhancing the communication of the ground-to-air link and degrading the quality of the BS-to-E_i channel. In contrast with the 2T&P scheme, we can observe that on its way to the final location, the UAV J flies in a big arc path to keep away from the UAV U in the 2T/NP scheme as shown in Figure 3c. This is because the BS' s transmit power and UAV J's jamming power in 2T/NP are fixed, and thus the UAV J has to fly as far as possible to avoid severe interference with the UAV U over the whole duration, T.

Table 1. Simulation parameters.

Notation	Physical Meaning	Simulation Value
$\mathbf{q}_{U,0}$	Initial horizontal location of the UAV U	$[-800, -200]^T$ m
$\mathbf{q}_{U,F}$	Final horizontal location of the UAV U	$[800, -200]^T$ m
$\mathbf{q}_{J,0}$	Initial horizontal location of the the UAV J	$[-800, 200]^T$ m
$\mathbf{q}_{J,F}$	Final horizontal location of the UAV J	$[800, 200]^T$ m
$\mathbf{w}_B$	Horizontal location of the BS	$[0, 0]^T$ m
$\mathbf{w}_{E_1}$	Location of the first eavesdropper	$[150, 0]^T$ m,
$\mathbf{w}_{E_2}$	Location of the second eavesdropper	$[-200, 0]^T$ m
H	Altitude of UAVs	100 m
H_B	Altitude of BS	10 m
$V_{\max}$	Maximum speed of UAVs	40 m/s
$d_{\min}$	Minimum safe distance between UAVs	20 m
δ_t	Time slot length	1 s
ρ_0	Channel power gain at the reference distance	-60 dB
σ^2	Noise power levels	-110 dBm
$\bar{P}$ and $\hat{P}$	Average and peak power of UAV U	20 dBm and 26 dBm
$\bar{Q}$ and $\hat{Q}$	Average and peak power of UAV J	10 dBm and 16 dBm
ϵ	Accuracy threshold	10^{-4}

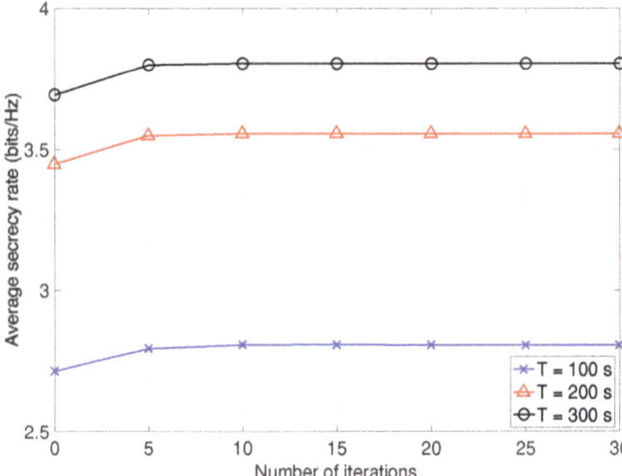

Figure 2. The convergence performance of the proposed Algorithm 1.

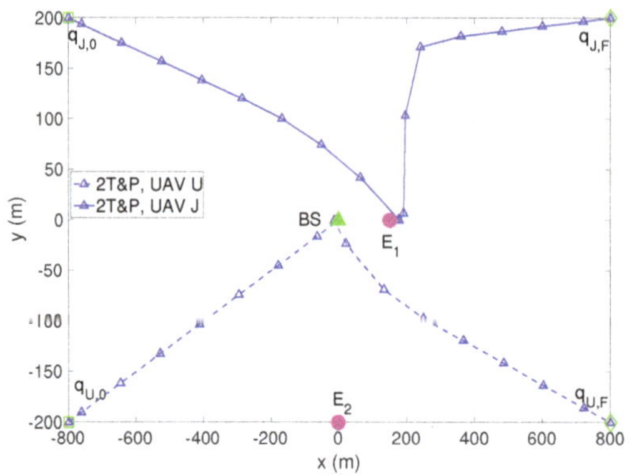

(**a**) UAVs' trajectories for the 2T&P scheme.

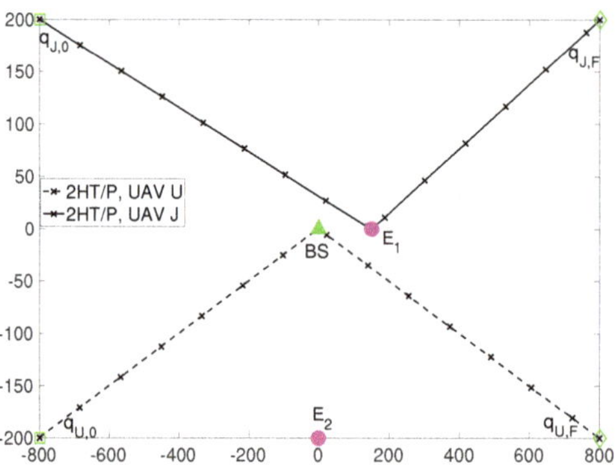

(**b**) UAVs' trajectories for the 2HT/P scheme.

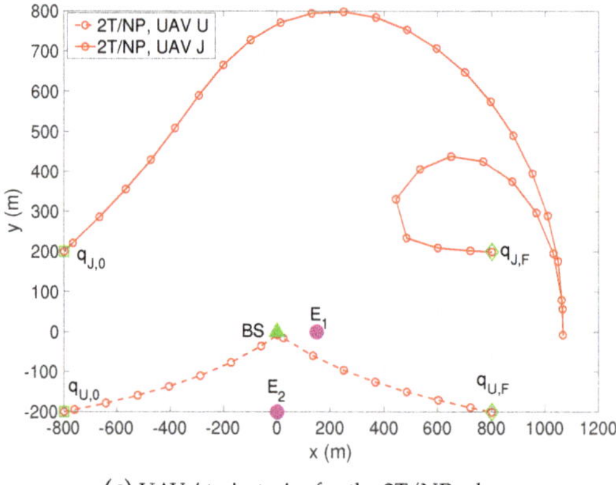

(**c**) UAVs' trajectories for the 2T/NP scheme.

Figure 3. UAVs' trajectories by different schemes for $T = 300$ s. All trajectories are sampled every 5 s. The horizontal locations of the BS, eavesdroppers, UAVs' initial and final locations are marked with ▲, ●, □ and ◇, respectively.

Note that there is a tradeoff between improving the average achievable secrecy rate of the UAV U and avoiding the interference induced by the UAV J. For 2HT/P in Figure 3b, with the pre-specified UAV trajectories, the BS and the UAV jammer can adjust their power allocations to enhance the secrecy performance. Specifically, the BS gradually increases its transmit power before the UAV U flies to its hovering location, while the UAV J properly decreases its jamming power when it reaches above E_1 to suppress the interference to the UAV U. In contrast, a secure communication-aware UAV trajectory design provides additional flexibility to avoid interference between UAVs in our 2T&P scheme. Thus, the UAV J adaptively adjusts its jamming power and trajectory according to the BS's transmit power and the location of the UAV U to further achieve the better secrecy rate.

Figure 4 illustrates the average secrecy rate versus T. It is expected that the average secrecy rates obtained by all schemes raise with T, and the proposed 2T&P scheme significantly outperforms other benchmark schemes owing to its joint optimization. Moreover, the proposed 2T&P scheme provides the significant gain as compared to the scheme in [18], i.e., 1T&P. This indicates that the advantage brought by the aerial cooperative jamming is more effective and important on notably improving the average secrecy rate. However, the 2T/NP presents the worst performance, which demonstrates that the power control also plays a key role in avoiding jamming from other UAVs, which is necessary in our cellular-enabled UAV communication networks with aerial cooperative jamming; otherwise, the secrecy rate can be significantly degraded as shown in Figure 4. Note that we expect that the proposed 2T&P algorithm can still achieve the best secrecy performance via the joint design, even if the number of the eavesdropper increases. This is because the joint design guarantees that the eavesdropper with the best channel condition can be effectively jammed by the UAV J; other eavesdroppers cannot wiretap confidential messages from the BS. The obtained results validate the advantages of introducing aerial cooperative jamming, and the joint optimization of UAV trajectories and power allocations.

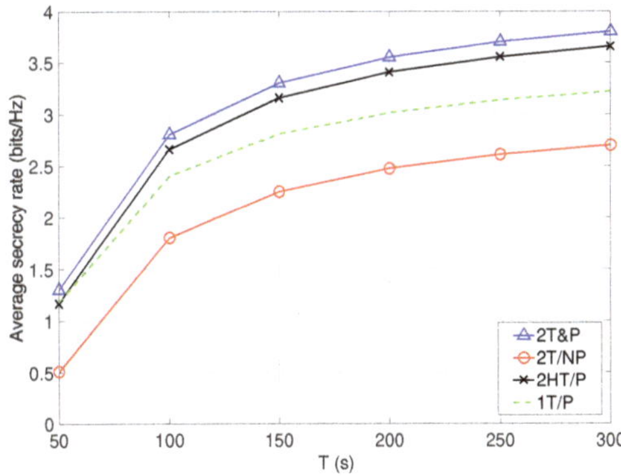

Figure 4. Average secrecy rate versus T with different trajectory and power control designs.

5. Conclusions

Integrating UAVs into the forthcoming 5G cellular networks faces new security challenges. Thus, a new type of cooperative aerial jamming scheme for the cellular-enabled UAV secure communication networks has been investigated in this paper. In particular, the UAV receiver and the UAV jammer cooperate closely with each other to maximize the worst-case average secrecy rate by jointly optimizing their trajectories and the BS/UAV transmit/jamming power. An efficient iterative solution has been proposed to approximately tackle the secrecy rate maximization problem over a given flight period, by means of the BCD and SCA methods. The proposed algorithm is guaranteed to converge to a locally optimal solution with suitable computational complexity. We have demonstrated, by numerical results, that the friendly UAV jammer provides flexible mobility for interference with the ground eavesdroppers, as well as effective power control of preventing it from jamming the UAV receiver, and thereby improves the system secrecy performance. Furthermore, the proposed scheme significantly outperforms the benchmark schemes with simple heuristic trajectories and pre-configured powers. The current scenario can also be extended to the general case with multiple legitimate UAVs, where optimal communication scheduling between the BS and each UAV should be considered. In this case, the design for UAV trajectories needs to avoid collision between UAVs more effectively, and reconciles a tradeoff between maximizing the minimum secrecy rate among multiple UAVs and suppressing the interference from the UAV jammer, which is an interesting problem to be resolved in the future.

Author Contributions: Conceptualization, methodology, and software, H.S. and B.D.; validation and investigation, Z.W. and X.L.; and writing—original draft preparation, B.D.; and writing—review and editing, H.S. and C.G.

Funding: This work was partially supported by the National Natural Science Foundation of China (Nos. 61701064, 71874027), partially by the China Scholarship Council (No. 201808510008), partially by the Sichuan Science and Technology Program (No. 2018GZ0454), partially by the Chengdu Science and Technology Program (No. 2017-RK00-00363-ZF), and partially by the Chongqing Natural Science Foundation (cstc2019jcyj-msxmX0264).

Acknowledgments: The authors the appreciate suggestions of the editors and reviewers, which are of great value to the paper.

Conflicts of Interest: The authors declare no conflicts of interest.

Appendix A. Appendix

We prove Lemma 1 via the definition of convex functions. First, since $f(x,y) = \log_2\left(1 + \frac{K_1}{x} + \frac{K_2}{y}\right)$ where $x > 0$, $y > 0$, $K_1 > 0$ and $K_2 > 0$, the first-order derivatives of $f(x,y)$ w.r.t. x and y are given by

$$f_x(x,y) = -\frac{K_1}{x^2 G \ln 2}, \qquad f_y(x,y) = -\frac{K_2}{y^2 G \ln 2}, \tag{A1}$$

where we let $G = 1 + \frac{K_1}{x} + \frac{K_2}{y}$ for brevity. Then, the Hessian of $f(x,y)$ is

$$\nabla^2 f(x,y) = \begin{bmatrix} \frac{2K_1 xy + 2K_1 K_2 x + K_1^2 y}{x^4 y G^2 \ln 2} & -\frac{K_1 K_2}{x^2 y^2 G^2 \ln 2} \\ -\frac{K_1 K_2}{x^2 y^2 G^2 \ln 2} & \frac{2K_2 xy + 2K_1 K_2 y + K_2^2 x}{xy^4 G^2 \ln 2} \end{bmatrix} \tag{A2}$$

For any $\mathbf{t} = [t_1, t_2]^T$, we have

$$\mathbf{t}^T \nabla^2 f(x,y)\mathbf{t} = \frac{F_1(x,y) + F_2(x,y) + F_3(x,y)}{x^4 y^4 G^2 \ln 2} \geq 0$$

where $F_1(x,y) = 2K_1 t_1^2 xy^3(y + K_2)$, $F_2(x,y) = 2K_2 t_2^2 x^3 y(x + K_1)$ and $F_3(x,y) = (t_1 K_1 y^2 - t_2 K_2 x^2)^2$ for $x > 0$, $y > 0$, $K_1 > 0$ and $K_2 > 0$, which finally leads to the convexity of the function $f(x,y)$.

References

1. Moon, J.; Lee, H.; Song, C.; Lee, S.; Lee, I. Proactive eavesdropping with full-duplex relay and cooperative jamming. *IEEE Trans. Wirel. Commun.* **2018**, *17*, 6707–6719. [CrossRef]
2. Lu, H.; Zhang, H.; Dai, H.; Wu, W.; Wang, B. Proactive eavesdropping in UAV-aided suspicious communication systems. *IEEE Trans. Veh. Technol.* **2019**, *68*, 1993–1997. [CrossRef]
3. Li, Y.; Zhang, Y.; Cai, L. Optimal location of supplementary node in UAV surveillance system. *J. Netw. Comput. Appl.* **2019**, *140*, 23–39. [CrossRef]
4. Zhao, N.; Lu, W.; Sheng, M.; Chen, Y.; Tang, J.; Yu, F.R.; Wong, K. UAV-assisted emergency networks in disasters. *IEEE Wirel. Commun.* **2019**, *26*, 45–51. [CrossRef]
5. Tuna, G.; Nefzi, B.; Conte, G. Unmanned aerial vehicle-aided communications system for disaster recovery. *J. Netw. Comput. Appl.* **2014**, *41*, 27–36. [CrossRef]

6. Peng, K.; Du, J.; Lu, F.; Sun, Q.; Dong, Y.; Zhou, P.; Hu, M. A hybrid genetic algorithm on routing and scheduling for vehicle-assisted multi-drone parcel delivery. *IEEE Access* **2019**, *7*, 49191–49200. [CrossRef]

7. You, C.; Zhang, R. 3D trajectory optimization in Rician fading for UAV-enabled data harvesting. *IEEE Wirel. Commun.* **2019**, *18*, 3192–3207. [CrossRef]

8. Zeng, Y.; Zhang, R.; Lim, T.J. Throughput maximization for UAV-enabled mobile relaying systems. *IEEE Wirel. Commun.* **2016**, *64*, 4983–4996. [CrossRef]

9. Zeng, Y.; Wu, Q.; Zhang, R. Accessing from the Sky: A Tutorial on UAV Communications for 5G and Beyond. 2019. Available online: https://arxiv.org/abs/1903.05289 (accessed on 13 March 2019).

10. Zeng, Y.; Zhang, R.; Lim, T.J. Wireless communications with unmanned aerial vehicles: Opportunities and challenges. *IEEE Commun. Mag.* **2016**, *54*, 36–42. [CrossRef]

11. Zeng, Y.; Lyu, J.; Zhang, R. Cellular-connected UAV: Potential, challenges, and promising technologies. *IEEE Wirel. Commun.* **2019**, *26*, 120–127. [CrossRef]

12. Wu, Q.; Zhang, R. Common throughput maximization in UAV-enabled OFDMA systems with delay consideration. *IEEE Trans. Commun.* **2018**, *66*, 6614–6627. [CrossRef]

13. Wu, Q.; Zeng, Y.; Zhang, R. Joint trajectory and communication design for multi-UAV enabled wireless networks. *IEEE Trans. Wirel. Commun.* **2018**, *17*, 2109–2121. [CrossRef]

14. Lyu, J.; Zeng, Y.; Zhang, R.; Lim, T.J. Placement optimization of UAV-mounted mobile base stations. *IEEE Commun. Lett.* **2017**, *21*, 604–607. [CrossRef]

15. Zhang, S.; Zeng, Y.; Zhang, R. Cellular-enabled UAV communication: A connectivity-constrained trajectory optimization perspective. *IEEE Trans. Commun.* **2019**, *67*, 2580–2604. [CrossRef]

16. Lin, X.; Yajnanarayana, V.; Muruganathan, S.D.; Gao, S.; Asplund, H.; Maattanen, H.; Bergstrom, M.; Euler, S.; Wang, Y.E. The sky is not the Limit: LTE for unmanned aerial vehicles. *IEEE Commun. Mag.* **2018**, *56*, 204–210. [CrossRef]

17. Der Bergh, B.V.; Chiumento, A.; Pollin, S. LTE in the sky: Trading off propagation benefits with interference costs for aerial nodes. *IEEE Commun. Mag.* **2016**, *54*, 44–50. [CrossRef]

18. Zhang, G.; Wu, Q.; Cui, M.; Zhang, R. Securing UAV communications via joint trajectory and power control. *IEEE Trans. Wirel. Commun.* **2019**, *18*, 1376–1389. [CrossRef]

19. Cui, M.; Zhang, G.; Wu, Q.; Ng, D.W.K. Robust trajectory and transmit power design for secure UAV communications. *IEEE Trans. Veh. Technol.* **2018**, *67*, 9042–9046. [CrossRef]

20. Li, A.; Wu, Q.; Zhang, R. UAV-enabled cooperative jamming for improving secrecy of ground wiretap channel. *IEEE Wirel. Commun. Lett.* **2019**, *8*, 181–184. [CrossRef]

21. Duo, B.; Wu, Q.; Zhang, R. Energy Efficiency Maximization for Full-Duplex UAV Secrecy Communication. 2019. Available online: https://arxiv.org/abs/1906.07346 (accessed on 18 June 2019).

22. Wang, Q.; Chen, Z.; Li, H.; Li, S. Joint power and trajectory design for physical-layer secrecy in the UAV-aided mobile relaying system. *IEEE Access* **2018**, *6*, 62849–62855. [CrossRef]

23. Zhong, C.; Yao, J.; Xu, J. Secure UAV communication with cooperative jamming and trajectory control. *IEEE Commun. Lett.* **2019**, *32*, 286–289. [CrossRef]

24. Cai, Y.; Cui, F.; Shi, Q.; Zhao, M.; Li, G.Y. Dual-UAV enabled secure communications: Joint trajectory design and user scheduling. *IEEE J. Sel. Areas Commun.* **2018**, *36*, 1972–1985. [CrossRef]

25. Beck, A.; Tetruashvili, L. On the convergence of block coordinate descent type methods. *SIAM J. Optim.* **2013**, *23*, 2037–2060. [CrossRef]

26. Hong, M.; Razaviyayn, M.; Luo, Z.; Pang, J. A unified algorithmic framework for block-structured optimization involving big data: With applications in machine learning and signal processing. *IEEE Sign. Process. Mag.* **2016**, *33*, 57–77. [CrossRef]

27. Razaviyayn, M.; Hong, M.; Luo, Z. A unified convergence analysis of block successive minimization methods for nonsmooth optimization. *SIAM J. Optim.* **2013**, *23*, 1126–1153. [CrossRef]

28. Zeng, Y.; Zhang, R. Energy-efficient UAV communication with trajectory optimization. *IEEE Trans. Wirel. Commun.* **2017**, *16*, 3747–3760. [CrossRef]

29. Mukherjee, A.; Fakoorian, S.A.A.; Huang, J.; Swindlehurst, A.L. Principles of physical layer security in multiuser wireless networks: A survey. *IEEE Commun. Surv. Tuts* **2014**, *16*, 1550–1573. [CrossRef]

30. You, C.; Zhang, R. Hybrid Offline-Online Design for UAV-Enabled Data Harvesting in Probabilistic LoS Channel. 2019. Available online: https://arxiv.org/abs/1907.06181v1 (accessed on 14 July 2019).

31. Gopala, P.K.; Lai, L.; El Gamal, H. On the secrecy capacity of fading channels. *IEEE Trans. Inf. Theory* **2008**, *54*, 4687–4698. [CrossRef]

32. Boyd, S.; Boyd, S.; Vandenberghe, L.; Press, C.U. *Convex Optimization*; Cambridge University Press: Cambridge, UK, 2004.

33. Grant, M.; Boyd, S. CVX: Matlab Software for Disciplined Convex Programming, Version 2.1. 2018. Available online: http://cvxr.com/cvx (accessed on 14 October 2019).

34. Ben-Tal, A.; Nemirovskiaei, A.S. *Lectures on Modern Convex Optimization: Analysis, Algorithms, and Engineering Applications*; Society for Industrial and Applied Mathematics: Philadelphia, PA, USA, 2001.

Article

A Sensor-Driven Analysis of Distributed Direction Finding Systems Based on UAV Swarms

Zhong Chen [1], Shihyuan Yeh [1], Jean-Francois Chamberland [1] and Gregory H. Huff [2,*]

[1] Department of Electrical and Computer Engineering, Texas A&M University, College Station, TX 77843-3128, USA; zhongchen@tamu.edu (Z.C.); steven.yeh66@gmail.com (S.Y.); chmbrlnd@tamu.edu (J.-F.C.)

[2] School of Electrical Engineering and Computer Science, Pennsylvania State University, University Park, PA 16802, USA

* Correspondence: ghuff@psu.edu; Tel.: +1-814-863-2226

Received: 29 April 2019; Accepted: 8 June 2019; Published: 12 June 2019

Abstract: This paper reports on the research of factors that impact the accuracy and efficiency of an unmanned aerial vehicle (UAV) based radio frequency (RF) and microwave data collection system. The swarming UAVs (agents) can be utilized to create micro-UAV swarm-based (MUSB) aperiodic antenna arrays that reduce angle ambiguity and improve convergence in sub-space direction-of-arrival (DOA) techniques. A mathematical data model is addressed in this paper to demonstrate fundamental properties of MUSB antenna arrays and study the performance of the data collection system framework. The Cramer–Rao bound (CRB) associated with two-dimensional (2D) DOAs of sources in the presence of sensor gain and phase coefficient is derived. The single-source case is studied in detail. The vector-space of emitters is exploited and the iterative-MUSIC (multiple signal classification) algorithm is created to estimate 2D DOAs of emitters. Numerical examples and practical measurements are provided to demonstrate the feasibility of the proposed MUSB data collection system framework using iterative-MUSIC algorithm and benchmark theoretical expectations.

Keywords: direction-of-arrival estimation; unmanned aerial vehicles; UAV swarm; aperiodic arrays; MUSIC; Cramer–Rao bound

---

1. Introduction

DOA estimation of source using array sensors plays an important role in the field of array signal processing. DOA estimation includes one-dimensional (1D) DOA estimation (azimuth) and 2D DOA estimation (azimuth and elevation). The accuracy of DOA estimation is mainly impacted by the algorithm, the geometry of sensor array structure, signal-to-noise (SNR), and snapshots, etc. Algorithms and array geometry are two essential research topics.

The DOA estimation method has been studied for decades and it is still an active research topic in recent years. Initially, DOA estimation based on sensor array structures used the Bartlett beamforming method, but it cannot offer high-resolution due to the Rayleigh limit [1,2]. Then, Burg came up with the maximum entropy (ME) method, which is a high-resolution method, but it has a low robustness and a considerable computation [3]. Later, a series of high-resolution spatial spectrum estimation methods based on decomposition of matrix eigenvectors for array signal processing came out and created a new era. All those spatial spectrum estimation methods were represented by MUSIC and estimation signal parameters via rotational invariance technique (ESPRIT) [4–7]. These two methods have greater resolution and accuracy than other classical methods. The simulations in [8] indicate that MUSIC algorithm is more accurate and stable than ESPRIT algorithm for uniform linear array (ULA). Furthermore, ESPRIT can only be used for invariant geometry, while MUSIC can be applied for arbitrary non-uniform sensor arrays and multiple-source estimation. The MUSIC algorithm attracted more and more attention since it came out, which was the milestone of spatial spectrum estimation methods. In certain conditions, MUSIC algorithm is 1D implementation of maximum likelihood (ML) method, which shares the same characteristic with ML [9,10]. Then, due to the relatively high calculation complexity of MUSIC, some search-free algorithms, such as root-MUSIC, manifold separation based on root-MUSIC, use the root-solving technique to reduce the computational complexity [11,12]. Nevertheless, it can only be used for ULA. In order to deal with arbitrary non-uniform arrays, the Fourier domain root-MUSIC (FD root-MUSIC) algorithm was addressed [13], but the FD root-MUSIC algorithm is mainly used for 1D DOA estimation.

Most of algorithms above are still in simulation and theoretical stages. Recently, practical implementation of the DOA estimation system using field programmable gate array (FPGA) and digital signal processing (DSP) are reported [14–16], those implementations achieve real-time application of DOA estimation based on the MUSIC algorithm. Different from other well-known subspace-based techniques like ESPRIT, MUSIC algorithm has many advantages in the real implementation owing to its simplicity and suitability for parallel processing [16].

Apart from DOA estimation methods, the geometry of the antenna array is also very important. Most of the investigations on the array structure are limited in either 1D linear array or 2D planar array configurations [17]. The 1D array can only detect 1D DOA, and 2D array structures such as uniform rectangular array (URA) and uniform circular array (UCA) can well estimate azimuth angles but cannot well estimate elevation angles due to its small antenna aperture in the elevation direction. In order to improve the elevation angle estimation accuracy, one may develop a three-dimension (3D) array structure by putting more elements in the vertical direction to make a large elevation aperture [18,19]. However, it requires additional hardware and computational cost. Furthermore, the uniform linear and planar array will cause the ambiguity problem (angle aliasing) due to the symmetry of array structures [20–22]. Xia et al. proposed that cubic arrays still have ambiguity problem and the spherical array can significantly reduce this problem [22]. Recently, 3D antenna array configurations have attracted much more research interest in array signal processing [22–28]. Most of those 3D arrays above are constructed from regular structure (i.e., cubic, cylinder), extending the planar array (i.e., URA, UCA) or configuring the virtual 3D array based on the planar array. Even though those special 3D arrays increase the elevation angle estimation accuracy, their array apertures are still relatively small since the physical size of static arrays are restricted. Moreover, conventional investigations in the DOA estimation require that the number of sensors is more than the number of receiving signals, which increase hardware cost and system complexity.

In contrast, in order to investigate the compromise between hardware cost and signal processing time, time-variant arrays whose element positions are changed over time are examined by time-divided sampling rather than simultaneously sampling as static array. Many researchers have reported using time-variant arrays to improve DOA estimation performance [29–35]. Instead of using a set of different elements to process the incident signals, the time-variant array can only use one or a small number of moving elements to construct virtual antenna arrays. Wan et al. proposed a method of combining the characteristics of arbitrary virtual baseline to construct virtual 3D array [29]. However, the number of sub-array elements is too small so that it cannot have high resolution. Liu examined a rotating long baseline interferometer whose length is much larger than one wavelength to estimate 2D DOAs by constructing virtual 2D circular arrays [30]. However, the 2D circular array has limited elevation aperture and still cannot well estimate the elevation angle.

The MUSB aperiodic array reconstructed from swarming UAVs proposed herein has aperture dimensions in both azimuth and elevation directions, which increase the accuracy of both azimuth and elevation angle estimation. Corner and Lamont proposed a parallel simulation of UAV swarm scenarios [36], and Saad et al. reported a testbed of vehicle swarm rapid prototyping [37]. Recently, many researchers investigated the methods and impact factors of designing the robust MUSB antenna arrays for signal collection platforms [38–41]. However, those works are limited in MUSB array constructing investigations including UAV positional precision, turbulence of the environment, micro-UAV swarm algorithm, and swarm-based real-time data collection. In this paper, 2D DOA estimation using the

MUSB aperiodic array is provided, a mathematical model of the MUSB data collection system for signal processing is offered firstly and the impact of associated parameters on DOA estimation accuracy and convergence in this model are analyzed. The MUSB arrays have advantages of large aperture, large interelement spacing, no shadowing effect, low mutual coupling effect and large spatial sampling data from different locations in free space.

In practice, the MUSB system for DOA estimation requires low snapshots and might be applied in low SNR scenarios. However, the subspace-based techniques require adequate SNR and snapshots to guarantee good performance. We utilize the iterative method to lower the noise floor by multiplying the current MUSIC spectrum and previous spectrum for each iteration. The details of the iterative-MUSIC algorithm will be presented in Section 5.

This paper mainly contributes to the array signal processing area in three aspects. First, the mathematical model of the MUSB aperiodic array data collection system is introduced to demonstrate the fundamental DOA estimation impact factors and performance. Second, the CRB associated with DOAs in the presence of the gain and phase coefficient in the system is derived to reveal some direction-finding properties such as the global convergence, snapshots, SNR, and the number of arrays. Third, a successive DOA refinement procedure with iterative-MUSIC algorithm is provided based on the reconstructed arrays and spectrum to meet the requirement of high-precision DOA estimation. The rest of this paper mainly consists of six sections. Section 2 introduces the MUSB mathematical model; Section 3 derives the CRB associated with source DOAs in the presence of the sensor gain and phase coefficient; Section 4 proposes performance analysis of the MUSB system for the single-emitter case; Section 5 gives the algorithm applied in this paper; Section 6 provides simulation and measurement results in different scenarios and Section 7 concludes the paper.

Glossary of notations is listed below: $C^{k \times p}$ = the space of $k \times p$ complex-valued matrices; E = expectation operator; A_{ij} = the i, j element of a general matrix $A \in C^{k \times p}$; A^T = the transpose of $A \in C^{k \times p}$; A^H = the conjugate transpose of $A \in C^{k \times p}$; Re (A) = the real part of $A \in C^{k \times p}$; Im (A) = the image part of $A \in C^{k \times p}$; tr (A) = the trace of $A \in C^{k \times k}$; det (A) = the determinant of $A \in C^{k \times k}$; $A \odot B$ = the Schur–Hadamard matrix product of $A, B \in C^{k \times p}$, defined by $[A \odot B]_{ij} = A_{ij}B_{ij}$; $A \otimes B$ = the Kronecker matrix product of $A, B \in C^{k \times p}$, defined by

$$A \otimes B = \begin{bmatrix} a_{11}B & \cdots & a_{1j}B \\ \vdots & \ddots & \vdots \\ a_{i1}B & \cdots & a_{ij}B \end{bmatrix} \tag{1}$$

$z \sim cN(\mu(\alpha), \zeta(\alpha))$ = the complex Gaussian distribution of the complex random vector z with mean μ and variance ζ, and α is a real-valued parameter vector that completely and uniquely specifies the distribution of z (see [42]).

2. Problem Formulation

2.1. Swarming UAV Synthetic Aperture

A swarming UAV synthetic aperture was presented in our early published paper [43]. Figure 1 shows a graphical representation of a UAV swarm as it morphs in time (iteration I in this paper). Each of the M agents in the swarm has a location, orientation, and trajectory. Notionally, these have position $P_{m,i}(r, \theta, \phi)$, where m is the agent's index and i is the iteration. During swarming, the agents undergo rotations and translations, where a dual quaternion framework provides a convenient mechanism to handle this behavior. This motion rotates the agents' local (u, v, w) coordinate systems that describe the spatial orientation of their antenna radiation pattern with respect to (w.r.t.) the global coordinate system and the direction to the incoming signal of interest $S(\theta_n, \phi_n)$, which is the nth source. The collection of these measurements over iteration creates a synthetic aperture that can be used to calculate the parameters of interest (θ_n, ϕ_n). Notionally, K is independent data sampled for each agent in each iteration and K is usually called a "snapshot".

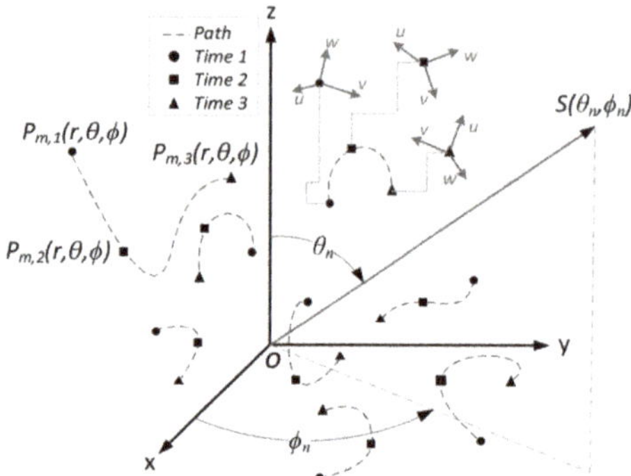

Figure 1. Morphing micro-UAV swarm-based (MUSB) antenna array configuration.

2.2. Signal Model

Friedlander and Weiss presented a mutual coupling model in the presence of sensor mutual coupling, gain, and phase uncertainties [44]. We neglect the mutual coupling effect in the signal model since the spacing of aperiodic array reconstructed from swarming UAVs is expected to be much larger than a wavelength. Consider that an arbitrary array of M elements receive N uncorrelated incident signals in the far-field demonstrated in part 1. Thus, the received signal for the i-th iteration can be represented as

$$X_i(k) = \Gamma_i \cdot \widetilde{\gamma}_i \cdot S_i(k) + W_i(k) \quad k = 1,2,\cdots,K; \quad i = 1,2,\cdots,I \tag{2}$$

where $X_i(k) = [X_{1,1}(k),\cdots, X_{1,M}(k),\cdots,X_{i,1}(k), \cdots, X_{i,M}(k)]^T$, $S_i(k) = [S_{1,1}(k),\cdots,S_{1,N}(k),\cdots,S_{i,1}(k),\cdots,S_{i,N}(k)]^T$, $W_i(k) = [W_{1,1}(k), \cdots, W_{1,M}(k),\cdots,W_{i,1}(k),\cdots, W_{i,M}(k)]^T$, $\Gamma_i = diag\{g_{1,1}e^{-j\omega_0\psi_{1,1}},\cdots, g_{1,M}e^{-j\omega_0\psi_{1,M}},\cdots, g_{i,1}e^{-j\omega_0\psi_{i,1}}, \cdots, g_{i,M}e^{-j\omega_0\psi_{i,M}}\}$, $\widetilde{A}_{i,mn} = e^{-j\omega_0\tau_{i,mn}}$ and $m = 1,2,\cdots, M; n = 1,2,\cdots, N; i = 1,2,\cdots, I$. Then,

$$X(k) = \sum_{i=1}^{I} X_i(k) = \sum_{i=1}^{I}[\Gamma_i \cdot \widetilde{\gamma}_i \cdot S_i(k) + W_i(k)] \tag{3}$$

therefore, $S(k) = \sum_{i=1}^{I} S_i(k), W(k) = \sum_{i=1}^{I} W_i(k), \Gamma = \sum_{i=1}^{I} \Gamma_i$, and $\widetilde{A} = \sum_{i=1}^{I} \widetilde{A}_i$. Note that the sensor gain $g_{i,m}$ and phase $\psi_{i,m}$ change w.r.t. element location based on orientation of UAV; $\tau_{i,mn}$ changes w.r.t. location; S_i is constant; W_i may change w.r.t. location, velocity of UAV, and environment.

Since the sources we consider here are in the far field from the observing array. It is easy to find that $\tau_{i,mn}$ can be represented by $\tau_{i,mn} = -d_{i,mn}/c$, and then

$$d_{i,mn} = x_{i,m} \sin\theta_n \cos\phi_n + y_{i,m} \sin\theta_n \sin\phi_n + z_{i,m} \cos\theta_n \tag{4}$$

where $d_{i,mn}$ is the distance from origin (reference sensor) of the coordinate to the m-th sensor in the direction of the nth source for the i-th iteration, c is the propagating velocity in free space, $(x_{i,m}, y_{i,m}, z_{i,m})$ are the coordinates of the m-th sensor for the i-th iteration, (θ_n, ϕ_n) are the DOAs of the n-th source in the sphere coordinate. From Equations (5), the matrix $\widetilde{A}$ can be obtained by

$$\widetilde{A}_{i,mn} = e^{j(\omega_0/c)(x_{i,m}\sin\theta_n\cos\phi_n + y_{i,m}\sin\theta_n\sin\phi_n + z_{i,m}\cos\theta_n)} = e^{j(2\pi/\lambda)(x_{i,m}\sin\theta_n\cos\phi_n + y_{i,m}\sin\theta_n\sin\phi_n + z_{i,m}\cos\theta_n)} \tag{5}$$

where λ is wavelength.

3. The CRB

3.1. Swarming UAV Synthetic Aperture

In theory, for a static array, the steering vector $\widetilde{A}$ is considered invariant over different snapshots since the array geometry is invariant. Assume $z \sim cN(\mu(\alpha), \zeta(\alpha))$, then, the *m,n*-th general formula of the Fisher information matrix (FIM) on the covariance matrix of any unbiased estimate of α is:

$$F_{mn} = tr\left[\zeta^{-1}(\alpha)\frac{\partial\zeta(\alpha)}{\partial\alpha_m}\zeta^{-1}(\alpha)\frac{\partial\zeta(\alpha)}{\partial\alpha_n}\right] + 2\mathrm{Re}\left\{\frac{\partial\mu(\alpha)}{\partial\alpha_m}\zeta^{-1}(\alpha)\frac{\partial\mu(\alpha)}{\partial\alpha_n}\right\} \tag{6}$$

where α_m denotes the *m*-th component of α. The general formula has been presented in [45] and proved in [46].

Petre and Nehorai presented the deterministic and stochastic CRB in [46]. For deterministic CRB, the parameters, mean, and variance of the complex distribution are given by $\alpha = \left\{\theta, \{\mathrm{Re}[s[t]], \mathrm{Im}[s[t]]\}_{k=1}^{K}, \sigma^2\right\}$, $\mu(\alpha) = [As[k]]_{k=1}^{K}$, $\zeta(\alpha) = block - diag(\sigma^2 I)$. Then, the *m,n*-th FIM is

$$F_{mn} = K\frac{\mu}{\sigma^4}\frac{\partial\sigma^2}{\partial\alpha_m}\frac{\partial\sigma^2}{\partial\alpha_n} + \frac{2}{\sigma^2} \cdot \sum_{k=1}^{K}\left[\frac{\partial}{\partial\alpha_m}As[k]\right]^{H}\left[\frac{\partial}{\partial\alpha_n}As[k]\right] \tag{7}$$

For stochastic CRB, the parameters, mean and variance of the complex distribution are given by $\alpha = \left\{\theta, \{\mathrm{Re}\{P_{mn}\}, \mathrm{Im}\{P_{mn}\}\}_{m,n=1}^{K}, \sigma^2\right\}$, $\mu(\alpha) = 0$, $\zeta(\alpha) = block - diag(R)$. Then, the *m,n*-th FIM is

$$F_{mn} = K \; tr\left[R^{-1}(\alpha)\frac{\partial R(\alpha)}{\partial\alpha_m}R^{-1}(\alpha)\frac{\partial R(\alpha)}{\partial\alpha_n}\right] \tag{8}$$

Since the signals estimated cannot be known completely and the signals in the practical are stochastic, this paper considers the stochastic CRB.

3.2. CRB for the UAV Swarming System

Before deriving the CRB, we assume that both incident signals and noise are stationary and the ergodic complex Gaussian random process with zero mean and nonsingular covariance matrix is uncorrelated with each other. The columns of $A = \Gamma\widetilde{A}$ are linearly independent. An additional assumption is that the number of array elements reconstructed from swarming UAVs is greater than the number of sources. Therefore, the matrix of the steering vector has a full column rank.

The covariance matrices of signal, noise and observation vectors for *i*-th iteration are given by

$$\begin{aligned}
P_i &= E\left[S_i S_i^{H}\right], \sigma_i^2 I_0 = E\left[W_i W_i^{H},\right] \\
R_i &= E\left[X_i(k)X_i^{H}(k)\right] = \Gamma_i\widetilde{A}_i P_i\widetilde{A}_i^{H}\Gamma_i^{H} + \sigma_i^2 I_0 = A_i P_i A_i^{H} + \sigma_i^2 I_0
\end{aligned} \tag{9}$$

where $\widetilde{A}_i$ is the steering vector for the *i*-th iteration and $A_i \triangleq \Gamma_i\widetilde{A}_i$. It is useful to observe that if we let the sample data covariance matrix $\hat{R}_i = \frac{1}{K}\sum_{k=1}^{K}X_i(k)X_i^{H}(k)$, and $\hat{R} = \frac{1}{I}\sum_{i=1}^{I}\hat{R}_i$, then $\lim_{K\to\infty}\hat{R}_i = R_i$, and $\lim_{I\to\infty}\hat{R} = R$.

The log-likelihood function of K independent samples in *i*-th iteration of a zero-mean complex Gaussian random process $X_i(k)$ whose statistics depend on a parameter vector α is given by

$$\begin{aligned}
L_i(\alpha) &= -K\ln[\det[R_i\pi]] - \sum_{k=1}^{K}x_i^{H}(k)R_i^{-1}x_i(k) \\
&= Z - K\ln[\det[R_i]] - \sum_{k=1}^{K}x_i^{H}(k)R_i^{-1}x_i(k)
\end{aligned} \tag{10}$$

where Z denotes the constant term of the log-likelihood function, det(R) represents the determinant of the matrix R, and R_i is the time-varying covariance matrix w.r.t. the iteration. Thus, the log-likelihood function of $X(k)$ is:

$$
\begin{aligned}
L(\alpha) &= -K \sum_{i=1}^{I} \ln[\det(R_i \pi)] - \sum_{i=1}^{I} \sum_{k=1}^{K} x_i^{H}(k) R_i^{-1} x_i(k) \\
&= \sum_{i=1}^{I} \left\{ Z - K \ln[\det(R_i)] - \sum_{k=1}^{K} x_i^{H}(k) R_i^{-1} x_i(k) \right\} = \sum_{i=1}^{I} L_i(\alpha)
\end{aligned}
\tag{11}
$$

Therefore, the *m,n*-th elements of FIM are given by

$$
F_{mn} = -E\left[\frac{\partial^2 L(\alpha)}{\partial \alpha_m \partial \alpha_n} \right] = -\sum_{i=1}^{I} E\left[\frac{\partial^2 L_i(\alpha)}{\partial \alpha_m \partial \alpha_n} \right] = \sum_{i=1}^{I} \left\{ K \, tr\left[R_i^{-1}(\alpha) \frac{\partial R_i(\alpha)}{\partial \alpha_m} R_i^{-1}(\alpha) \frac{\partial R_i(\alpha)}{\partial \alpha_n} \right] \right\} = \sum_{i=1}^{I} F_{i,mn}
\tag{12}
$$

It follows that the FIM's submatrix F_{mn} for the UAV swarming data collecting system can be obtained by summing the single-iteration $F_{i,mn}$ of FIM over the iterations. Furthermore, the FIM's submatrix $F_{i,mn}$ can be obtained by multiplying the single-snapshot $F_{i,mn}$ and the number of snapshots. Thus, we only need to know the single-snapshot $F_{i,mn}$ for the i-th iteration. The problem of the major interest is the estimation of the incident angles of the sources. Expression of CRB for 1D DOA of each iteration for the present problem is listed in [47]. The CRB of 2D DOA estimation with arbitrary array for the i-th iteration, which can be considered as an arbitrary static array, presented in this paper is given in the Appendix A.

4. Analysis of Single-Emitter Case

In this section, we investigate more details of the MUSB data collecting system with single-emitter case. The unknown parameters we consider here are the 2D DOAs (θ, ϕ). Assume the source variance is P, the noise variance is σ^2, the snapshot for each iteration is K, and the iteration is I. From the Appendix A, we have the formula of FIM w.r.t. 2D DOAs in the i-th iteration of the MUSB system.

$$
F_{i,s,2} = \begin{bmatrix} F_{i,\theta\theta} & F_{i,\theta\phi} \\ F_{i,\phi\theta} & F_{i,\phi\phi} \end{bmatrix} = \frac{2K}{\sigma^2} \left\{ \mathrm{Re}\left[D_i^{H} A_i^{\perp} D_i \odot 1_2 1_2^{T} \otimes U_i \right] \right\}
\tag{13}
$$

The *m*th element of the steering vector for the i-th iteration is given by

$$
a_{i,m}(\theta, \phi) = \gamma_{i,m} \cdot \exp\left[j\frac{2\pi}{\lambda}(x_{i,m} \sin\theta \cos\phi + y_{i,m} \sin\theta \sin\phi + z_{i,m} \cos\theta) \right]
\tag{14}
$$

where $\gamma_{i,m}$ is the gain and phase parameter and can be represented as $\gamma_{i,m} = g_{i,m} e^{-j\omega_0 \psi_{i,m}}$. Taking the derivative of $a_i(\theta, \phi)$ w.r.t. θ and ϕ for the i-th iteration, we obtain

$$
d_{i,\theta} = \frac{d}{d\theta} a_i(\theta, \phi) = j\frac{2\pi}{\lambda} b_{i,m} \odot a_i(\theta, \phi), \quad d_{i,\phi} = \frac{d}{d\phi} a_i(\theta, \phi) = j\frac{2\pi}{\lambda} q_{i,m} \odot a_i(\theta, \phi)
\tag{15}
$$

where $b_{i,m} = x_{i,m} \cos\theta \cos\phi + y_{i,m} \cos\theta \sin\phi - z_{i,m} \sin\theta, q_{i,m} = -x_{i,m} \sin\theta \sin\phi + y_{i,m} \sin\theta \cos\phi$. Note

$$
A_i^{H} A_i = \sum_{m=1}^{M} g_{i,m}^2 = G_{i,M}
\tag{16}
$$

Therefore, it is straightforward to verify that

$$
D_i^{H}(\theta) D_i(\theta) = \left(-j\frac{2\pi}{\lambda} b_i A_i^{H} \right)\left(j\frac{2\pi}{\lambda} b_i A_i \right) = \frac{4\pi^2}{\lambda^2} \sum_{m=1}^{M} b_{i,m}^2 g_{i,m}^2
\tag{17}
$$

$$D_i^H(\phi)D_i(\phi) = \left(-j\frac{2\pi}{\lambda}q_i A_i^H\right)\left(j\frac{2\pi}{\lambda}q_i A_i\right) = \frac{4\pi^2}{\lambda^2}\sum_{m=1}^{M} q_{i,m}^2 g_{i,m}^2 \tag{18}$$

$$D_i^H(\theta)D_i(\phi) = \left(-j\frac{2\pi}{\lambda}b_i A_i^H\right)\left(j\frac{2\pi}{\lambda}q_i A_i\right) = \frac{4\pi^2}{\lambda^2}\sum_{m=1}^{M} b_{i,m}q_{i,m}g_{i,m}^2 = D_i^H(\phi)D_i(\theta) \tag{19}$$

Substituting Equations (14), (15), and (17), we can obtain

$$\begin{aligned}
D_i^H(\theta)A_i^{\perp}D_i(\theta) &= D_i^H(\theta)\left(I_0 - A_i\left(A_i^H A_i\right)^{-1}A_i^H\right)D_i(\theta) = D_i^H(\theta)D_i(\theta) - \frac{D_i^H(\theta)A_i A_i^H D_i(\theta)}{G_{i,M}} \\
&= \frac{4\pi^2}{\lambda^2}\sum_{m=1}^{M} b_{i,m}^2 g_{i,m}^2 - \frac{4\pi^2}{\lambda^2}\frac{1}{G_{i,M}}\left(\sum_{m=1}^{M} b_{i,m}g_{i,m}^2\right)^2 = \frac{4\pi^2}{\lambda^2}\left\{\sum_{m=1}^{M} b_{i,m}^2 g_{i,m}^2 - \frac{1}{G_{i,M}}\left(\sum_{m=1}^{M} b_{i,m}g_{i,m}^2\right)^2\right\}
\end{aligned} \tag{20}$$

Using the same derivative procedure, we can obtain

$$D_i^H(\phi)A_i^{\perp}D_i(\phi) = \frac{4\pi^2}{\lambda^2}\left\{\sum_{m=1}^{M} q_{i,m}^2 g_{i,m}^2 - \frac{1}{G_{i,M}}\cdot\left(\sum_{m=1}^{M} q_{i,m}g_{i,m}^2\right)^2\right\} \tag{21}$$

$$D_i^H(\theta)A_i^{\perp}D_i(\phi) = \frac{4\pi^2}{\lambda}\left\{\sum_{m=1}^{M} b_{i,m}q_{i,m}g_{i,m}^2 - \frac{1}{G_{i,M}}\cdot\sum_{m=1}^{M} b_{i,m}g_{i,m}^2\sum_{m=1}^{M} q_{i,m}g_{i,m}^2\right\} = D_i^H(\theta)A_i^{\perp}D_i(\phi) \tag{22}$$

Furthermore,

$$U_i = PA_i^H R_i^{-1}A_i P = P^2 A_i^H\left(A_i^H PA_i + \sigma^2 I_0\right)^{-1}A_i = P^2 G_{i,M}\left(PG_{i,M} + \sigma^2\right)^{-1} = \frac{P^2 G_{i,M}}{PG_{i,M} + \sigma^2} \tag{23}$$

Therefore,

$$F_{i,\theta\theta} = \frac{8K\pi^2}{\lambda^2\sigma^2}\frac{P^2 G_{i,M}^2}{PG_{i,M} + \sigma^2}\left\{\frac{1}{G_{i,M}}\sum_{m=1}^{M} b_{i,m}^2 g_{i,m}^2 - \left\{\frac{1}{G_{i,M}}\sum_{m=1}^{M} b_{i,m}g_{i,m}^2\right\}^2\right\} \tag{24}$$

$$F_{i,\phi\phi} = \frac{8K\pi^2}{\lambda^2\sigma^2}\frac{P^2 G_{i,M}^2}{PG_{i,M} + \sigma^2}\left\{\frac{1}{G_{i,M}}\sum_{m=1}^{M} q_{i,m}^2 g_{i,m}^2 - \left\{\frac{1}{G_{i,M}}\sum_{m=1}^{M} q_{i,m}g_{i,m}^2\right\}^2\right\} \tag{25}$$

$$F_{i,\theta\phi} = \frac{8K\pi^2}{\lambda^2\sigma^2}\frac{P^2 G_{i,M}^2}{PG_{i,M} + \sigma^2}\left\{\frac{1}{G_{i,M}}\sum_{m=1}^{M} b_{i,m}q_{i,m}g_{i,m}^2 - \left\{\frac{1}{G_{i,M}^2}\sum_{m=1}^{M} b_{i,m}g_{i,m}^2\sum_{m=1}^{M} q_{i,m}g_{i,m}^2\right\}\right\} = F_{i,\phi\theta} \tag{26}$$

Assume

$$B_i = \frac{1}{G_{i,M}}\sum_{m=1}^{M} b_{i,m}^2 g_{i,m}^2 - \left(\frac{1}{G_{i,M}}\sum_{m=1}^{M} b_{i,m}g_{i,m}^2\right)^2, \; Q_i = \frac{1}{G_{i,M}}\sum_{m=1}^{M} q_{i,m}^2 g_{i,m}^2 - \left(\frac{1}{G_{i,M}}\sum_{m=1}^{M} q_{i,m}g_{i,m}^2\right)^2 \tag{27}$$

$$V_i = \frac{1}{G_{i,M}}\sum_{m=1}^{M} b_{i,m}q_{i,m}g_{i,m}^2 - \left(\frac{1}{G_{i,M}^2}\sum_{m=1}^{M} b_{i,m}g_{i,m}^2\sum_{m=1}^{M} q_{i,m}g_{i,m}^2\right), \; C_i = \frac{G_{i,M}^2 SNR^2}{G_{i,M}SNR + 1} \tag{28}$$

where $SNR = P/\sigma^2$. Thus, summing over iteration, we obtain the CRB

$$CRB = \begin{bmatrix} F_{\theta\theta} & F_{\theta\phi} \\ F_{\phi\theta} & F_{\phi\phi} \end{bmatrix}^{-1} \tag{29}$$

where $F_{\theta\theta} = \frac{8K\pi^2}{\lambda^2}\sum_{i=1}^{I} C_i B_i$, $F_{\phi\phi} = \frac{8K\pi^2}{\lambda^2}\sum_{i=1}^{I} C_i Q_i$, $F_{\theta\phi} = \frac{8K\pi^2}{\lambda^2}\sum_{i=1}^{I} C_i V_i = F_{\phi\theta}$. If we ignore the sensor orientation (i.e., $\Gamma = 1$) and let $K = 1$, then $A_i^H A_i = M$ and the FIM w.r.t. 1-D DOA is given by

$$F_\theta = \frac{8\pi^2}{\lambda^2}\frac{M^2 SNR^2}{M SNR + 1}B \tag{30}$$

where $B = \sum_{i=1}^{I} B_i$ and $B_i = \frac{1}{M}\sum_{m=1}^{M} h_{i,m}^2 - \left(\frac{1}{M}\sum_{m=1}^{M} b_{i,m}\right)^2$, which coincides with the results in [4].

5. The Algorithm

5.1. UAV Parameters

Assume there are M swarming UAVs and each UAV swarms in a cylinder region (radius r, $h = 2r$). Each UAV has an initial location (x, y, z) in the swarming region with a vector velocity $\vec{V}_{i,m}$ in each iteration and the initial locations of the M UAVs are considered as the first iteration. The swarming short distance in each iteration is represented as a vector $\vec{d}_{i,m}$. The scalar quantity can be represented as d, which uses the following relation:

$$d = \beta_t \cdot \lambda / \rho \tag{31}$$

where β_t is a uniformly distributed random number matrix between zero and one, λ is wavelength and p is the coefficient determining the distribution mean of swarm distance. Figure 2 shows the distribution of short distance with mean $\mu = 0.25$ wavelength. Here, the range of d also depends on the speed of UAV and data sampling interval and can be configured by the customer. We can obtain

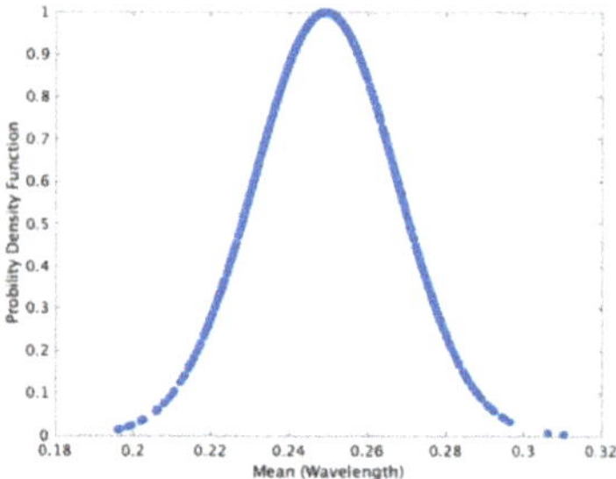

Figure 2. Distribution of short distance that one unmanned aerial vehicle (UAV) swarms.

$$\vec{d}_{i,m} = \vec{V}_{i,m} \cdot t \tag{32}$$

5.2. Data Processing and Algorithm

When UAV swarms, a significant amount of data/information may be sampled due to the number of swarming UAVs and number of iterations. Herein, the location of each UAV is considered as one data point. Thus, when M UAVs morph I times, we have $M * I$ data points which will be used to reconstruct virtual (e.g., synthetic) 3D aperiodic arrays to calculate the MUSIC spectrum and estimate DOAs. The number of data points (represented as N_d) is equivalent to the number of elements for a static array.

One challenge is how to choose N_d for computing the MUSIC spectrum at each signal processing iteration (represented as p-iteration). We would like to use more N_d in each p-iteration since more N_d used for MUSIC spectrum calculation each time means that more array elements are used for data processing in a static antenna array, and are more accurate for DOA estimation. However, more data processing points cause higher calculation cost. Thus, it is necessary to compromise N_d at each p-iteration and computational complexity.

Another problem is how to well choose data collecting and processing methods. Wide range of methods can be deployed to collect, concatenate, and process data collected from time-dependent measurements. MUSIC algorithm requires a minimum of three unique samples collected from these time-dependent measurements to provide DOA estimation. The required data samples can be obtained from agents in the swarm (e.g., one UAV locally collecting three samples at different times, or three spatially distributed UAVs collecting one sample simultaneously). We need to consider spatiotemporal distribution of samples (such as sampling rate w.r.t. wavelength, trajectory and velocity of UAVs, orientation of sensors, etc.) and iterative processing of measurements (use all data collected, truncate or applying a moving window, etc.). The iterative processing method with a moving window is called the iterative-MUSIC used in this paper.

Figure 3 shows the data processing schematic for three data points in each iteration within the MUSB system. The left part in Figure 3 shows a UAV swarm and sampling method, and the right part shows the algorithm and data processing method. When UAV swarms to a certain location, we will sample K times and each snapshot takes time d_t. After taking K snapshots, the program sets up a data point. When $N_d = 3$, the program computes the stand MUSIC spectrum using three data points and stores the result. Then when UAV swarms again, we accumulate the current data point and two previous data points to calculate the MUSIC spectrum. Then we multiply the current and previous MUSIC spectrum at each p-iteration where we obtain the iterative-MUSIC spectrum to reduce noise level and improve DOA estimation performance. Note that if the p-iteration is too big, the value of spectral points will be very small and might be taken as zero. If so, DOA cannot be estimated and we may use dB instead of a number at that situation. Then, we refine and estimate DOAs and use predefined DOA estimation precision criteria to stop the process.

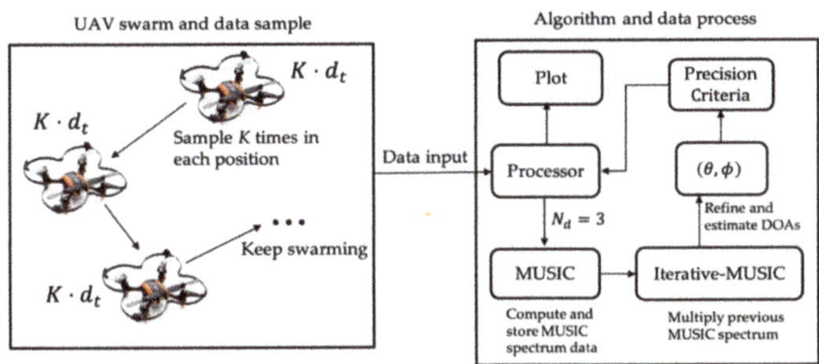

Figure 3. Data processing schematic for MUSB system.

5.3. MUSIC Algorithm for MUSB Array

As stated in part 2 of this section, the number of reconstructed MUSB arrays is N_d. Here, we rewrite the covariance of signals listed in Part 2 of Section 3 and give the first p-iteration (slightly different from UAV swarming iteration) of the iterative-MUSIC algorithm.

Rewriting the data model (1), we obtain

$$X(k) = \Gamma \widetilde{A} \cdot S(k) + W(k) \quad = A \cdot S(k) + W(k) \qquad k = 1, 2, \cdots, K \qquad (33)$$

where $\left\{ X\{k\} \in C^{N_d \times 1} \right\}$ are the vectors of sampled data, $\left\{ S\{k\} \in C^{N \times 1} \right\}$ are the source signals, $\left\{ \Gamma \in C^{N_d \times N_d} \right\}$ are the gain and phase of sensors and $\left\{ \widetilde{A} \in C^{N_d \times N} \right\}$ are the regular steering vectors. Thus, the covariance of $X(k)$ is

$$R = E\left[X(k)X^H(k) \right] = \Gamma \widetilde{A} P \widetilde{A}^H \Gamma^H + \sigma^2 I_0 = A P A^H + \sigma^2 I_0 \qquad (34)$$

Define

$$R_s = APA^H \tag{35}$$

R_s is $N_d \times N_d$ matrix with rank N. Therefore, it has $N_d - N$ repeated eigenvectors corresponding to the minimum eigenvalues σ^2. Let e_i be such an eigenvector so that $R_s e_i = 0$ or $A^H e_i = 0$, thus, $N_d - N$ eigenvectors e_i corresponding to the minimal eigenvalues are orthogonal to each of N signal columns of $A = \Gamma \widetilde{A}$, proved in [5]. $N_d - N$ dimensional subspace spanned by the noise eigenvectors is defined as noise subspace and N dimensional subspace spanned by incident signal mode vectors is defined as signal subspace.

Let Q_n be $N_d \times (N_d - N)$ matrix of noise eigenvectors, then the MUSIC spatial spectrum function is given by

$$P_{MU}(\theta, \phi) = \frac{1}{\widetilde{a}(\theta, \phi)^H \Gamma^H Q_n Q_n^H \Gamma \widetilde{a}(\theta, \phi)} = \frac{1}{\left\| Q_n^H \Gamma \widetilde{a}(\theta, \phi) \right\|^2} = \frac{1}{\left\| Q_n^H a(\theta, \phi) \right\|^2} \tag{36}$$

Then, the search spectrum peaks in the range of θ and ϕ, and the peak spectrum points we obtain are the estimation of arrival angles of incident waves.

5.4. Convergence Check

The algorithm performs the calculation until the system converges. The convergence can be guaranteed since the estimated DOA is a convergent series.

When the signal is covered by a high noise level, the estimated DOAs might be far from the ground truth and cannot be judged for convergence. However, as the iteration increases, the noise level is reduced and estimated DOAs are converged gradually. The Equation for judging convergence is given by

$$DOA_{i+1} - DOA_i \leq \varepsilon, \qquad i = 1, 2, \cdots, I_m \tag{37}$$

where I_m is the number of p-iteration of the iterative-MUSIC algorithm and ε is the preset threshold. The numerical simulation is given in part 1 of Section 6.

5.5. Computational Complexity Analysis

The orders of computational complexities of conventional spectral MUSIC [5], iterative-MUSIC, and eigenstructure-based algorithm with array interpolation (denoted by array interpolation) [48] are compared in Table 1. In this table, we assume the interpolated number of the array interpolation technique is equal to the number of actual sensors ($M_1 = N_d$) and the total number of angular sectors is denoted as $I_\theta I_\phi$. Those methods in Table 1 include the eigendecomposition step represented by the term $O(N_d^2 N)$ and the computation of $J_\theta J_\phi$ samples of the MUSIC null-spectrum function represented by $O(J_\theta J_\phi (N_d + 1)(N_d - N))$, where J_θ and J_ϕ stand for the search numbers along the directions of θ and ϕ. $\widetilde{J}_\theta \widetilde{J}_\phi$ stands for the total search numbers for each sector of array interpolation algorithm. The iterative-MUSIC at each iteration may have very low snapshot ($K = 1$) different from the traditional spectral MUSIC algorithm which requires high snapshots. The computation complexity of the iterative-MUSIC algorithm is a little bit lower than the array interpolation at each iteration ($J_\theta J_\phi \approx I_\theta I_\phi \widetilde{J}_\theta \widetilde{J}_\phi \gg N_d$), but it is higher in the whole UAV swarming period. However, the array interpolation algorithm will be very complex when applied for 3D random arrays.

Table 1. The orders of computational complexities of real-valued operations.

Algorithm	Primary Computations
Spectral MUSIC	$O\left(N_d^2 N + J_\theta J_\phi (N_d + 1)(N_d - N)\right)$
Iterative-MUSIC	$O\left(N_d^2 N + J_\theta J_\phi (N_d + 1)(N_d - N)\right)$ for each iteration
Array Interpolation	$O\left(I_\theta I_\phi \left(N_d^2 N + \widetilde{J}_\theta \widetilde{J}_\phi (N_d + 1)(N_d - N)\right)\right)$

6. Simulation and Measurement Results

In this section, several groups of simulations will be carried out to demonstrate the performance of the presented distributed directional finding system in this paper. As the framework of the MUSB system established in this paper is mentioned for the first time, we focus mainly on analyzing the impact of various factors on the feasibility of the MUSB system. Practical measurements in the lab are also given in part 3 to show the DOA estimation performance in practice.

The wavelength of signal is fixed at 1 meter (m) and the simulation in each scenario is repeated 500 times. The elevation angle of the source emitter is 85° and the azimuth angle is 270°. As one UAV swarms till $N_d = 3$, the iterative-MUSIC algorithm begins to search (0°, 179°) space for elevation angle and (0°, 359°) space for azimuth angle with 1° interval to form the overcomplete MUSIC spectrum for each p-iteration. Then, UAV keeps swarming and the iterative-MUSIC algorithm keeps computing the MUSIC spectrum before the precision of DOA estimation is satisfied. When the preset threshold is satisfied, UAV stops swarming and the reconstructed process of phased arrays is terminated. The refined DOA estimations are obtained by scanning the reconstructed signal peaks from the iterative-MUSIC algorithm with 0.1° step during the refinement procedure 10 times.

Moreover, the speed of UAV will influence the snapshot at each location where the system samples the source emitter. The snapshot at each location should be very low if the UAV swarms very fast and the snapshot can be high when the swarming rate of UAV is low. The speed of UAV in this paper will be represented by the distance between two iterations of swarming UAVs shown in Figure 2.

The joint root-mean-square error (RMSE) of incident signals is used for statistical DOA estimation precision evaluation, which is defined as

$$RMSE_{\theta,\phi} = \sqrt{\sum_{w=1}^{W}\sum_{n=1}^{N}\left[\left[\hat{\theta}_n^w - \theta_n^w\right]^2 + \left[\hat{\phi}_n^w - \phi_n^w\right]^2\right]/2WN} \tag{38}$$

where W is the number of Monte Carlo simulations, (θ_n^w, ϕ_n^w) represent the actual DOAs of the nth signal, and $\left(\hat{\theta}_n^w, \hat{\phi}_n^w\right)$ represent the estimated DOAs of the nth signal in the w-th simulation.

First, the system convergence will be studied in a typical scenario. Second, the DOA estimation performance using the iterative-MUSIC algorithm will be compared with CRB in various scenarios. Finally, DOA estimation performance in practice will be investigated.

6.1. System Convergence

Figure 4 provides and example of the MUSB distributed directional finding system gradually converging to the ground truth as iteration increases by using iterative-MUSIC algorithm.

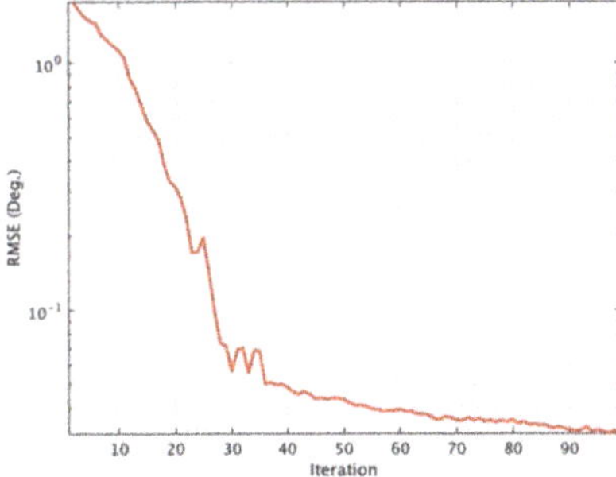
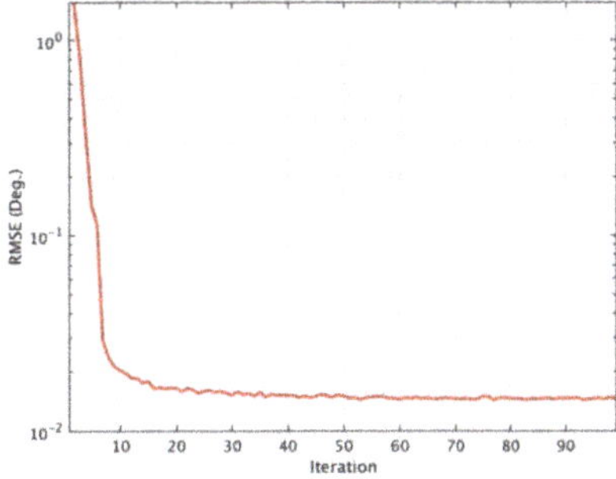

Figure 4. Direction-of-arrival (DOA) estimation convergence for the MUSB system. (**a**) $N_d = 3$, SNR = 0 dB, $K = 1$. (**b**) $N_d = 3$, SNR = 0 dB, $K = 16$.

6.2. DOA Estimation Performance

The performance of DOA estimation depends on multiple factors such as SNR, K, N_d, velocity of UAV and number of iterations. Furthermore, the performance also depends on the distinctness of the array geometries due to the diversity of different observations at different time instants. In this section, we only consider the single-emitter case. When a single source is present, a typical scenario is set in which UAV swarming short distance mean $\mu = 2\lambda$, the Monte Carlo simulation number and angles of the source are the same as before. Figure 5 depicts the results predicted by the stochastic CRB derived in Section 4. The joint RMSE of elevation and azimuth angles obtained from the iterative-MUSIC algorithm together with CRB from Section 4 are shown in Figures 6 and 7.

The extreme case is $K = 1$, $N_d = 3$ (2D DOA estimation requires a minimum of three unique samples). The fixed settings and changed settings are listed as: Figure 5a varies snapshots K from 1 to 96 for different SNR with iteration $i = 100$, $N_d = 3$; Figure 5b varies SNR from -30 to 0 dB for different UAV iterations with $K = 1$, $N_d = 3$; Figure 5c varies SNR from -30 to 0 dB for different data points N_d in each iteration with $K = 1$, $i = 100$; Figure 5d varies SNR for different speed of UAV with $K = 1$, $N_d = 3$, $i = 100$; Figure 6a varies SNR of one signal from -20 to 20 dB with $K = 1$, $N_d = 3$; Figure 6b varies K from 1 to 99 with SNR = 0 dB, $N_d = 3$; Figure 6c varies speed of UAV with $K = 1$, $N_d = 3$, SNR = 0 dB; Figure 6d varies incident elevation angles from 5° to 90° with SNR = 0 dB, $N_d = 3$, $K = 1$; Figure 7a varies sensor gain deviation with SNR = 0 dB, $N_d = 3$, $K = 1$; Figure 7b varies sensor phase deviation with SNR = 0 dB, $N_d = 3$, $K = 1$. Scenarios (a)–(d) in Figure 5 give the lower bounds of the MUSB system. Scenarios (a)–(d) in Figure 6 show the performance of MUSB array without element rotation (sensor gain $g = 1$ and sensor phase $\psi = 0$ and scenarios (a)–(b) in Figure 7 show the impact factors with the element rotation of MUSB array (sensor gain and phase coefficients have certain deviations).

From Figures 6 and 7, we can find that Figure 6a shows the DOA estimation performance of the MUSB system will increase with increasing SNR; Figure 6b shows that the system can estimate DOAs even when snapshot $K = 1$; Figure 6c shows that when the UAV speed increases, the precision of DOA estimation increases; Figure 6d shows that the precision of DOA estimation increases with increasing elevation angles. Figure 7 shows the performance of the MUSB array with UAV rotating associated with sensor gain and phase varying. Figure 7a shows that the DOA estimation is not significantly impacted as the standard deviation of sensor gain increases; Figure 7b shows the precision of DOA estimation decreases significantly when the standard deviation of the sensor phase increases to a certain value.

In Figure 7b, we find that the stochastic CRB is flat for the single-emitter case as the phase standard phase varies because Equation 16 shown in Section 4 cancels the sensor phase errors by multiplying the steering vector A and the conjugate transpose of A, while the iterative-MUSIC algorithm does not have the advantages CRB has utilized. Equation 39 shows the advantages of CRB for the single-emitter case and Equations 16–21 give the derivation process with advantages of sensor phase error cancel.

$$A_i^H A_i = \left[g_{i,1} \cdot e^{-j\varphi_{1,1}}, \cdots, g_{i,m} \cdot e^{-j\varphi_{1,m}} \right] \begin{bmatrix} g_{i,1} \cdot e^{j\varphi_{1,1}} \\ \vdots \\ g_{i,m} \cdot e^{j\varphi_{1,m}} \end{bmatrix} = \sum_{m=1}^{M} g_{i,m}^2 \tag{39}$$

where $\varphi_{i,m}$ includes the sensor phase deviation and phase difference from sensor m to relative phase center in the i-th iteration.

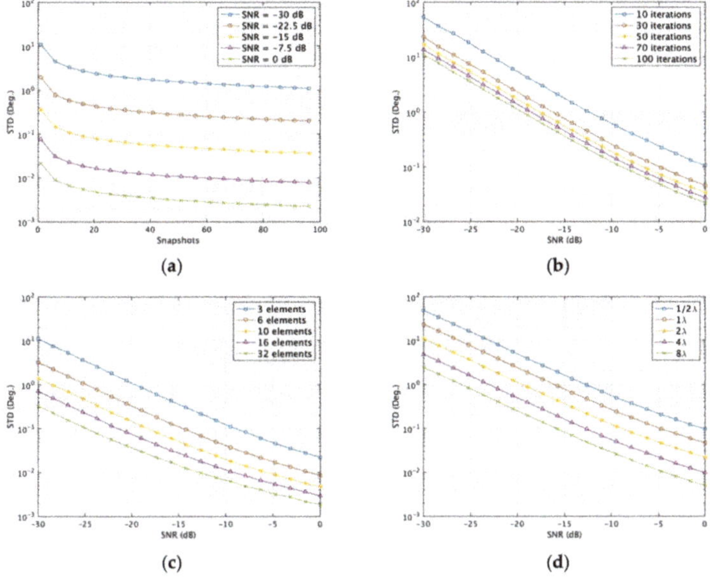

Figure 5. DOA standard deviation predicted by the stochastic Cramer–Rao bound (CRB). (**a**) Varying snapshots with different signal-to-noise ratio (SNR); (**b**) varying SNR with different iterations of UAV swarming; (**c**) varying SNR with different array element number; (**d**) varying SNR with different average speed.

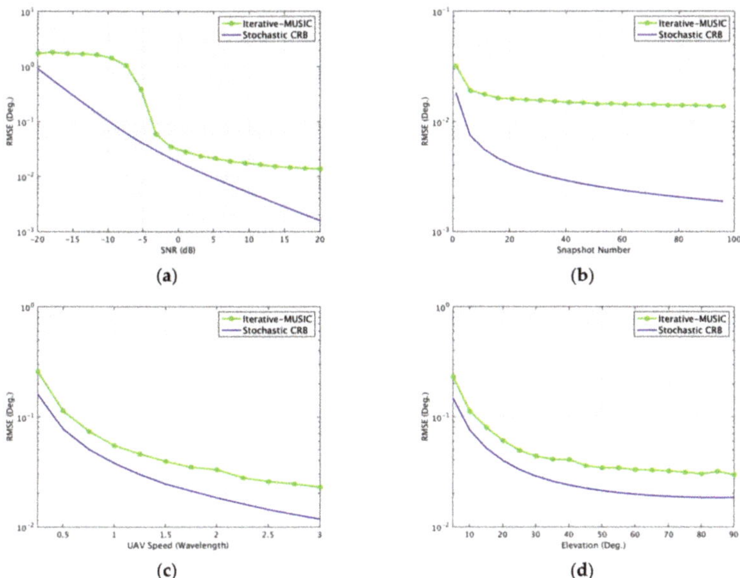

Figure 6. DOA estimation root mean square error (RMSE) of iterative-MUSIC (multiple signal classification) and CRB in different scenarios. (**a**) Varying SNR ($N_d = 3$); (**b**) varying K; (**c**) varying the speed of UAV w.r.t. wavelength; (**d**) varying incident elevation angles.

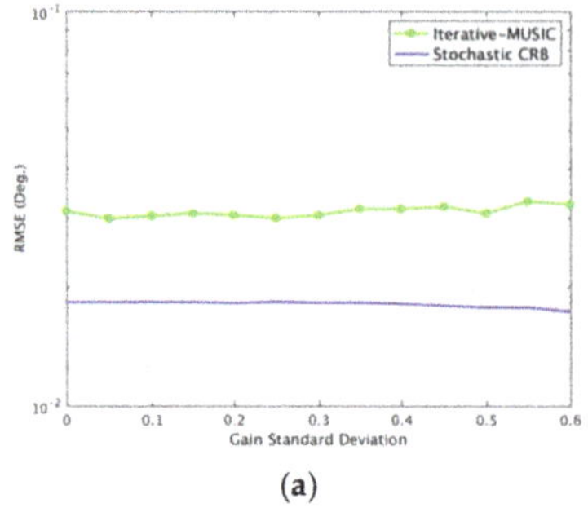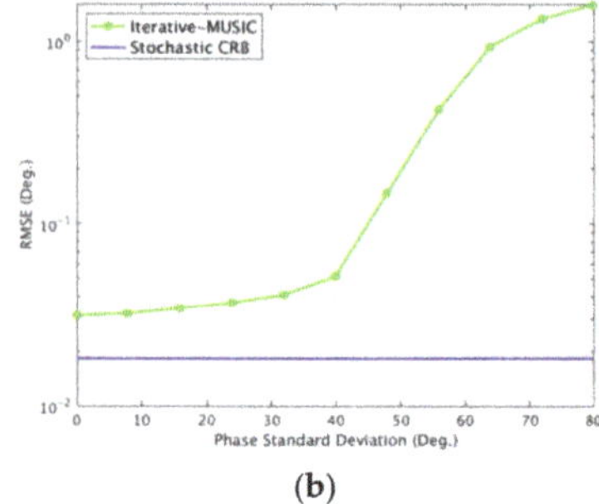

Figure 7. DOA estimation RMSE of iterative-MUSIC and CRB in different scenarios. (**a**) Varying sensor gain standard deviation; (**b**) varying sensor phase standard deviation.

6.3. Measurement

In the experiment, the test fixture provides a convenient platform to study this morphing in time (using sixteen elements). Randomly positioned rectangular patch antennas designed for 2.45 GHz are used with a randomly morphing volume provided by a moving platform named as "Medusa" to estimate DOAs. Figure 8 shows that one source in the far-field transmits and receiving sensors receive the signal with the vector network analyzer (VNA) measuring the S_{21} of transmitting and receiving antennas. Figure 9 shows the measured MUSIC spectrum with 16 spatially distributed elements as a static volumetric random array in Medusa with element rotation (Rotate 0–45 degrees around x, y, z axis randomly, no UAV swarming). Figure 10 shows the measured azimuth and elevation errors gradually converge from around 10–20 degrees to around 1–2 degrees as the iteration increases with UAV swarming. In Figure 10, we take the extreme case ($N_d = 3$, $K = 1$).

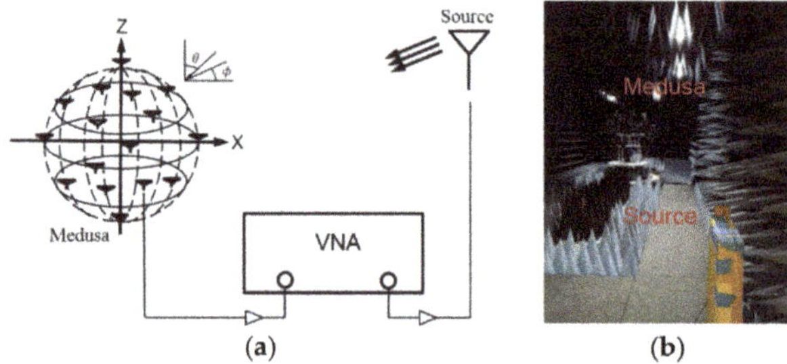

Figure 8. Test diagram. (**a**) Schematic diagram; (**b**) practical test diagram.

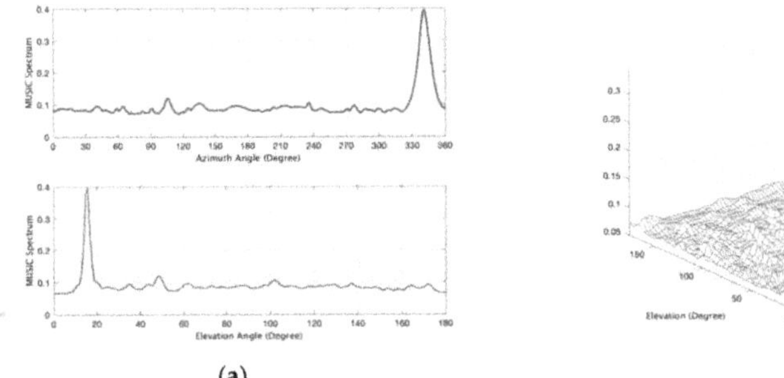
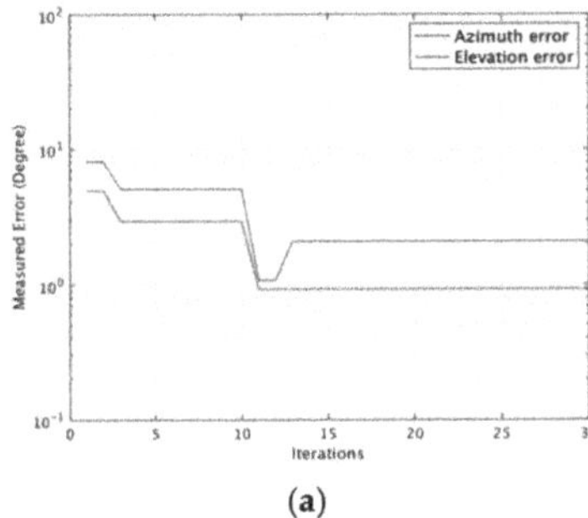
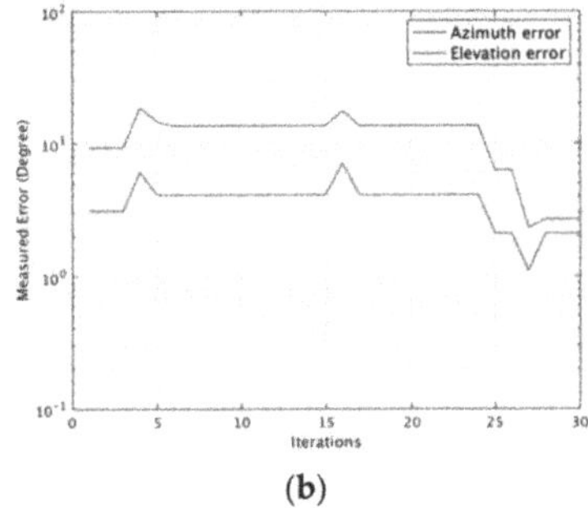

Figure 9. Measured MUSIC spectrum with iteration $i = 1$, $K = 1$, $N_d = 16$, an incident signal of azimuth 356.3° and elevation 18°. (**a**) 2D MUSIC spectrum; (**b**) 3D MUSIC spectrum.

Figure 10. Measured errors compared to iterations with $N_d = 3$, $K = 1$. (**a**) Element without rotation; (**b**) element with rotation (element rotates 0–45 degrees around x, y, z axis randomly to simulate UAV flying).

7. Conclusions

This paper establishes a MUSB data collection framework for the first time, which makes it possible to realize source estimation with low snapshot under low SNR environment in a UAV swarming period. Theoretical and experiment results are given to reveal the performance of the MUSB phased array system used for 2D DOA estimation, which supports the feasibility of the system. The iterative-MUSIC algorithm is applied for the framework and it can estimate the DOAs efficiently only with one snapshot in each iteration when UAV swarms very fast. The UAV speed controls the structure of the reconstructed aperiodic phased arrays from the MUSB system and the DOA estimation precision is increased when the distance between the two iterations of swarming UAV is increased. The impact of known sensor gain errors and phase errors from the UAV rotation for the DOA estimation performance are also investigated. Practical experiment results match the theoretical expectation of the MUSB system using the iterative-MUSIC algorithm. Our results will benefit future research on performance analysis and optimal design of time-varying antenna arrays based on the UAV swarm. It is also interesting to extend the results when position errors are present in the future.

Author Contributions: Z.C. and G.H.H. proposed the main idea; Z.C., G.H.H., S.Y., G.-F.C. conceived and designed the simulations; Z.C. wrote the paper.

Funding: This research received no external funding.

Conflicts of Interest: The authors declare no conflict of interest.

Appendix A

Referred to [49], the stochastic FIM's F_θ w.r.t. 1D incident angle θ of static array is available:

$$F_\theta = CRB^{-1} = \frac{2K}{\sigma^2}\{\mathrm{Re}[H \odot U]\} \tag{A1}$$

where $H = D^H\left[I_0 - A\left[A^H A\right]^{-1} A^H\right]D$, $D = [d_1, d_2, \cdots d_N]$, $d_n = \frac{da(\theta)}{d\theta}\big|_{\theta=\theta_n}$, and $U = PA^H R^{-1} AP$. For each iteration, the FIM's $F_{i,\theta}$ for the presented problem in this paper is given by

$$F_{i,\theta} = CRB_{i,\theta}^{-1} = \frac{2K}{\sigma^2}\{\mathrm{Re}[H_i \odot U_i]\} \tag{A2}$$

where $H_i = D_i{}^H\left[I_0 - A_i\left[A_i{}^H A_i\right]^{-1} A_i{}^H\right]D_i$, $D_i = [d_{i,1}, d_{i,2}, \cdots d_{i,N}]$, $d_{i,n} = \frac{da_i(\theta)}{d\theta}\big|_{\theta=\theta_n}$, $U_i = P_i A_i{}^H R_i^{-1} A_i P_i$. As already stated,

$$F_\theta = \sum_{i=1}^{I} F_{i,\theta} \tag{A3}$$

Omitting to the parameters $\mathrm{Re}(P_{mn})$, $\mathrm{Im}(P_{mn})$, σ^2, gain and phase of signals, only consider the estimation of elevation and azimuth angles (θ_n, ϕ_n) hereby is given by $\alpha = \begin{bmatrix} \theta^T, & \phi^T \end{bmatrix}^T$, $\theta = [\theta_1, \theta_2, \cdots \theta_N]^T$, and $\phi = [\phi_1, \phi_2, \cdots \phi_N]^T$. Thus, the submatrix of FIM associated with 2D DOA is

$$F_{i,s,2} = \begin{bmatrix} F_{i,\theta\theta} & F_{i,\theta\phi} \\ F_{i,\phi\theta} & F_{i,\phi\phi} \end{bmatrix} = \frac{2K}{\sigma^2}\{\mathrm{Re}[D_i^H A_i^\perp D_i \odot 1_2 1_2^T \otimes U_i]\} \tag{A4}$$

$$A_i^\perp = I_0 - A_i\left(A_i{}^H A_i\right)^{-1} A_i{}^H \tag{A5}$$

where "$i, s, 2$" denotes the 2D stochastic bounds for the i-th iteration, 1_2 represents 2x1 vectors of ones and

$$\begin{aligned} F_{i,\theta\theta} &= \frac{2K}{\sigma^2}\{\mathrm{Re}[D_i^H[\theta]A_i^\perp D_i[\theta] \odot U_i]\}, \\ F_{i,\theta\phi} &= \frac{2K}{\sigma^2}\{\mathrm{Re}[D_i^H[\theta]A_i^\perp D_i[\phi] \odot U_i]\} \end{aligned} \tag{A6}$$

$$\begin{aligned} F_{i,\phi\theta} &= \frac{2K}{\sigma^2}\{\mathrm{Re}[D_i^H[\phi]A_i^\perp D_i[\theta] \odot U_i]\}, \\ F_{i,\phi\phi} &= \frac{2K}{\sigma^2}\{\mathrm{Re}[D_i^H[\phi]A_i^\perp D_i[\phi] \odot U_i]\} \end{aligned} \tag{A7}$$

where $A_i(\theta, \phi) = [a_i[\theta_1, \phi_1], a_i[\theta_2, \phi_2], \cdots, a_i[\theta_N, \phi_N]]$. Thus, the system stochastic FIM is given by

$$F_{s,2} = \sum_{i=1}^{I} F_{i,s,2} = \begin{bmatrix} F_{\theta\theta} & F_{\theta\phi} \\ F_{\phi\theta} & F_{\phi\phi} \end{bmatrix} \tag{A8}$$

where $F_{\theta\theta} = \sum_{i=1}^{I} F_{i,\theta\theta}$, $F_{\theta\phi} = \sum_{i=1}^{I} F_{i,\theta\phi}$, $F_{\phi\theta} = \sum_{i=1}^{I} F_{i,\phi\theta}$, $F_{\phi\phi} = \sum_{i=1}^{I} F_{i,\phi\phi}$.

References

1. Bartlett, M.S. Smoothing periodograms from time series with continuous spectra. *Nature* **1948**, *161*, 686–687. [CrossRef]
2. Bartlett, M.S. Periodogram analysis and continuous spectra. *Biometrika* **1950**, *37*, 1–16. [CrossRef] [PubMed]
3. Burg, J.P. Maximum Entropy Spectral Analysis. Ph.D. Thesis, Stanford University, Stanford, CA, USA, 1975.
4. Gavish, M.; Weiss, A.J. Performance analysis of the VIA ESPRIT algorithm. *IEE Proc. F-Radar Signal Process.* **1993**, *140*, 123–128. [CrossRef]

5.   Schmidt, R.O. Multiple emitter location and signal parameter estimation. *IEEE Trans. Antennas Propag.* **1986**, *34*, 276–280. [CrossRef]

6.   Dongarsane, C.R.; Jdhav, A.N. Simulation study on DOA estimation using MUSIC algorithm. *Int. J. Technol. Eng. Syst.* **2011**, *2*, 54–57.

7.   Tang, H. DOA Estimation Based on MUSIC Algorithm. Master's Thesis, Linnaeus University, Kalmar, Sweden, May 2015.

8.   Oumar, O.A.; Siyau, M.F.; Sattar, T.P. Comparison between MUSIC and ESPRIT direction of arrival estimation algorithms for wireless communication systems. In Proceedings of the First International Conference on Future Generation Communication Technologies, London, UK, 12–14 December 2012.

9.   Stoica, P.; Nehorai, A. MUSIC, maximum likelihood, and Cramer-Rao bound. *IEEE Trans. Acoust. Speech Signal Process.* **1989**, *37*, 1783–1795. [CrossRef]

10.   Lee, H.B. Resolution threshold beamspace MUSIC for two closely spaced emitters. *IEEE Trans. Acoust. Speech Signal Process.* **1990**, *38*, 723–738. [CrossRef]

11.   Barabell, A.J. Improving the resolution performance of eigenstructure-based direction-finding algorithms. In Proceedings of the IEEE International Conference on Acoustics, Speech, and Signal Processing, Boston, MA, USA, 14–16 April 1983.

12.   Rao, B.D.; Hari, K.V. Performance of analysis of root-MUSIC. *IEEE Trans. Acoust. Speech Signal Process.* **1989**, *37*, 1939–1949. [CrossRef]

13.   Rubsamen, M.; Gershman, A.B. Direction-of-arrival estimation for nonuniform sensor arrays: From manifold separation to Fourier domain MUSIC methods. *IEEE Trans. Signal Process.* **2009**, *57*, 588–599. [CrossRef]

14.   Liu, Y.; Liu, H. Antenna array signal direction of arrival estimation in digital signal processing (DSP). *Procedia Comput. Sci.* **2015**, *55*, 782–791. [CrossRef]

15.   Xie, Y.; Peng, C.; Jiang, X.; Shan, O. Hardware design and implementation of DOA estimation algorithms for spherical array antennas. In Proceedings of the 2014 IEEE International Conference on Signal Processing, Communications and Computing (ICSPCC), Guilin, China, 5–8 August 2014.

16.   Kim, M.; Ichige, K.; Arai, H. Implementation of FPGA based fast DOA estimator using unitary MUSIC algorithm. In Proceedings of the 2003 IEEE 58th Vehicular Technology Conference, Orlando, FL, USA, 6–9 October 2003.

17.   Abouda, A.A.; El-Sallabi, H.M.; Haggman, S.G. Impact of antenna array geometry on MIMO channel eigenvalues. In Proceedings of the 2005 IEEE 16th International Symposium on Personal, Indoor and Mobile Radio Communications, Berlin, Germany, 11–14 September 2005.

18.   Vu, D.; Renaux, R.B.; Marcos, S. Performance Analysis of 2D and 3D Antenna Arrays for Source Localization. In Proceedings of the 18th European Signal Processing Conference, Aalborg, Denmark, 23–27 August 2010.

19.   Hotta, H.; Teshima, M.; Amano, T. Estimation of radio propagation parameters using antennas arrayed 3-dimensional space. *TEICE Tech. Rep.* **2005**, *105*, 23–28.

20.   Godara, L.C.; Cantoni, A. Uniqueness and linear independence of steering vectors in array space. *J. Acoust. Soc. Am.* **1981**, *70*, 467–475. [CrossRef]

21.   Gazzah, H.; Karim, A.M. Optimum ambiguity-free directional and omnidirectional planar antenna arrays for DOA estimation. *IEEE Trans. Signal Process.* **2009**, *57*, 3942–3953. [CrossRef]

22.   Xia, Z.; Huff, G.H.; Chamberland, J.F.; Pfister, H.; Bhattacharya, R. Direction of arrival estimation using canonical and crystallographic volumetric element configurations. In Proceedings of the 2012 6th European Conference on Antennas and Propagation (EUCAP), Prague, Czech Republic, 26–30 March 201.

23.   Kiani, S.; Pezeshk, A.M. A Comparative Study of Several Array Geometries for 2D DOA Estimation. *Procedia Comput. Sci.* **2015**, *58*, 18–25. [CrossRef]

24.   Moriya, H. Cuboid array: A novel 3-d array configuration for high resolution 2-D DOA estimation. In Proceedings of the IEEE Workshop on Signal Processing Systems, Taipei City, Taiwan, 16–18 October 2013.

25.   Moriya, H. High-resolution 2-D DOA estimation by 3-D array configuration based on CRLB formulation. In Proceedings of the 2012 IEEE 11th International Conference on Signal Processing, Beijing, China, 21–25 October 2012.

26.   Poormohammad, S.; Farzaneh, F. Precision of direction of arrival (DOA) estimation using novel three-dimensional array geometries. *Int. J. Electron. Commun.* **2017**, *75*, 34–45. [CrossRef]

27. Asgari, S. Theoretical and experimental study of DOA estimation using AML algorithm for an isotropic and non-isotropic 3D array. In Proceedings of the SPIE—The International Society for Optical Engineering, San Diego, CA, USA, 26–27 August 2007.

28. Zhang, R.; Wang, S.; Lu, X.; Duan, W.; Cai, L. Two-dimensional DoA estimation for multipath propagation characterization using the array response of PN-sequences. *IEEE Trans. Wirel. Commun.* **2016**, *15*, 341–356. [CrossRef]

29. Wan, L.; Liu, L.; Han, G.; Rodrigues, J.J. A low energy consumption DOA estimation approach for conformal array in ultra-wideband. *Future Internet* **2013**, *5*, 611–630. [CrossRef]

30. Liu, Z.; Guo, F.C. Azimuth and elevation estimation with rotating long-baseline interferometers. *IEEE Trans. Signal Process.* **2015**, *63*, 2405–2419. [CrossRef]

31. Liu, Z. Direction-of-arrival estimation with time-varying arrays via Bayesian multitask learning. *IEEE Trans. Veh. Technol.* **2014**, *63*, 2405–2419. [CrossRef]

32. Dvorkind, T.G.; Greenberg, E. DOA estimation and signal separation using antenna with time varying response. In Proceedings of the 2014 22nd European Signal Processing Conference (EUSIPCO), Lisbon, Portugal, 1–5 September 2014.

33. Naaz, A.; Rao, R. DOA estimation-a comparative analysis. *Int. J. Comp. Commun.* **2014**, *3*, 141–144. [CrossRef]

34. Demissie, B.; Oispuu, M.; Ruthotto, E. Localization of multiple sources with a moving array using subspace data fusion. In Proceedings of the 2008 11th International Conference on Information Fusion, Cologne, Germany, 30 June–3 July 2008.

35. Kendra, J.R. Motion-extended array synthesis-part I: Theory and method. IEEE Trans. *Geosci. Remote Sens.* **2017**, *55*, 2028–2044. [CrossRef]

36. Corner, J.J.; Lamont, G.B. Parallel simulation of UAV swarm scenarios. In Proceedings of the 2004 Winter Simulation Conference, Washington, DC, USA, 5–8 December 2004.

37. Saad, E.; Vian, J.; Clark, G.J.; Bieniawski, S. Vehicle swarm rapid prototyping testbed. In Proceedings of the IAAA Infotech at Aerospace Conference, Seattle, WA, DC, USA, 6–9 April 2009.

38. Petko, J.S.; Werner, D.H. Positional tolerance analysis and error correction of micro-UAV swarm-based antenna arrays. In Proceedings of the 2009 IEEE Antennas and Propagation Society International Symposium, Charleston, SC, USA, 1–5 June 2009.

39. Almeida, M.; Hildmann, H.; Solmaz, G. Distributed UAV-swarm-based real-time geomatic data collection under dynamically changing resolution requirements. *Int. Arch. Photogramm. Remote Sens. Spat. Inf. Sci.* **2017**, *W6*, 5–12. [CrossRef]

40. Namin, F.; Petko, J.S.; Werner, D.H. Design of robust aperiodic antenna array formations for micro-UAV swarms. In Proceedings of the 2010 IEEE Antennas and Propagation Society International Symposium, Toronto, ON, CA, 11–17 July 2010.

41. Namin, F.; Petko, J.S.; Werner, D.H. Analysis and design optimization of robust aperiodic micro-UAV swarm-based antenna arrays. *IEEE Trans. Antennas Propag.* **2012**, *60*, 2295–2308. [CrossRef]

42. Goodman, N.R. Statistical analysis based on a certain multivariate complex distribution (an itroduction). *Ann. Math. Stat.* **1963**, *34*, 152–177. [CrossRef]

43. Chen, Z.; Chamberland, J.F.; Huff, G.H. Impact of UAV swarm density and heterogeneity on synthetic aperture DoA convergence. In Proceedings of the IEEE USNC-URSI Radio Science Meeting (Joint with AP-S Symposium), San Diego, CA, USA, 9–14 July 2017.

44. Friedlander, B.; Weiss, A.J. Direction finding in the presence of mutual coupling. *IEEE Trans. Antennas Propag.* **1991**, *39*, 273–284. [CrossRef]

45. Hung, H.; Kaveh, M. On the statistical sufficiency of the coherently averaged covariance matrix for the estimation of the parameters of wide-band sources. In Proceedings of the IEEE International Conference on Acoustics, Speech, and Signal Processing, Dallas, TX, USA, 6–9 April 1987.

46. Stoica, P.; Nehorai, A. MODE, maximum likelihood, and Cramer-Rao bound: Conditional and unconditional results. In Proceedings of the International Conference on Acoustics, Speech, and Signal Processing, Albuquerque, NM, USA, 3–6 April 1990.

47. Zeira, A.; Friedlander, B. Direction finding with time-varying arrays. *IEEE Trans. Signal Process.* **1995**, *43*, 927–937. [CrossRef]

48. Friedlander, B.; Weiss, A.J. Eigenstructure-based algorithms for direction finding with time-varying arrays. *IEEE Trans. Aerosp. Electron. Syst.* **1996**, *39*, 689–701. [CrossRef]
49. Stoica, P.; Nehorai, A. Performance study of conditional and unconditional direction-of-arrival estimation. *IEEE Trans. Acoust. Speech Signal Process.* **1990**, *38*, 783–1795. [CrossRef]

Article

Energy-Aware Management in Multi-UAV Deployments: Modelling and Strategies

Victor Sanchez-Aguero [1,2,*], Francisco Valera [2], Ivan Vidal [2], Christian Tipantuña [3,4] and Xavier Hesselbach [3]

1   IMDEA Networks Institute, Avda. del Mar Mediterráneo, 22, 28918 Madrid, Spain
2   Departamento de Ingeniería Telemática, Universidad Carlos III de Madrid, Avda. Universidad, 30, 28911 Leganés, Madrid, Spain; fvalera@it.uc3m.es (F.V.); ividal@it.uc3m.es (I.V.)
3   Department of Network Engineering, Universitat Politècnica de Catalunya, Calle Jordi Girona 1-3, E-08034 Barcelona, Spain; christian.jose.tipantuna@upc.edu (C.T.); xavier.hesselbach@upc.edu (X.H.)
4   Escuela Politécnica Nacional, Ecuador, Avda. Ladrón de Guevara, E11-253 Quito, Ecuador
*   Correspondence: victor.sanchez@imdea.org

Received: 30 March 2020; Accepted: 12 May 2020; Published: 14 May 2020

Abstract: Nowadays, Unmanned Aerial Vehicles (UAV) are frequently present in the civilian environment. However, proper implementations of different solutions based on these aircraft still face important challenges. This article deals with multi-UAV systems, forming aerial networks, mainly employed to provide Internet connectivity and different network services to ground users. However, the mission duration (hours) is longer than the limited UAVs' battery life-time (minutes). This paper introduces the UAV replacement procedure as a way to guarantee ground users' connectivity over time. This article also formulates the practical UAV replacements problem in moderately large multi-UAV swarms and proves it to be an NP-hard problem in which an optimal solution has exponential complexity. In this regard, the main objective of this article is to evaluate the suitability of heuristic approaches for different scenarios. This paper proposes betweenness centrality heuristic algorithm (BETA), a graph theory-based heuristic algorithm. BETA not only generates solutions close to the optimal (even with 99% similarity to the exact result) but also improves two ground-truth solutions, especially in low-resource scenarios.

Keywords: UAV; UAV fleet; UAV swarm; energy consumption; self-organization; algorithms; optimization; UAV replacement

1. Introduction

The unstoppable growth of the Unmanned Aerial Vehicles (UAVs) (commonly known as drones) ecosystem during these last years, has been proven to be just the beginning of a near-future global phenomenon. The US Federal Aviation Administration predicts [1] that UAVs providing commercial services will triple over the next five years, and will overtake consumer off-the-shelf UAVs by the year 2024. UAVs will grow eightfold over the next decade and will become the largest segment of the civilian market.

The utilization of multi-UAV systems, because of their rapid deployment, mobility, and flexibility, has recently attracted attention to support/extend the fifth-generation cellular network technology (5G) in extraordinary situations (e.g., massified events, natural disasters, infrastructure failures). The 5G networks will certainly bring faster uploading and downloading speeds in combination with a dramatic decrease of the network latency. However, in exceptional or emergency circumstances, the deployment of 5G terrestrial infrastructure may not be economically viable. In addition, the deployment times of these extraordinary on-demand 5G network services should meet the Key Performance Indicators [2] (KPI) defined by the 5G-PPP, which states that new deployments must finish within

90 min. Accordingly, it is here where UAVs are expected to play a crucial role. If properly deployed and configured, UAV networks can provide fast-ubiquitous 5G access (which is also within the 5G KPIs) employing wireless communications solutions in a diversity of real-world scenarios.

5G UAV missions, e.g., to complement existing cellular networks in high-density environments, deliver network coverage in hard to reach rural areas (Remote Access Networks), or in IoT scenarios, may require the management of moderately large UAV fleets. UAVs mainly act as aerial communication platforms such as (i) aerial base stations (BS) (to support existing 5G infrastructure in high traffic demand) [3], or (ii) aerial WiFi access points (AP) forming a Flying Ad Hoc Network (FANET) (to create new networks) [4]. So that UAV research area aims to extend the 5G network (where it has no range) or support the existing 5G network (when it is not enough) using radio solutions as payload [5], e.g., 5G or LTE microcells, multi-hop solutions based on commodity WiFi. When compared to terrestrial antennas, aerial units may have some advantages since they can change their altitude with the possibility of avoiding obstacles, including no geographical restriction on the antenna location. However, these advantages turn into crucial design challenges, such as the optimal positioning, the limited flight time, or the optimal trajectories calculation and network planning [6].

In particular, multi-UAV environments may give rise to long-endurance missions that require uninterrupted service provisioning (performing UAV replacements) that are not achievable using a single UAV due to battery capacity constraints (at most around 20 min flight [7]). A UAV replacement means that a UAV that is waiting in the Ground Control Station (GCS) becomes active and goes into the scenario to substitute one of the UAVs that are on service, so as to provide the same functionality. This is only possible by providing a fleet exceeding the number of UAVs that have to be active on service at the same time, i.e., there is a reasonable number of fresh UAVs for replacement.

Nevertheless, developing an appropriate replacement strategy of UAVs, is one of the critical hurdles that have not yet been properly addressed by the research community. The replacement strategy enables to optimize the cost in terms of required aerial infrastructure resources while keeping the provided level of service. To guarantee that long-term (beyond battery life) services can be deployed, some of the UAVs that are at the GCS must be able to successfully replace the UAVs that provide the actual service on the stage whenever necessary (for example, when a UAV has low battery or fails). However, the economic cost of oversizing the fleet is enormous, as these devices commonly have high prices. For this reason, it is necessary to develop a resource optimization mechanism in order to allow intelligent and autonomous UAV systems to be managed with the lowest possible number of UAVs.

Figure 1 illustrates a representative use case of UAVs delivering network coverage. As it can be appreciated, some UAVs provide connectivity to several end-users. A Controller entity, located in the GCS, is in charge of scheduling when UAV replacements will take place. Once the replacement procedure is started for a certain UAV, that UAV directly goes back to the GCS to change its battery ①, while another one comes back in its place ②. As soon as it has a charged battery installed, it is available again in the replacement pool for other UAVs to be changed when required. By following this methodology, an uninterrupted service may be provided. Reducing the UAV fleet while ensuring a reasonable quality of service is not a straightforward procedure (Further details about the methodology depicted in Figure 1 can be found in Section 4, Sections 4.1 and 4.4).

This article presents the practical UAV replacement problem and analyses its complexity. Next, different scenarios, as well as the methodology to evaluate the service performance, are presented. A sub-optimal heuristic algorithm is proposed that guarantees the proper modeling and control of a fleet of UAVs that are used to provide Internet connectivity minimizing the fleet size. This algorithm is validated comparing it with the optimal solution, using an improved version of the brute-force search combinatorial algorithm developed in our previous work [8]. Finally, the practical limitations of the proposal are analyzed, while the possible alternatives solutions are considered.

The rest of the article is organized as follows: the related work and background are reviewed in Section 2. Section 3 states the proposed problem and analyses its complexity. Section 4 describes the methodology to operate with large multi-UAV fleets. Then, Section 5 details the suggested scenarios and remarks the results obtained from the simulation. Finally, Section 6 concludes the article and depicts some future research lines.

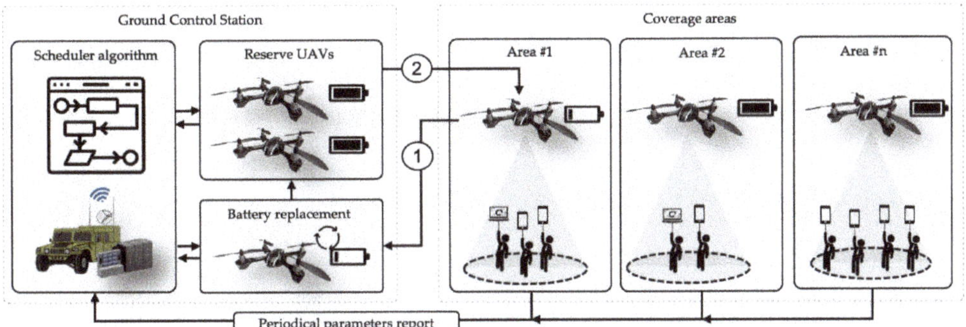

Figure 1. Typical Unmanned Aerial Vehicles (UAV) use case using the proposed methodology.

2. Related Work and Background

Due to the versatility of UAVs, these devices are used in a wide variety of fields. The following section introduces the evolution of management strategies used to overcome battery limitations and algorithmic solutions for UAV battery replacements which are the main topics of this article. It also explains how UAVs may complement 5G networks, and wireless communication solutions in large geographical areas using UAV swarms.

Ubiquitous connectivity is one of the current challenges of 5G networks and beyond 5G [6]. UAVs have appeared as a promising solution to provide reliable and flexible wireless communication services for ground users in a wide variety of scenarios [3]. The usage of UAVs promises to provide cost-effective wireless connectivity for devices without infrastructure coverage. Concretely, UAVs are considered as flying BSs for coverage extension and capacity enhancement of the existing 5G cellular networks. In this paper [9], authors explore the use of UAV-BSs to provide coverage during natural disasters. In this work [5], an evolved packet core (EPC) inside a UAV is introduced, to orchestrate the LTE RAN in the presence of multiple BSs. This EPC can also interoperate with commercial BSs as well as commodity user equipment. In [10], the authors provide an overview of UAV-aided networks, introducing the underlying architecture and wireless channel characteristics.

One of the most critical design challenges in multi-UAV systems is the achievement of the all-to-all communication between UAVs, which is necessary for cooperation and collaboration [11,12]. If every UAV is connected to existing network infrastructure such as a GCS, satellite network or base stations, swarm communications can be delivered via this infrastructure. This type of network scheme simplifies some problems that may be associated with UAVs ad hoc networks alternatives, like routing protocols or the distributed control of the network. However, it also brings as a consequence certain limitations such as the expensive equipment (long-range or satellite antennas) and obviously less flexibility since the deployment is fixed to existing infrastructure. An alternative solution is the usage of FANETs. In this type of system, UAVs have several roles, not only as functional devices to provide coverage, gathering sensor data, or video dissemination but also to be used as network relays to connect all UAVs through the UAV network itself. Commonly, only one (or a few) UAV (also known as backbone UAV) are required to be connected to the fixed infrastructure (GCS). The backbone UAV is generally equipped with two radios: (i) low-power radio (WiFi or Bluetooth, for instance) is used for communication between the UAVs and (ii) high power long-range radio to communicate with the GCS [13]. It is common to find quite a few examples of research works that use FANETs to support 5G networks [14]. For instance, Reference [4] extends a 5G network slice for video monitoring with a

FANET composed of small low-altitude UAVs with multi-access edge computing (MEC) facilities to allow high-speed transmission.

Although the development of UAV networks is receiving significant attention from the research community, some challenges must be solved before their proper deployment and consolidation. One of them is their limited battery capacity since normally a UAV source power mainly depends on small batteries (we are considering in this article small rotary-wing UAVs and not big fixed-wing UAVs with fuel engines). Consequently, these SUAVs (Small UAVs) are hardware-constrained devices that cannot be too heavy or carry heavy payloads. Besides, to the power consumption of the flight engines, it is essential to consider the additional energy required by onboarded computers, that may not be carrying their own external batteries and in case they were, extra weight would be added to the system. As a consequence, we find that the useful lifetime of a UAV system is undoubtedly limited by these restrictions. Different research works propose solutions to provide uninterrupted service on long endurance missions and overcome the reduced-battery challenge. For instance, Ref. [15] presents an algorithm to offer continuous structural inspection services using UAVs not only through simulation results, but also by using an implementation. In this solution, authors replace a UAV unit before its battery is drained. The replacement algorithm employed in this article will be used later in this paper to compare and contextualize the proposed solution. The modification of their methodology (since in this particular case, authors work with a single-UAV system while the proposed scenarios require moderately large fleets of UAVs) is explained and detailed in Section 5.2.2. In [16], authors consider UAV replacement (among other possible alternatives, such as refueling [17] or recharging) to maintain total surveillance of an area perimeter. Additionally, some articles propose an automatic battery replacement [18–20]. They offer a GCS capable of swapping UAV batteries without human interaction. Ground task automation not only reduces human interaction but also increases the multi-UAV system operation area, improving the coverage and enabling operation in hazardous environments. This trend makes us choose battery replacement as the preferred option in the solution proposed in this paper. Battery price is considerably lower than the cost of a UAV, and the time to replace the battery is remarkably shorter than the time to recharge it. Moreover, thanks to these studies and their practical experimentation, we use these results as input for our scheduling algorithms to provide accuracy to the design of UAV replacement strategies. Diverse works attempt to solve the limited battery life problem which is inherent to current SUAVs by proposing diverse alternatives. In [21], it is considered that the UAVs land to provide service (if possible and secure operation). The work in [22] summarizes different techniques to prolong the UAV operation time from Battery dumping [23] to Photovoltaic arrays [24,25]. Some other additional techniques have been proposed like wireless charging using lasers is in [26].

The optimization field, to improve the restricted communication performance of UAV networks while using the minimum amount of physical resources, is also an actual discussion topic in state of the art. In [27], the effective use of flight-time constrained devices is investigated, maximizing the average data service to ground users following a fair resource allocation policy. The solution of the cooperative allocation problem proposed in [28] significantly improves the performance of several network parameters. In [29], the authors try to minimize the number of vehicle-mounted BSs required to guarantee wireless coverage for a group of distributed ground users. Similar work in [30] proposes a placement algorithm for vehicle-mounted BSs that maximizes the number of covered ground users using the minimum UAVs. In [31], authors investigate the UAV coverage problem and propose a multi-UAV coverage model based on energy-efficient communication. The work in [32,33] focuses on the application of a multi-layout multi-subpopulation genetic algorithm achieving significantly better performance results than the other meta-heuristic algorithms also considered to improve the coverage deployment of multi-UAV networks. An explicit definition of the minimum-energy paths between a predefined initial and final configuration of a quadrotor by solving an optimal control problem concerning the angular accelerations of rotors is detailed in [34]. Their solution yielded minimum-energy and fixed-energy paths for the aerial vehicle.

3. Problem Statement

As it has just been mentioned, one of the main challenges in multi-UAV systems is to keep all the target geographic areas covered overtime by UAVs, since their battery life is limited (minutes) as compared to the typical mission timelines (hours). In order to face this problem, our approach is to use a fleet with the number of UAVs that are required to cover the whole scenario, and then maintain extra UAVs in a backup pool to serve as replacement units (as it can be seen in Figure 2). Once a replacement has been scheduled by the GCS, a fully recharged UAV enters the scenario while the replaced UAV goes back home to substitute its empty battery and be therefore ready to be changed by the next active UAV that requires a replacement. However, this procedure of identifying the minimum number of necessary (extra) UAVs and scheduling UAV replacements in the appropriate moment (to guarantee a minimum level of service availability) resembles a sophisticated approach and is the main problem that is treated in this paper.

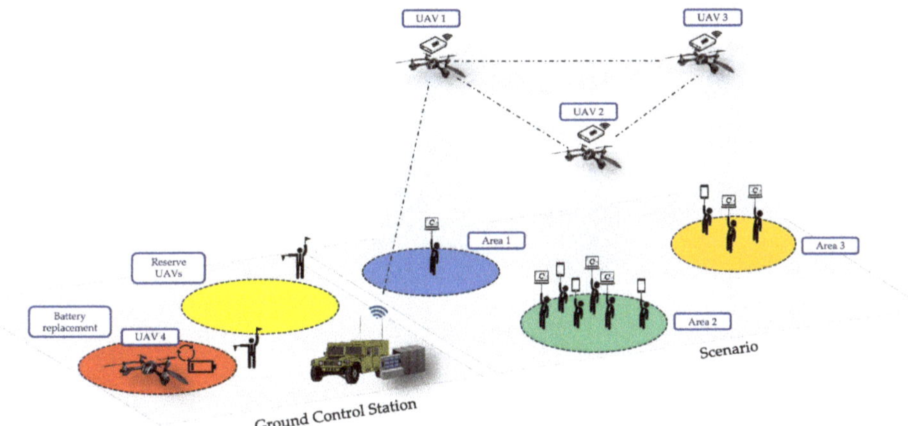

Figure 2. Multi-UAV system during a mission for three target areas ($j = 3$) and four UAVs ($i = 4$).

Figure 2 depicts the reference scenario considered in our analysis. In this scenario, different colors are used to represent different geographic areas, which encompass the target areas where UAVs are intended to provide network coverage to end-users, the geographic location where UAVs are directed for a battery replacement, and the specific area where the backup pool of UAVs is kept for subsequent use. These color patterns have been reproduced in Figure 3 to classify not only what task the UAVs are doing at a certain moment but also to show which UAVs are covering the target areas at any given time. The following subsection faces the practical UAV replacement problem from an optimization viewpoint stating a simplified and manageable procedure, checking its complexity, and solving it through different approaches (optimal brute force algorithm, heuristic algorithm).

Complexity Analysis

In this subsection, we prove that a simplified version of the proposed problem maps to an NP-hard problem (bin-packing problem [35] in this particular case) so that we are able to state its complexity. We denote a UAV using the index i and the target areas using the index j. Each UAV i has $C_{B_i}(t)$ battery level at instant t and the UAV might be in four different states: (i) battery replacement state (landed in the GCS), (ii) flying state (towards the GCS, or towards a region where it is intended to provide a network service), (iii) covering a region, or (iv) waiting in the reserve UAVs area to replace an active UAV. These four states can be appreciated in Figure 3. This diagram represents a hypothetical scenario with three regions ($j = 3$) and 4 UAVs ($i = 4$), also indicating when the replacements take place to guarantee system availability over time. Note that a region/area indicates where a UAV has to fly.

In the case (e.g., the number of users, a high volume of traffic) that two UAVs have to be geographically near, we consider two different areas.

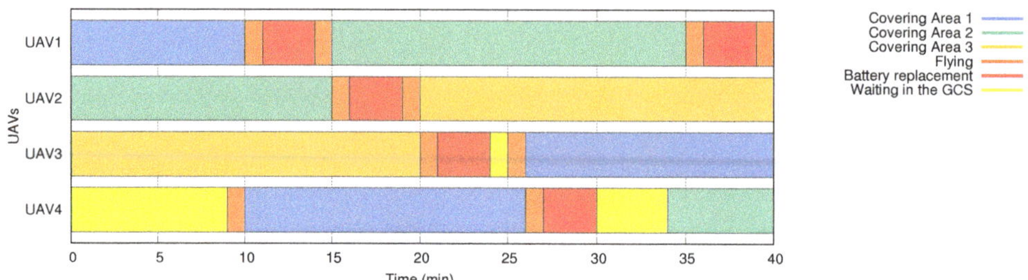

Figure 3. Multi-UAV system states during a mission for three target areas ($j = 3$) and four UAVs ($i = 4$).

This problem statement must guarantee each region j is always covered by a UAV, i.e.,

$$\sum_i x_{i,j}(t) = 1, \quad \forall j, \forall t \tag{1}$$

with $x_{i,j}(t) = 1$ whenever UAV i covers region j. Moreover, at any time t, the battery level of UAV i has to stay above a safe threshold a_j (e.g., 20%) for each region, so as to ensure the flight back to the GCS:

$$\sum_j a_j x_{i,j}(t) \leq C_{B_i}(t) y_i(t), \quad \forall i, \forall t, \tag{2}$$

here $y_i(t) = 1$ whenever UAV i is not in the GCS.

Additionally, battery levels keep on decreasing while UAV is covering a region. Otherwise, we consider its battery levels is set to 100% once it has returned to the GCS, and the operator has replaced the battery:

$$C_{B_i}(t+1) = \left(C_{B_i}(t) - c\right) y_i(t) + R_{T_d, T_r}\left(1 - y_i(t)\right), \quad \forall i, \forall t. \tag{3}$$

with c the battery consumption, and R_{T_d, T_r} the average battery charge ratio during the time spent in returning to the GCS T_d, and the operator replacement task T_r.

T_d remains constant in this simplified version of the problem, no matter how far a UAV i is from the region j it was covering, to the GCS.

The main goal of this problem is to minimize the number of active UAVs over time:

$$\min \sum_{i,t} y_i(t). \tag{4}$$

This optimization problem with objective function (4), and constraints (1) and (2), maps to the bin-packing problem. Notice that this simplified problem has as bins the drones and as items the areas. Battery levels $C_{B_i}(t)$ are just the bin capacities (Note that having time-dependent variables corresponds to having t repeated in such variables multiple times, i.e., with $t = \{1, 2\}$, $y_i(t)$ is expressed as two different variables $y_{i,1}$ and $y_{i,2}$.), and the battery threshold of each region j becomes the items' weights. Thus, constraint (2) is just the bin-packing restriction that prevents exceeding bin capacities. Furthermore, constraint (1) imposes that all items (our regions j) are fitted inside a bin.

Without considering (2), we already have an instance of the bin-packing problem. Since this makes some instances of our problem being NP-hard, our reduced problem automatically becomes NP-hard. Then, the next step is to generate a heuristic algorithm that will provide a sub-optimal solution. At the same time, it is required to develop a methodology that will enable the algorithm evaluation.

4. Methodology

This section describes the different elements depicted in Figure 1 and explains the steps to be followed by the mission planner to provide uninterrupted network services. It first describes the parameters that UAVs must report to the GCS in order to serve as input for the scheduler algorithms. Then, it details the diverse assumptions taken for system modeling, that enable simulations to evaluate the preliminary proposals. Later, it presents the metrics to assess the performance of the proposed solutions. Finally, it describes the different strategies used in this article to schedule UAVs replacements.

4.1. Reported Parameters

Current UAV systems regularly report to their control station their location (GPS coordinates if the UAV incorporates this type of navigation) and the remaining battery. However, this knowledge may not be enough to have a holistic view of the UAV network which enables the scheduler algorithm to satisfy the objective function (4) of minimizing the number of required UAVs to provide guaranteed service availability. This is the list of parameters periodically reported by the UAVs to the GCS that enable the calculation of essential inputs for the scheduler algorithms that will be defined later to make the appropriate replacement scheduling:

- GPS coordinates: longitude, latitude, and altitude enable the calculation of the distance between each UAV and the GCS. Consequently, taking into account the cruising speed of the UAV, it is possible to estimate the time, and also the required battery, needed to complete the replacement procedure.
- Remaining battery: the current value of the available battery, in combination with the historical battery values (last n values), allows calculating the average energy consumption. With these values, an approximation of the UAVs' lifetime can be determined.
- Network neighbors: the neighboring nodes enable us to generate a graph that represents the UAV network. With the GPS position and the theoretical wireless range, an overview of the network topology can be obtained. However, in certain circumstances, such as several packet collisions or high interferences, having a UAV nearby does not guarantee to have a proper communication channel established. Assuming that communications are bidirectional, for a network link to exist between two UAVs, both have to report each other to the controller, i.e., UAV_A reports UAV_B and UAV_B reports UAV_A. This functionality is deployed in the UAV payload equipment.
- Number of connected users: if a UAV acts as a BS or AP (it can also act as relay, video transmitter, telemetry/sensor transmitter), it must report the number of users that it is serving. This way, we can better determine the impact that a failure (disconnection) of this particular UAV causes in the network.

4.2. Assumptions

Using a discrete-time model and in order to provide a reasonable implementation of the UAV replacement strategies, it is required to make some assumptions (simplifications):

1. The UAVs in the fleet are all the same model. It also implies that all batteries have the same dimensions and therefore have the same capacity/duration. This approach is reasonable since handling UAVs that use the same battery model reduces the number of these in the GCS and also simplifies the battery exchange procedure.
2. As long as there is a topological path existing between two nodes, it is assumed that the network route is possible and it is configured, i.e., no time is needed to configure different routes when topology modifications happen. Unquestionably, the routing protocol used in the network may eventually affect the system but under normal circumstances, the convergence time is negligible [36].

3. The chosen path between two network nodes (UAVs or network users) is the shortest path based on the number of hops. A priori, this decision makes sense since taking the shortest path minimizes the delay (and actually reducing the delay is one of the main objectives of 5G). However, if the network has several users distributed heterogeneously, using different paths may be interesting to balance the network load and avoid packet collisions.

4. When a UAV is in flight, any non-flight related energy consumption (for instance due to wireless transmissions) is negligible [37]. Furthermore, UAVs usually incorporate two batteries [38]. The primary battery is in charge of supplying the flying engines while the secondary supplies the payload equipment. The secondary battery enables a static mode of operation and, in particular situations, UAVs may land to extend the life of the provided network service, stopping the flying engines while keeping the payload powered with its own battery [21]. Our laboratory experiments with the secondary battery (3.7 V and 3800 mAh) result in more than 2 h of duration, so the battery limiting the UAV operation is the primary one in any case. Moreover, the battery consumption model is linear and the same in all the UAVs (it does not depend on the flight conditions).

5. Because the price of batteries is remarkably lower than the cost of UAVs, it is assumed that the number of available batteries cells is huge. This way, the GCS never runs out of charged batteries. Batteries can also be recharged during the mission, and some of them could be reused.

6. UAV payloads have enough computing capacity, consequently not saturated under any conditions. In our previous work [38], we have carried out experiments using Raspberry Pi 3B single board computers to prove their correct functioning.

Although all these assumptions may affect the results of the simulations, the primary purpose of this article is not to achieve accurate results but to verify that it is worth using a replacement scheduler algorithm to manage moderately large UAV fleets. Once this hypothesis is demonstrated, the progressive replacement of each simplification opens interesting future work for the evaluation of more realistic results.

4.3. Performance Metric

The average number of users connected (over time) is used as the metric to evaluate the performance of UAV replacement strategies. Each sampling period (e.g., 5 s in our simulations), this metric is examined in order to calculate the percentage of end-users connected to the GCS, which is in charge of providing Internet connectivity (i.e., the number of end-users that through a path established across the UAVs network are actually connected to the GCS). The average value of all the partial results during the simulation time will be used as performance metric.

4.4. Scheduler Algorithm Proposals

The following subsection outlines the strategies that have been taken into account when performing the simulations. Obtaining the optimal solution and defining a heuristic algorithm is part of the optimization process. The optimal solution will not predictably serve for large scenarios (in a reasonable time), but it will validate the heuristic algorithm in small scenarios for its future application in real environments. A summary of the parameters that describe the proposed UAV replacement strategies is shown in Table 1.

4.4.1. Optimal Algorithm

To find the optimal UAV scheduling strategy that minimizes the number of UAVs used to cover a certain analysis region and a given number of users, the a brute-force algorithm has been proposed. This algorithm is an evolution of the strategy developed by us and presented in [8], which has incorporated positioning information, a number of users, and specific parameters related to the displacement between the GCS and the regions to be covered (i.e., landing time, take off time, and cruising speed of UAVs). In this regard, the proposed algorithm can be seen as an evolution of the approach addressed in [8]. The UAV scheduling strategy is explained in Algorithm 1, and its operation can be summarized in the following three stages.

Algorithm 1: Optimal UAV battery replacement strategy.

Input: GPS GCS and regions, connected users, j, i, C_B, d, L, v, T, T_l, T_w, T_s
Output: UAV ID, replacement time

1   Compute parameters for UAV allocation: T_B, T_r
2   Compute Region ranking   `// From first to last, based on the number of possible disconnected`
     `users`
3   **while** *Analysis time* $< T_w$ **do**
4   **if** *Topological change* **then**
5   Compute Region ranking
                                 `// Combinatorial analysis`
6   **foreach** *UAV* $\in i$ **do**
7   **foreach** *Region* $\in j$ **do**
8   Create *pair(UAV, Region)*
9   *SetPairsUAVRegion*
10   **foreach** *pairUAVRegion* $\in$ *SetPairsUAVRegion* **do**
11   Combinations of j elements between *pairUAVRegion* with other pairs $\in$ *SetPairsUAVRegion*
12   Compute service availability
13   *SetCombinationsUAVRegion*
14   Sorting based on a decreasing service availability level
15   Selection of the best combination: *UAVpreallocated*   `// Analysis of battery charge levels`
16   **foreach** *UAV* $\in$ *UAVpreallocated* **do**
17   **foreach** *Battery level* $\in \left\lceil \frac{T_B}{T_s} \right\rceil$ **do**
18   Create *infoUAVBattLevel*   `// Vector with battery levels for each UAV`
19   Compute service availability
20   Computer number of UAVs for next allocation process
21   *SetInfoUAVBattLevel*
22   Sorting based on a decreasing service availability and number of UAVs in reserve
23   Selection of the best battery levels for each UAV
24   Perform UAV allocation to designated regions   `// UAV allocation process`
25   Update available UAVs in reserve for replacements   `// Unused UAVs and UAVs to which the battery`
     `has been replaced`

- Computation of parameters for UAV allocation. This procedure consists of calculating the lifetime of each UAV (i.e., the battery lifetime to exclusively provide the service in the designated region) and its corresponding replacement time, considering the information about the locations of the GCS and the service regions (GPS coordinates) as well as the parameters v, T_o and T_l. In this step, a priority level or ranking is also assigned to each region according to the number of users that can be affected (disconnected) directly or indirectly if the UAV allocated to that region, and acting as an AP, suffers a failure. Thus, a higher priority level corresponds to a region that, if is not covered (no UAV allocated), it produces a higher number of disconnected users directly or indirectly (AP in that location with a link or links to other locations). This information is used in the process of allocation of UAVs to each region (next step), and ensures that fewer users are affected if $i < j$ at a given time.
- Optimal distribution of UAVs to cover the service regions. Through a brute-force analysis, all possible combinations of available UAVs to cover the different service regions are explored. The best distribution (combination) of UAVs, which consists of those whose characteristics (battery duration) allow for the highest service availability time, is systematically selected. If necessary,

a detailed example of the combinatorial analysis of UAVs to cover services is described in Table 4 in [8].

- Analysis of the percentage of battery charge to perform the replacement. With the information from the previous step (UAV allocation per service region), the algorithm analyses the optimal charge level for each UAV in which the corresponding replacement must be performed. This procedure is carried out by means of an exhaustive exploration of each level of charge for every UAV, and seeks to guarantee the highest service availability time and efficient use of the available resources (minimization of the number of UAVs for replacements). In a traditional approach, as shown in Figure 4a, replacement is performed when the battery capacity reaches its minimum threshold ($T_B = 75$ s in the example). Although this procedure allows the full capacity of the battery to be used, the simultaneous discharge of several or all UAVs may cause a greater demand of resources (UAVs) for the subsequent allocations (in the worst case $i = j$) and unavailability of one or several regions if there are no UAVs available for replacements. On the contrary, a desynchronization in the replacement time, as shown in Figure 4b, allows not only a greater availability of services but also minimization of the number of UAVs in the system. In the example presented in Figure 4a, four UAVs are required (two UAVs in services and two in the reserve) to guarantee a service availability equal to 100%, whereas in Figure 4b, only three UAVs are necessary to reach the same availability level. Once the algorithm has determined the charge levels for a replacement that allow the maximum service availability and the maximum number of UAVs available for the next allocation, these UAVs are allocated to their corresponding regions. The allocation process continues iteratively (i.e., execution of step two and step three) until reaching the maximum time horizon T_w, as shown in the example in Figure 4b with $T_w = 100$ s.

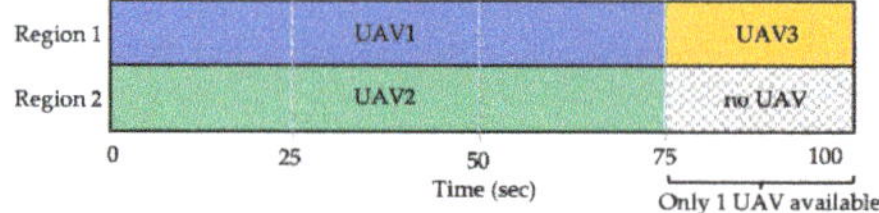

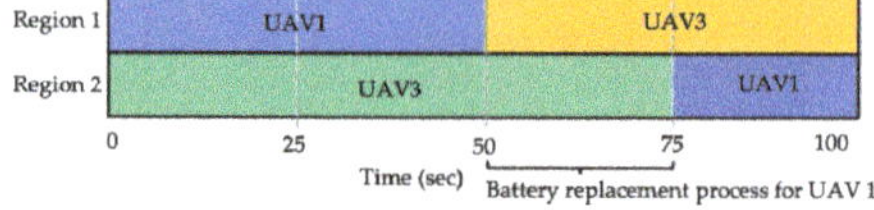

(a) UAV replacement considering all battery consumption

(b) UAV replacement to offer the maximum service availability and the highest number of UAVs for the next allocation

Figure 4. Differences between the analysed scheduling procedures. Example for $j = 2$ and $i = 3$ (2 UAVs in services and 1 UAV for replacement).

The proposed optimal strategy is an offline exhaustive search mechanism whose complexity is given by (5)

$$f(i,j,T_B,T_s) = C(i \times j, j) + \left(\left\lceil \frac{T_B}{T_s} \right\rceil \right)^j \tag{5}$$

where the first term represents the combinatorial analysis for the allocation of UAVs and the second term corresponds to the analysis of the charge levels for replacements. Both terms in (5) are non-polynomial, the first term is the dominant and, according to the Big-O classification [39], the order of growth of the algorithm is $O(C(ixj, j))$, i.e., non-polynomial. Based on preliminary tests we can report that if the UAVs have the same characteristics (i.e., equal battery capacity) the analysis in step three (exploration of the charge levels) is only necessary for the first allocation process, because desynchronization is maintained all other allocations, as shown in Figure 5. While this mechanism can partially reduce the complexity of the algorithm, obtaining an optimal solution using exhaustive search limits it to real-time applications and reveals the drawbacks in selecting the number of regions and UAVs (at most $i = 12$ and $j = 6$). In this regard, this strategy can be used in planning stages to estimate the number of UAVs needed for a mission, such as an emergency or rescue scenario.

However, in these cases a suitable alternative is a strategy described in [8], because it is a more generic and less complex approach.

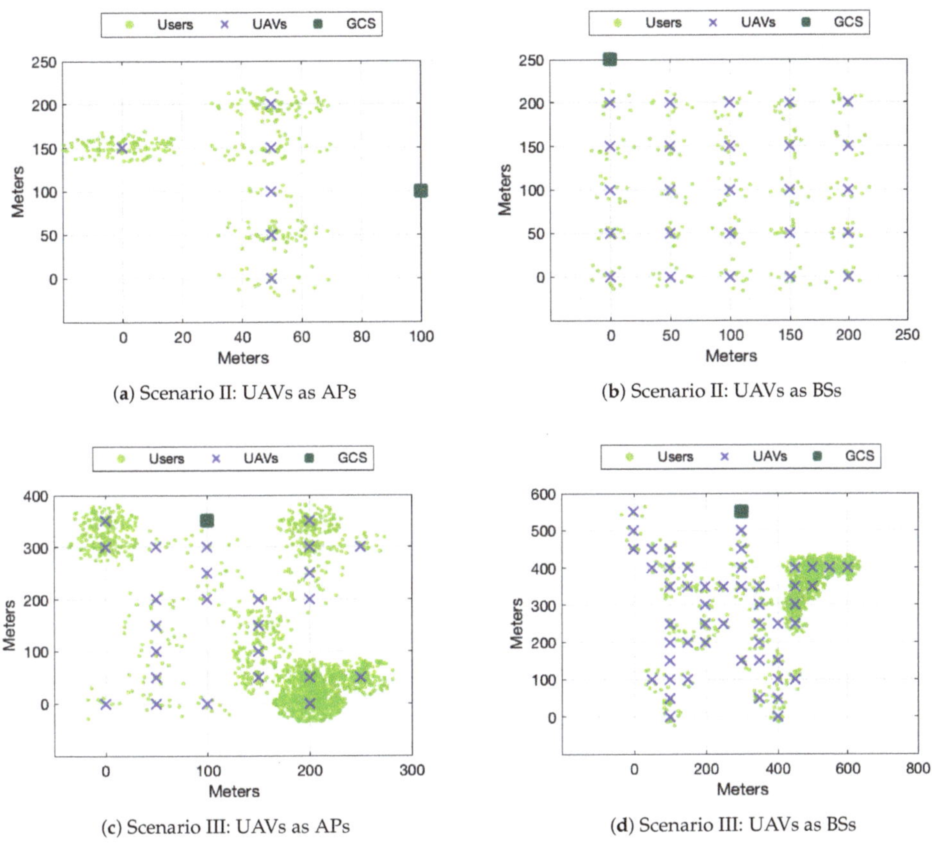

(**a**) Scenario II: UAVs as APs

(**b**) Scenario II: UAVs as BSs

(**c**) Scenario III: UAVs as APs

(**d**) Scenario III: UAVs as BSs

Figure 5. Proposed scenarios for algorithm performance evaluation.

Therefore, the hardness of the problem analysed in Section 3 (NP-Hard) and the complexity of the optimal solution shown in (5) (exponential) demonstrate the need for less complex heuristic mechanisms that can be used in real-time implementations. These strategies are described in the following sections and represent the major contributions of this paper.

4.4.2. BETA: Betweenness Centrality Heuristic Algorithm

Heuristic algorithms are employed to solve optimization problems that are out of scope in reasonable times by optimal algorithms. In this particular case, it is also essential that this heuristic algorithm has a fast execution time because it must be run in real-time. BETA schedules the replacements based on the relevance of each participant within the network. To determine the relevance of an area in a network scenario, we apply graph theory fundamentals. Each area/UAV (an area is covered by a UAV) would correspond to the graph vertices (also called nodes), while the links among UAVs correspond to the graph edges (also called links or lines). One of the most well-known metrics to identify which are the most significant vertices in a graph is centrality, more specifically the betweenness centrality, which resembles the number of times a vertice acts as a connection along the shortest path between two other nodes. However, in the proposed multi UAV networks, nodes do not communicate with other nodes randomly since they do it with those that have Internet connectivity to the public network (either the GCS or a 5G-enabled UAV), as this provides the ground users with Internet connectivity.

To formulate this custom metric, we have divided the graph into two sub-graphs: (i) sub-graph which is composed by those UAVs that do not have Internet connectivity and (ii) sub-graph which is formed not only by the GCS, but also by the UAVs that may eventually have connectivity to the core network. Therefore, due to these specifications, the centrality metric has been calculated in the following way:

$$g'(v) = \sum_{\substack{s \neq v \\ s \in A \\ t \in B}} \left(\frac{\sigma_{st}(v)}{\sigma_{st}} U_s \right), v \in A \tag{6}$$

where $\sigma_{st}(v)$ is the number of shortest paths from UAV s to UAV t that traverse UAV v, and σ_{st} the total number of shortest paths from UAV s to UAV t. U_s is the number of users connected to UAV U_s. The amount of users is crucial since if there are no users connected to UAV U_s, there is no impact on the network. This statement (6) (which is quite versatile) despite being designed for FANETs is also suitable for BS scenarios (where UAVs are directly connected to the core network).

A ranking was computed using the $g'(v)$ metric as input. In case two UAVs have the same value $g'(v)$ the one closer to the GCS will be above the other in the ranking since this will minimize the total replacement procedure time, and the replaced UAV will be active sooner to perform another UAV replacement. Now it is possible to assume which scheduling strategy to follow. BETA attends the following strategy (it can be appreciated in Algorithm 2): (i) if there is any topological change or the ground users move around the scenario, the algorithm must compute the ranking again; (ii) whenever there is a UAV available in the reserve, the algorithm schedules a replacement to the UAV with less remaining battery. However, this replacement takes place only if it does not affect UAVs that have a higher position in the ranking, i.e., that means that the remaining lifetime of the top UAVs is shorter than the time needed to make a UAV available again (after flying towards the GCS and battery replacement). In the case that this UAV replacement cannot be performed, the same analysis is repeated for the next UAV with the lower battery until the algorithm finds a UAV to make the replacement. For this algorithm to work correctly, it has to be executed periodically. In our case BETA runs every 5 s which coincides with the sampling period.

Algorithm 2: UAV replacement methodology.

    Input: GPS, remaining battery, neighbors, connected users
    Output: UAV ID, replacement time

1	Update UAV lifetime	`// based on remaining batteries and GPS`
2	Update battery ranking	`// based on remaining batteries`
3	**if** *Topological change* **then**	
4	Compute UAV ranking	`// based on neighbors and connected users`
5	**for** *UAVs battery ranking* **do**	
		`// From lowest to higher`
6	**if** *UAVs in reserve > 0* **then**	
7	time = time to replaced UAV to be available again	
8	warnings = find (time > lifetime)	
9	**if** *warnings < reserve UAVs* **then**	
10	Schedule replacement	

---

5. Simulation Details and Results

In order to validate these algorithms we have used different scenarios with different properties that will be discussed in this section. The following subsections detail *(i)* the simulation parameters and the justification of their selection, *(ii)* the ground-truth solutions with which the BETA algorithm is also compared, *(iii)* the simulation setup, and finally *(iv)* the simulated scenarios in combination with achieved results.

5.1. Simulation Parameters

This subsection details the parameters that have been taken into account to carry out the simulations and the selection criteria. This data, together with the related notation, can be seen in Tables 1 and 2.

Table 1. System parameters.

Parameter	Notation	Units/Coments
Number of regions	j	Integer number
Number of UAVs	i	Integer number
Location GCS	P_{GCS}	x,y coordinates
Location UAV$_i$	P_{UAV_i}	x,y coordinates
Number of users per region	u_j	Integer number
Total number of users	U	Integer number
Battery replacement time	T_r	Time units, e.g., seconds
Battery capacity	C_B	Electric current per time units, e.g., mAh
Device consumption	d	Electric current, e.g., mA
Link distance	L	Length units, e.g., meters
UAV cruising speed	v	Speed units, e.g., meters/seconds
Take-off time	T_o	Time units, e.g., seconds
Landing time	T_l	Time units, e.g., seconds
Simulation time	T_w	Time units, e.g., seconds
Sampling time	T_s	Time units, e.g., seconds

Table 2. Simulation parameters.

Scenario	Parameters											
	j	i	U	T_r	C_B	d	L	v	T_o	T_l	T_w	T_s
I	6	6–12	300									
II	25	25–50	250	180 s	2700 mAh	5670 mA	70 m	5 m/s	60 s	60 s	3600 s	5 s
III	25	25–50	300									
IV	50	50–100	500									

The time needed to perform a battery replacement is based on [15]. The battery capacity is based on the Parrot Bebop 2 specifications [7] (It is chosen because we have performed several tests using this model, and it is the selected unit in the technical validations we have worked previously [38,40,41], since it has demonstrated that it is able to carry a single board computer onboard like a Raspberry Pi for a reasonable time without problems and a reasonable cost). To calculate the device consumption, we have assumed that the UAV flies for 20 min (also specified in the technical characteristics). For WiFi range and although the standards state that the range is quite large, in practice, we have found that the WiFi range is relatively short for an acceptable received signal level [42]. The cruising speed has been calculated based on its maximum speed (also on technical specifications). Meanwhile, takeoff and landing times have also been calculated by our own measurements since we have not found accurate information. The simulations are iterated assuming a fixed number of areas to be served (and obviously one UAV per area) and then increasing the number of replacement UAVs (starting by 0 and increasing until the number of UAVs in reserve equals the number of UAVs in the scenario, which would mean doubling the size of the fleet).

5.2. Ground-Truth Solutions

To provide context to the BETA and optimal algorithms performance, they will both be compared with two alternative solutions (with smaller complexity). The primary purpose of the article is not to measure how far the heuristic solution is from the optimal but to highlight that the use of this type of solution is worthwhile and under which conditions and in which scenarios. In order to do that, the four scheduling techniques will be compared (BETA, optimal, baseline and simple scheduling) and different conclusions will be obtained

5.2.1. Baseline

This is the simplest strategy. UAVs are assumed to periodically send their current battery level and GPS position, however, no further calculations are made from the GCS. When an active UAV reaches a minimum battery threshold, i.e., only the required battery to return to the GCS plus a safety threshold, e.g., 20%, a replacement is scheduled (if UAVs are available), i.e., the drained UAV flights to the GCS, and at that moment (when the drained UAV starts flying to the GCS), a fresh UAV takes off and flies to the uncovered target area to provide the service. If no fresh UAVs are available, there will be no service in that area until a UAV is ready to go and cover it again. The lack of intelligence in this baseline solution prevents us from reaching 100% service provisioning in any case because even with infinite UAVs to serve as fresh replacements there will always be a gap without network service corresponding to the time that passes since the drained UAV leaves the stage towards the GCS until the moment the new UAV enters the stage and starts operating.

5.2.2. Simple Scheduling

This strategy is inspired by [15]. UAVs are also assumed to send their current battery level and GPS periodically. However, in this case, the controller is required to estimate a battery threshold (based on battery reports and other parameters such as the UAV speed and the takeoff and landing times) that includes not only the minimum battery needed to return to GCS but also includes the time needed for the fresh UAV (in case there are available units) to reach the target area. That way the new UAV will start serving the area just after the old one leaves and predictably, if there are enough UAVs in reserve, all the areas can be covered for the whole mission time (or at least a high percentage of the time). In case the active UAV reaches the threshold and there is no fresh UAV to perform the replacement, the UAV can still continue providing service until it reaches the battery level needed to reach the GCS enlarging the service time.

5.3. Simulation Setup

A Matlab (Matlab R2017b) event-based simulator achieves all the paper results. To calculate the BETA, simple scheduling, and baseline solutions in all the scenarios, a computer equipped with a 2.6 GHz Intel Core i5 processor and 8 GB RAM was used. Meanwhile, the optimal algorithm has been run on a computer equipped with a 3.33 GHz x 12 cores Intel Core i7 Extreme processor and 24 GB RAM. If the reader is interested in reproducing the experiment, the code is available in this repository [43].

5.4. Validation Scenarios

5.4.1. Scenario I: Proof of Concept

To start the analysis, we have defined a basic scenario (it can be appreciated in Figure 5a) as a proof of concept. In this stage, there are a total of 6 coverage areas and 300 ground users heterogeneously distributed. In a scene with these reduced dimensions, it is possible (in terms of reasonable computation time) to run the algorithm that provides the optimal solution so it will be possible to compare all the alternatives. Figure 6a depicts the average connected users for the four algorithms when the UAVs act as a FANET which means that UAVs onboard commodity WiFi equipment and use the created UAV WiFi adhoc network itself to connect to the GCS (which in turn provides the Internet connectivity). Therefore if one of the UAVs that are geographically closer to the GCS (and hence connecting part of the topology to the GCS) runs out of battery and there is no possible UAV replacement, some parts of the network may get disconnected even though the rest of UAVs may be successfully covering other target areas. Following this logic, whenever there is a failure in the backbone UAV, the system gets completely divided. On the other hand, Figure 6b depicts the average connected users for the four algorithms when UAVs act as BSs (they are directly connected to the public network without the need of a hop by hop network like a FANET). These scenarios are usually employed in massified events where the existing cellular network is operating correctly, but may be insufficient. As expected, these results are better than the FANETs results since each UAV is only responsible for its own end users. However these on-boarded BS solutions are usually more expensive and it is not always viable (when the infrastructure does not exist or is temporarily damaged for instance).

Figure 6a shows that both BETA and the simple scheduling strategies perform similarly and are close to the optimal solution. To reach 100% of connected users with the simple scheduling approach, it is required to double the UAV fleet (12 UAVs) but in any case in reduced scenarios, the simple scheduling solution is enough to provide an adequate service. On the other hand, the baseline algorithm provides erratic and unintuitive results considering that the performance decreases as the fleet increases. This phenomenon happens because although the time that UAVs are covering the target areas is higher, the network is disconnected for longer, i.e., having more UAVs does not guarantee overall connectivity if the backbone UAV is not working. If there are no reserve UAVs in reserve (the fleet size is equal to the number of target areas), the return and battery replacement process (of all the UAVs in the scenario) is almost synchronized (and operate simultaneously). However, if there are some UAVs in reserve, this process may be unsynchronized. For this reason, the baseline results decrease and, in consequence, are worse and inconstant.

The results in Figure 6b (UAVs acting as BSs) are better as we commented and again BETA and simple scheduling strategies are close to the optimal solution. The baseline solution performance improves in this case, as the size of the fleet increases. All the strategies in fact stabilize with a fleet of 8 UAVs, two in reserve (fleet 25% oversize), and both BETA and simple scheduling achieve acceptable values.

In scenarios with reduced dimensions this 25% of fleet oversize (having two UAVs in reserve) seems quite reasonable. However, in a scenario with numerous areas, e.g., 25 areas, 50 areas, this oversize may imply a rather expensive operation. It is then important to validate the solutions in much bigger scenarios and see the performance of the algorithms there.

5.4.2. Scenario II: Grid

Figure 5b shows a scenario with 25 coverage areas and 250 ground users homogeneously distributed, i.e., ten ground users per area. We have selected a grid topology which is fail-tolerant since there are multiple alternative paths to reach the GCS from each area. Moreover, all the areas have the same number of users, which makes the difference in the UAV ranking insignificant in the FANET scenario and almost nonexistent in the BSs scenario.

Figure 6c shows that the performance of the heuristic strategy is better than the simple scheduling solution (when the fleet is formed by 30 UAVs, 5 UAVs in reserve, the results improve by more than 10%). Both strategies reach acceptable levels from 35 UAVs fleet. The heuristic solution reaches 100% of users connected with a 38 UAVs fleet while the simple scheduling solution, as in the first scenario, needs to double the fleet size to reach 100% of connected users. On the other hand, the baseline solution has similar behavior to the previous scenario. This outcome highlights that if no strategy (however simple) is used to schedule the UAV replacements, the results can be harmful, and even over-dimensioning the resources does not guarantee favorable results.

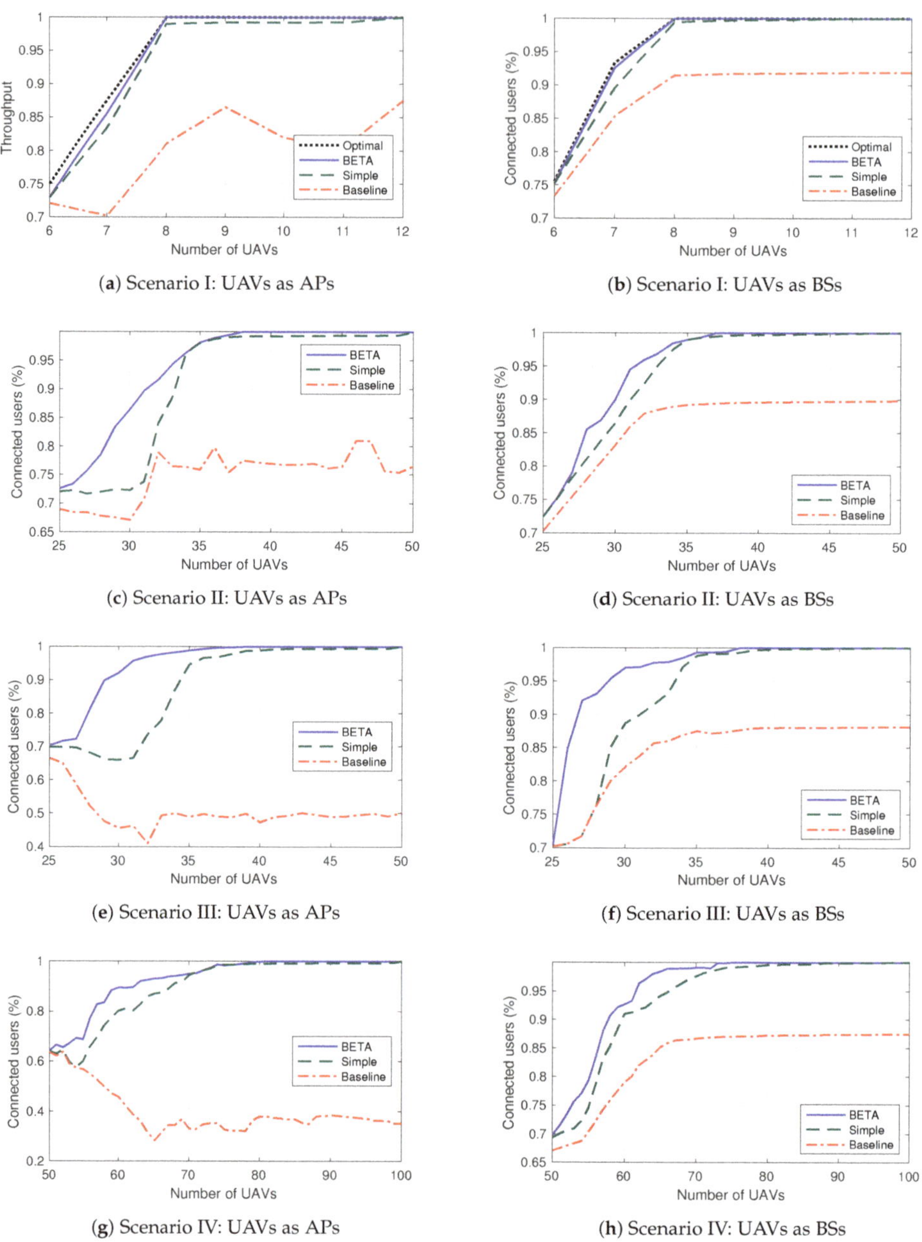

Figure 6. Average number of users connected in different scenarios increasing the fleet size.

Figure 6d presents the results of UAVs acting as BSs. In this case, the heuristic algorithm behaves similarly to the simple scheduling solution. The heuristic algorithm schedules the UAV replacements based on $g'(v)$ metric (6), which is determined using graph theory. In this scenario, the nodes representing the UAV network have the same $g'(v)$ since they are all directly connected to the infrastructure and provide connectivity to the same ground users; therefore, all UAVs connect the same number of users to the network. For this reason, scheduling the UAV replacements using the heuristic strategy has no advantage other than that they are performed as soon as there is an available fresh UAV. This phenomenon reveals that the heuristic solution makes the difference in scenarios where UAVs have different relevance within the network.

5.4.3. Scenario III and Scenario IV: Tree

Finally, we have designed two tree type scenarios with the users distributed very heterogeneously. This type of scheme makes some UAVs much more relevant, and scheduling replacements effectively seems to have a substantial impact on the final performance. The first scenario has 25 coverage areas and 300 ground users. The second scenario has 50 coverage areas, 500 ground users. The areas and user distribution can be appreciated in Figure 5c,d.

Figure 6e reveals that the difference between the BETA solution and the simple scheduling solution is significant in these scenarios. For a 30 UAVs fleet (5 UAVs in reserve), we achieve a 20% improvement by using the heuristic scheduler, which is an important variation when providing a network service. The heuristic strategy obtains 100% of users connected from 36 UAVs. As in previous scenarios, the baseline solution produces insufficient results. It is interesting to observe that as the UAV fleet increases (which have high economic cost), the users are not connected longer.

Similarly, when UAVs act as BSs, Figure 6f, we obtain better results using the heuristic strategy. This variation is because, in this scenario, the ground users are heterogeneously distributed, and consequently, UAVs have different $g'(v)$ since they connect diverse numbers of ground users, which implies that performing the correct replacement has more impact.

The conclusions of scenario IV are similar to the ones of scenario III, although the performance (Figure 6g,h) is worse because of the greater complexity of the UAV network topology and the greater failure possibility.

5.5. Comparison of the UAV Replacement Strategies

Once the results have been obtained (Figure 6) and discussed (Section 5.4), in this section we will perform a comparison of the results by computing: *(i)* for scenario I, the distance from the optimal solution (*Opt*) to the suboptimal or approximate solutions (*SubOpt*), in order to verify the quality of the results, and *(ii)* for the succeeding scenarios, the difference between the simple scheduling and baseline approaches against the Heuristic strategy. To this end, the criterion of approximation ratio (ρ) has been used [23]. This parameter, in the context of the proposal, estimates how many times lower the approximate solution is as compared to the exact or optimal result; it is defined as the average value of the ratio between the suboptimal and optimal solutions, as shown in

$$\rho = \frac{1}{i} \sum_i \frac{SubOpt_i}{Opt_i}, \tag{7}$$

where $SubOpt_i$ and Opt_i are the results for all the variation of UAVs in reserve (from zero to fleet size) for the optimal and suboptimal strategies, respectively. The ρ factor ranges from 0 (whether Opt and $SubOpt$ have completely different values) to 1 (if Opt and $SubOpt$ produce the same solution); an intermediate value (i.e., $0 < \rho < 1$) represents the similarity or closeness factor to the optimal solution ($SubOpt = \rho x Opt$) [23]. For a better understanding of the calculation of this parameter, (8) presents an example for the simple scheduling approach of Scenario I when UAVs act as APs (Figure 6a).

$$\rho = \frac{1}{7} \times \left(\frac{72.95}{74.99} + \frac{83.38}{87.51} + \frac{99.02}{100} + \frac{99.24}{100} + \frac{99.27}{100} + \frac{99.28}{100} + \frac{100}{100} \right) = 0.98. \tag{8}$$

The result of (8) shows that the suboptimal solution (simple scheduling approach) is similar to the optimal solution in a factor equal to 0.98 (98% similarity between solutions). The rest of the ρ factors for Scenario I (Figure 6a,b) are summarized in Figure 7, while the ρ values for other scenarios are presented in Figure 8. The comparison between the optimal solution and the approximate solutions in Scenario I, based on ρ factor, reveals that all the proposed strategies produce not only near-optimal solutions, but also a stable performance (i.e., high-quality feasible solutions). In all cases, as illustrated in Figure 6a,b and then corroborated in Figure 7, the approximate algorithms (BETA, simple scheduling and baseline) generate solutions very close to the optimal, even with 99% similarity to the exact result (1% of error), which is achieved by the BETA approach. Then, this strategy is used as a baseline to evaluate the performance of the other strategies (simple scheduling and baseline) for Scenario II, Scenario III, and Scenario IV. In summary, from the information analyzed in Section 5.4, the average number of connected users allows us to appreciate where one algorithm improves another, while the distance to the optimal solution computed in this section allows us to quantify this variation. To provide a higher level of detail, ρ has been analyzed by ranges depending on the fleet size (all the analysis has been carried out by increasing the number of UAVs gradually). In addition, this value has also been computed at a global level. The main reason is to analyze in which areas the solution improves and quantify it. Since from a particular value, the solutions provide similar results, computing this metric at a global level makes it difficult to recognize the areas of improvement.

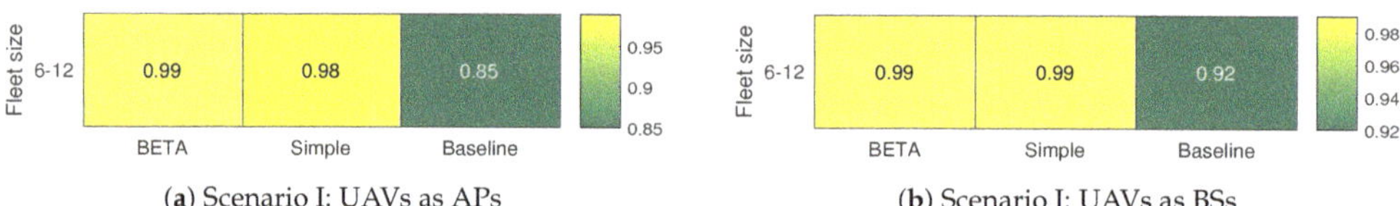

(**a**) Scenario I: UAVs as APs

(**b**) Scenario I: UAVs as BSs

Figure 7. Approximation ratio ρ: optimal strategy vs. heuristic strategies for Scenario I.

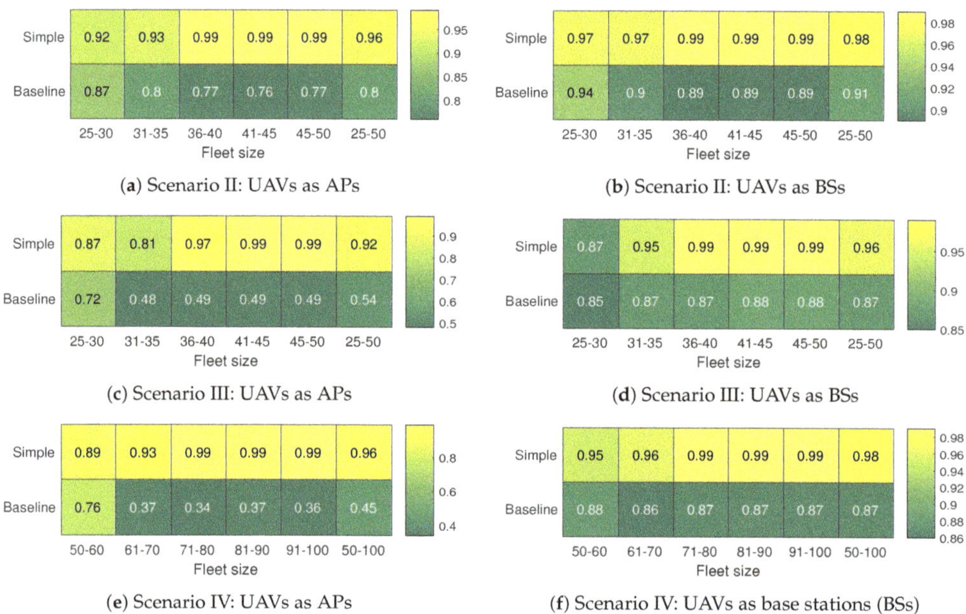

(**a**) Scenario II: UAVs as APs

(**b**) Scenario II: UAVs as BSs

(**c**) Scenario III: UAVs as APs

(**d**) Scenario III: UAVs as BSs

(**e**) Scenario IV: UAVs as APs

(**f**) Scenario IV: UAVs as base stations (BSs)

Figure 8. Approximation ratio ρ: betweenness centrality heuristic algorithm (BETA) vs. other suboptimal strategies.

It can be seen in Figure 8 that the central area of improvement is in the first two ranges, i.e., from 25 to 35 UAVs. This result is positive because, as expected, if there is a reasonable amount of UAVs (with their corresponding cost), a typical solution can perform adequately. However, in scenarios with limited resources using the heuristic strategy improves in all cases the simple scheduling solution.

Analyzing the above metrics, we can conclude that using strategies to make replacements is worthwhile. However, the heuristic strategy designed in this article is considerably aggressive since it schedules a replacement whenever UAVs are available. This strategy can result in the number of replacements skyrocketing over time, as well as the number of batteries to be used, which would bring a high economic cost. It should be noted that the price of a battery is much lower than that of a UAV but in any case it is not negligible.

Figure 9 displays the number of UAV replacements as a function of the number of UAVs forming the fleet and the mission time for scenario III using the BETA strategy. Approximately 1200 replacements are needed to provide a service of 10 h and obtain 100% of users connected. Furthermore, due to the aggressive nature of BETA, the replacement grows exponentially after reaching a fleet size that guarantees 100% of connected users. In addition, it should be mentioned that the main advantage of small UAVs is that deployments are generally done very quickly and very flexibly, but as we have seen it is both economically and logistically difficult, to achieve reasonable solutions when service time largely exceeds battery lifetime. Other alternatives should be used in these cases like bigger UAVs with more battery capacity (or even using fuel), land the UAVs on the ground to improve their autonomy, or even deploying of a fixed infrastructure if service is expected to be maintained for a long.

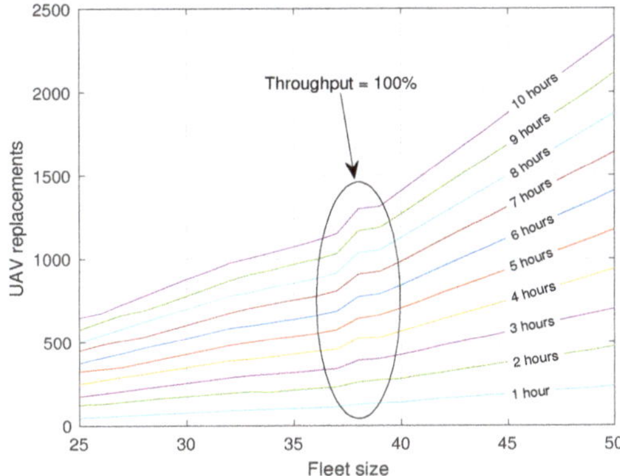

Figure 9. Number of UAV replacements using BETA in scenario III.

6. Conclusions and Future Work

This article states the practical UAV replacement problem, where a multi-UAV network is expected to provide long-endurance network services (in the order of hours) using constrained devices with limited autonomy (in the order of minutes). It is verified that the optimal UAV scheduling to minimize the number of UAVs for replacements while providing a guaranteed service availability, is NP-hard and that its optimal solution has exponential complexity. In this regard, some heuristics approaches have been analyzed and evaluated.

Secondly, the article details a methodology, including the simulation environment and the parametrization required to perform a preliminary evaluation of these heuristic strategies. The simulator code is available in [43] to reproduce the experiment and evaluate upcoming future strategies.

The article also introduces BETA, a heuristic replacement strategy that performs the replacements as soon as possible based on the relevance of each UAV within the network. BETA is presented as an example in order to verify if it is worthwhile using a heuristic replacement approach or not.

BETA is capable of running in real-time with a 99% similarity with the optimal solution in some simple scenarios (scenario I). In heterogeneous scenarios, BETA improves the basic solutions, achieving the most significant improvement in instances where the scenarios are heterogeneous, and the resources are limited. Furthermore, we conclude that it is far better to have a replacement strategy (no matter how simple it is) than having no strategy at all. BETA is compared with the optimal algorithm in order to evaluate the distance whenever possible and with other alternatives in some other scenarios and it has been possible to see that in some situations the advantages are not so relevant as in other ones.

The article opens several lines of future research, such as to be able to provide priority of replacements for UAVs serving users in emergency/disaster scenarios. The application of replacement strategies in disaster scenarios includes an uncertainty degree caused by several factors caused by moving UAVs while they are operating (that may change the topology), or extreme conditions that may force the engines and battery consumption. Some other futures lines include the combination of FANETs and BSs in the same scenario, to test UAVs models with different battery capacity, to model the energy consumption according to more realistic consumption patterns based on experimentation, or to schedule UAV replacements without making them through the GCS.

Author Contributions: Conceptualization, V.S.-A., F.V., I.V., C.T. and X.H.; methodology, V.S.-A., F.V., I.V., C.T. and X.H.; software, V.S.-A. and C.T.; validation, V.S.-A. and C.T.; formal analysis, V.S.-A., F.V. and C.T.; investigation, V.S.-A., F.V., I.V., C.T. and X.H.; data curation, V.S.-A. and C.T.; writing–original draft preparation, V.S.-A., F.V. and C.T.; writing–review and editing, V.S.-A., F.V., I.V., C.T. and X.H.; visualization, V.S.-A. and C.T.; supervision, F.V. and X.H.; project administration, F.V. and X.H.; funding acquisition, F.V. and X.H. All authors have read and agreed to the published version of the manuscript.

Funding: This article has been partially supported by the 5G-City project (TEC2016-76795-C6-1/3-R) funded by the Spanish Ministry of Economy and Competitiveness, and the H2020 5GRANGE project (grant agreement 777137).

Acknowledgments: The authors wish to thank Jorge Martín-Pérez for his guidance and support in the problem statement section. Christian Tipantuña acknowledges the support from Escuela Politécnica Nacional and from SENESCYT for his doctoral studies at UPC.

Conflicts of Interest: The authors declare no conflict of interest.

Abbreviations

The following abbreviations are used in this manuscript:

UAV	Unmanned Aerial Vehicle
5G	Fifth-generation cellular network
KPI	Key Performance Indicator
BS	Base Station
AP	Access Point
FANET	Flying Ad Hoc Network
GCS	Ground Control Station
SUAV	Small Unmanned Aerial Vehicle
GPS	Global Positioning System
BETA	Betweenness centrality heuristic algorithm
IoT	Internet of Things
LTE	Long Term Evolution
EPC	Evolved packet core
RAN	Radio access network
MEC	Multi-access edge computing
SUAV	Small Unmanned Aerial Vehicles
NP	Nondeterministic polynomial
RAM	Random Access Memory

References

1. Federal Aviation Administration. FAA Aerospace Forecast. Available online: https://www.faa.gov/ (accessed on 25 March 2020).
2. 5GPPP. Key Performance Indicators. Available online: https://5g-ppp.eu/kpis/ (accessed on 25 March 2020).
3. Hayat, S.; Yanmaz, E.; Muzaffar, R. Survey on unmanned aerial vehicle networks for civil applications: A communications viewpoint. *IEEE Commun. Surv. Tutor.* **2016**, *18*, 2624–2661. [CrossRef]
4. Grasso, C.; Schembra, G. A fleet of mec uavs to extend a 5g network slice for video monitoring with low-latency constraints. *J. Sens. Actuator Netw.* **2019**, *8*, 3. [CrossRef]
5. Moradi, M.; Sundaresan, K.; Chai, E.; Rangarajan, S.; Mao, Z.M. SkyCore: Moving core to the edge for untethered and reliable UAV-based LTE networks. In Proceedings of the 24th Annual International Conference on Mobile Computing and Networking, New Delhi, India, 29 October–2 November 2018; pp. 35–49.
6. Li, B.; Fei, Z.; Zhang, Y. UAV communications for 5G and beyond: Recent advances and future trends. *IEEE Internet Things J.* **2018**, *6*, 2241–2263. [CrossRef]
7. Parrot. Parrot Bebop 2. 2019. Available online: https://www.parrot.com/es/drones/parrot-bebop-2 (accessed on 25 March 2020).
8. Tipantuña, C.; Hesselbach, X.; Sánchez-Aguero, V.; Valera, F.; Vidal, I.; Nogales, B. An NFV-based energy scheduling algorithm for a 5G enabled fleet of programmable unmanned aerial vehicles. *Wirel. Commun. Mob. Comput.* **2019**, *2019*, 4734821. [CrossRef]
9. Merwaday, A.; Guvenc, I. UAV assisted heterogeneous networks for public safety communications. In Proceedings of the 2015 IEEE Wireless Communications and Networking Conference Workshops (WCNCW), New Orleans, LA, USA, 9–12 March 2015; pp. 329–334.
10. Zeng, Y.; Zhang, R.; Lim, T.J. Wireless communications with unmanned aerial vehicles: Opportunities and challenges. *IEEE Commun. Mag.* **2016**, *54*, 36–42. [CrossRef]
11. Bekmezci, I.; Sahingoz, O.K.; Temel, Ş. Flying ad-hoc networks (FANETs): A survey. *Ad Hoc Netw.* **2013**, *11*, 1254–1270. [CrossRef]
12. Sánchez-García, J.; García-Campos, J.; Arzamendia, M.; Reina, D.G.; Toral, S.; Gregor, D. A survey on unmanned aerial and aquatic vehicle multi-hop networks: Wireless communications, evaluation tools and applications. *Comput. Commun.* **2018**, *119*, 43–65. [CrossRef]
13. Khan, M.A.; Qureshi, I.M.; Khanzada, F. A hybrid communication scheme for efficient and low-cost deployment of future flying ad-hoc network (FANET). *Drones* **2019**, *3*, 16. [CrossRef]
14. Gupta, L.; Jain, R.; Vaszkun, G. Survey of important issues in UAV communication networks. *IEEE Commun. Surv. Tutor.* **2015**, *18*, 1123–1152. [CrossRef]
15. Erdelj, M.; Saif, O.; Natalizio, E.; Fantoni, I. UAVs that fly forever: Uninterrupted structural inspection through automatic UAV replacement. *Ad Hoc Netw.* **2019**, *94*, 101612. [CrossRef]
16. Burdakov, O.; Kvarnström, J.; Doherty, P. Optimal scheduling for replacing perimeter guarding unmanned aerial vehicles. *Ann. Oper. Res.* **2017**, *249*, 163–174. [CrossRef]
17. Fravolini, M.L.; Ficola, A.; Campa, G.; Napolitano, M.R.; Seanor, B. Modeling and control issues for autonomous aerial refueling for UAVs using a probe–drogue refueling system. *Aerosp. Sci. Technol.* **2004**, *8*, 611–618. [CrossRef]
18. Fujii, K.; Higuchi, K.; Rekimoto, J. Endless flyer: A continuous flying drone with automatic battery replacement. In Proceedings of the 2013 IEEE 10th International Conference on Ubiquitous Intelligence and Computing and 2013 IEEE 10th International Conference on Autonomic and Trusted Computing, Vietri sul Mere, Italy, 18–21 December 2013; pp. 216–223.
19. Ure, N.K.; Chowdhary, G.; Toksoz, T.; How, J.P.; Vavrina, M.A.; Vian, J. An automated battery management system to enable persistent missions with multiple aerial vehicles. *IEEE/ASME Trans. Mechatron.* **2014**, *20*, 275–286. [CrossRef]
20. Suzuki, K.A.; Kemper Filho, P.; Morrison, J.R. Automatic battery replacement system for UAVs: Analysis and design. *J. Intell. Robot. Syst.* **2012**, *65*, 563–586.

[CrossRef]

21. Sanchez-Aguero, V.; Valera, F.; Nogales, B.; Gonzalez, L.F.; Vidal, I. VENUE: Virtualized Environment for Multi-UAV Network Emulation. *IEEE Access* **2019**, *7*, 154659–154671. [CrossRef]

22. Lu, M.; Bagheri, M.; James, A.P.; Phung, T. Wireless charging techniques for UAVs: A review, reconceptualization, and extension. *IEEE Access* **2018**, *6*, 29865–29884. [CrossRef]

23. Chang, T.; Yu, H. Improving electric powered UAVs' endurance by incorporating battery dumping concept. *Procedia Eng.* **2015**, *99*, 168–179. [CrossRef]

24. Nugent, T.J., Jr.; Kare, J.T. Laser power beaming for defense and security applications. In *Unmanned Systems Technology XIII*; International Society for Optics and Photonics: Bellingham, WA, USA, 2011; Volume 8045, p. 804514.

25. Duncan, K.J. Laser based power transmission: Component selection and laser hazard analysis. In Proceedings of the 2016 IEEE PELS Workshop on Emerging Technologies: Wireless Power Transfer (WoW), Knoxville, TN, USA, 4–6 October 2016; pp. 100–103.

26. Galkin, B.; Kibilda, J.; DaSilva, L.A. UAVs as mobile infrastructure: Addressing battery lifetime. *IEEE Commun. Mag.* **2019**, *57*, 132–137. [CrossRef]

27. Mozaffari, M.; Saad, W.; Bennis, M.; Debbah, M. Wireless communication using unmanned aerial vehicles (UAVs): Optimal transport theory for hover time optimization. *IEEE Trans. Wirel. Commun.* **2017**, *16*, 8052–8066. [CrossRef]

28. Sharma, V.; Srinivasan, K.; Chao, H.C.; Hua, K.L.; Cheng, W.H. Intelligent deployment of UAVs in 5G heterogeneous communication environment for improved coverage. *J. Netw. Comput. Appl.* **2017**, *85*, 94–105. [CrossRef]

29. Lyu, J.; Zeng, Y.; Zhang, R.; Lim, T.J. Placement optimization of UAV-mounted mobile base stations. *IEEE Commun. Lett.* **2016**, *21*, 604–607. [CrossRef]

30. Alzenad, M.; El-Keyi, A.; Lagum, F.; Yanikomeroglu, H. 3-D placement of an unmanned aerial vehicle base station (UAV-BS) for energy-efficient maximal coverage. *IEEE Wirel. Commun. Lett.* **2017**, *6*, 434–437. [CrossRef]

31. Ruan, L.; Wang, J.; Chen, J.; Xu, Y.; Yang, Y.; Jiang, H.; Zhang, Y.; Xu, Y. Energy-efficient multi-UAV coverage deployment in UAV networks: A game-theoretic framework. *China Commun.* **2018**, *15*, 194–209. [CrossRef]

32. Reina, D.; Tawfik, H.; Toral, S. Multi-subpopulation evolutionary algorithms for coverage deployment of UAV-networks. *Ad Hoc Netw.* **2018**, *68*, 16–32. [CrossRef]

33. Reina, D.; Camp, T.; Munjal, A.; Toral, S. Evolutionary deployment and local search-based movements of 0th responders in disaster scenarios. *Future Gener. Comput. Syst.* **2018**, *88*, 61–78. [CrossRef]

34. Morbidi, F.; Cano, R.; Lara, D. Minimum-energy path generation for a quadrotor UAV. In Proceedings of the 2016 IEEE International Conference on Robotics and Automation (ICRA), Stockholm, Sweden, 16–21 May 2016; pp. 1492–1498.

35. Martello, S.; Vigo, D. Exact solution of the two-dimensional finite bin packing problem. *Manag. Sci.* **1998**, *44*, 388–399. [CrossRef]

36. Nogales, B.; Sanchez-Aguero, V.; Vidal, I.; Valera, F. Adaptable and automated small uav deployments via virtualization. *Sensors* **2018**, *18*, 4116. [CrossRef]

37. Zeng, Y.; Zhang, R. Energy-efficient UAV communication with trajectory optimization. *IEEE Trans. Wirel. Commun.* **2017**, *16*, 3747–3760. [CrossRef]

38. Nogales, B.; Vidal, I.; Sanchez-Aguero, V.; Valera, F.; Gonzalez, L.F.; Azcorra, A. Automated Deployment of an Internet Protocol Telephony Service on Unmanned Aerial Vehicles Using Network Functions Virtualization. *JoVE (J. Vis. Exp.)* **2019**, *153*, e60425. [CrossRef]

39. Garey, M.R.; Johnson, D.S. *Computers and Intractability: A Guide to the Theory of NP-Completeness*; W. H. Freeman: New York, NY, USA, 1990; pp. 37–79.

40. Telemadrid. La Universidad Carlos III junto al Instituto Imdea Networks Desarrolla un dron Antincendios. Available online: http://www.telemadrid.es/programas/telenoticias-1/Universidad-Carlos-III-Instituto-Networks-2-2208699163--20200228032649.html (accessed on 25 March 2020).

41. RTVE. Zoom Net—5GRange. 2020. Available online: https://www.rtve.es/m/alacarta/videos/zoom-net/zoom-net-5g-dive-entrevista-shou-zi-chew-dreams/5526638/?media=tve (accessed on 25 March 2020).

42. Sanchez-Aguero, V.; Nogales, B.; Valera, F.; Vidal, I. Investigating the deployability of VoIP services over wireless interconnected micro aerial vehicles. *Internet Technol. Lett.* **2018**, *1*, e40. [CrossRef]

43. Sanchez-Aguero, V.; Tipantuña, C.; Valera, F. Energy-Aware Management Strategies in Multi-UAV Fleets Repository. Available online: https://github.com/vsaguero/energyUAV (accessed on 25 March 2020).

Article

A Decentralized Low-Chattering Sliding Mode Formation Flight Controller for a Swarm of UAVs

Thiago F. K. Cordeiro [1], João Y. Ishihara [2] and Henrique C. Ferreira [2,*

[1] Gama Campus, University of Brasília, Brasília 72444-240, Brazil; thiagocordeiro@unb.br
[2] Department of Electrical Engineering, Darcy Ribeiro Campus, University of Brasília, Brasília 70910-900, Brazil; ishihara@ene.unb.br
[*] Correspondence: henrique@ene.unb.br

Received: 30 March 2020; Accepted: 7 May 2020; Published: 30 May 2020

Abstract: In this paper, a nonlinear robust formation flight controller for a swarm of unmanned aerial vehicles (UAVs) is presented. It is based on the virtual leader approach and is capable of achieving and maintaining a formation with time-varying shape. By using a decentralized architecture, the local controller in each UAV uses information only from the UAV itself, its neighbors, and from the virtual leader. Also, a synchronization control objective provides a mechanism to weight between the fleet achieving the desired formation shape, that is, achieving the desired relative position between the UAVs, and each UAV achieving its desired absolute position. The use of a combination of a sliding mode controller and a low pass filter reduces the usual chattering effect, providing a smooth control signal while maintaining robustness. Simulation results show the effectiveness of the proposed decentralized controller.

Keywords: unmanned aerial vehicle; synchronized multi-agent formation; decentralized sliding mode control

---

1. Introduction

The use of an unmanned aerial vehicle (UAV) swarm brings several advantages in search and rescue, disaster monitoring, aerial mapping, traffic monitoring, reconnaissance missions, and surveillance [1–3]. A swarm of UAVs provides system redundancy, reconfiguration ability, and structure flexibility, being more effective, flexible, robust, and reliable than single vehicles [4,5]. The formation control is a critical task of attempting cooperation among UAVs. In general, a formation control problem is to find a coordination scheme to enable UAVs to reach and maintain some desired, possibly time-varying formation or group configuration [6].

In the view of communication networks, the existing formation control approaches can be classified into the centralized method, where a single controller is used to control the whole team based on the information from the whole team [7] and the decentralized method, where each team member generates its own control based on local information from its neighbors [1,2,4,8–11]. Centralized formation control can be a good strategy for a small team of UAVs. When considering a team with a large number of UAVs, the need for greater computational capacity and a large communication bandwidth would mandate a decentralized formation control [4].

The main structures considered for formation control of UAVs swarm are leader-follower, behavioral, and virtual structure/virtual leader [12,13]. In the leader-follower approach [3,6,10,14], a common leader is chosen and the rest of the agents are assigned as followers. The group leader broadcasts its position information to the followers who then begin to follow the leader at an offset. In the behavioral approach [15,16], several desired behaviors are prescribed for agents in this approach. Such desired behaviors may include cohesion, collision avoidance, obstacle avoidance. In the virtual structure/virtual leader approach [4,9,11], the entire formation is treated as a single rigid body.

The virtual structure can evolve as a whole in a given direction with some given orientation and can maintain a rigid geometric relationship among multiple vehicles. In the virtual leader approach, the leader is a known virtual entity and its information can be made available for each agent software.

There are several control techniques used in UAV formation control, based on distinct premises and aiming to achieve distinct objectives. A common approach is to use a nonlinear dynamic inversion (NLDI) which, via nonlinear functions, encapsulates the nonlinear system in a box with virtual inputs/outputs that behaves as a linear system. This linearized system act as a set of double integrators, that then is controlled by any linear or nonlinear technique, such as pole placement [14,17], $\mathcal{H}_\infty$ control [9], differential game approach [10] or sliding mode control (SMC) [8,18]. There are two main approaches when using the NLDI fixed-wing UAV formation flight control, related to in which frame the whole formation is described. One is to choose a global frame, such as north-east-down [9,10], and the virtual inputs are accelerations in north/south, east/west and up/down directions. Other is to use a leader related frame [8,14,17,18], and the virtual inputs accelerates toward forward/backward, left/right and above/below the leader.

In References [14,17], classic controllers were designed after applying of NLDI procedure in the nonlinear dynamics of UAVs formation flight. In Reference [10], a differential game approach is used to achieve an optimal controller that weights between minimizing the terminal position and velocity error of each UAV and minimizing the control effort. Another option, as in Reference [2], is the model predictive control (MPC), which can be used to compute an optimal control output to achieve formation control while avoiding obstacle and dealing with actuator saturation. It is however computationally expensive, since it reevaluates, at each time instant, the optimal control output over a finite time horizon. In Reference [2], the computational cost is partially reduced by maintaining the previously computed control output and reevaluating only when certain trigger events indicates that the control output must be changed, which works well in steady maneuvers, such as straight level flight or in constant-rate turns. In all of these approaches [2,10,14,17], the project does not account for the effect of disturbances or model uncertainty. Robust [8,9,11,18,19] and adaptive [1] approaches are appropriate for tackling this problem, where robust approaches usually has fast response but has a high control effort and/or chattering, whereas adaptive approach has a slower convergence, but uses a smoother control signal. In this way, the robust approach is recommended if precision of the formation is more important than control effort. In Reference [9], the proposed $\mathcal{H}_\infty$ linear controller is robust to noises, disturbances, and delays in communication between the UAVs. The SMC is an interesting technique since it, ideally, can completely compensates the effects of model uncertainties and bounded disturbances. As disadvantage, it provides a discontinuous, chattering control signal whose source is a signum (sign) function [11,19,20]. A possible solution is to change this signum function to a saturation (sat) function, as in our previous work for UAV formation flight [18], but this generates a trade-off between precision and chattering. Another solution is the use of a second-order SMC (SOSMC) [21,22], which uses the integral of the chattering signal as control input of a plant. A generalization of the second-order SMC is the low-pass filter (LPF) [23–26]. The integral, or the more general low-pass filter, is included as part of the plant model, which improves precision. For example, in Reference [23] an architecture with sliding mode control and low pass filter is proposed for synchronous position control for multiple robotic manipulator systems. However, its control law involves the computation of derivatives whose order is higher than the plant order, which can difficult implementation for example in an embedded system. In Reference [24] an attitude controller for the reentry of a space vehicle based on low pass filter SMC architecture is proposed. Differently from, for example, Reference [23], the LPF in Reference [24] is used to filter only the signal component that contains a signum function, while bypassing the smooth component of the control signal directly to the plant. This approach, compared to the approach in Reference [23], avoids the computation of higher-order derivatives.

In this paper, using an LPF-based SMC approach, a decentralized controller for a time-varying synchronized formation of multiple UAVs with a virtual leader is proposed. It is considered that each UAV is subject to unknown bounded disturbance. The computation of higher-order derivatives is not required. This is achieved by decomposing the control signal in smooth and in chattering components and filtering only the chattering component. Compared with Reference [24], which considers a single space vehicle, the proposed controller considers a synchronized and decentralized formation of multiple UAVs. In a synchronized formation, multiple UAVs simultaneously converge to desired positions. In comparison with Reference [23], which uses LPF SMC for synchronized position control for multiple robotic manipulator systems in a ring-link communication topology, the proposed controller not require computation of higher-order derivatives, a more general information exchange topology is adopted, and the problem of UAV formation flying is considered. Different from our previous work [18], the proposed controller uses an LPF for chattering attenuation. The finite-time convergence to a linear sliding surface is proven by introduction of an appropriated Lyapunov function candidate and simulation results show the effectiveness of the proposed control architecture.

To use the LPF-based SMC, the upper bound of the disturbance must be known. Most of this disturbance is better described in the wind frame of the aircraft. For example, a model uncertainty can affect the lift force computation. The difference between the true and computed lift forces is equivalent to a disturbance force applied in the lift direction. Similar discussion can be made of the thrust, drag and side forces. If the NLDI linearizes the system in the leader's wind frame [8,14,17,18], this upper bound can be used directly, under the assumption that the leader's wind frame is similar enough to the followers' wind frames, as the fleet in formation flies approximated to the same direction. If, however, the NLDI linearizes the system in a global frame [9,10], the wind-frame-described upper bound must be translated to the global frame. The equations to translate the disturbance upper bound to the global frame are developed in this paper.

The main contribution is now summarized. A formation flight controller is developed that includes, in a single controller, the following characteristics:

- uses the robust sliding mode control technique [8,11,18,19];
- has low-chattering with low degradation in performance by the use of a low-pass filter modelled as a plant component [23–26];
- uses a variant of the LPF SMC that is mathematically and computationally simpler than the usual approach, and removes computation of higher-order derivatives [24];
- is a multi-agent decentralized/synchronous approach [18,23].

It is worth noting that each individual characteristic of the controller listed above has already been developed in other papers but, to the best knowledge of the authors, there is no controller that includes all characteristics in a single controller. It is also worth noting that, to include all characteristics in a single controller, an appropriate Lyapunov that unifies theses characteristics is developed.

As a second contribution, a set of equations that translate the disturbance upper bound and is derivative upper bound from the wind frame to the global frame is proposed. These equations are used in the proposed LPF-SMC, but can be used, with minor modifications, in most fixed-wing formation SMC or SOSMC that are described in a global frame.

The remainder of this paper is organized as follows. Section 2 defines the problem, presents the mathematical models for the UAVs, formation flight, and communication graph. Section 3 presents the proposed controller, equations to compute the disturbance's upper bound, and proves the stability of the controller. Section 4 evaluates the proposed controller by simulation against an unfiltered SMC, where it is shown that the controller significantly reduces its chattering without significantly reducing its performance, and Section 5 concludes this paper.

2. Preliminaries

In this section, models for an individual UAV and for a fleet formation are presented.

2.1. UAV Model

The dynamics relating the input and output of the *i*-th vehicle in a fleet of n UAVs can be described by using the so-called point-mass aircraft model. It assumes a non-rotating flat Earth with a constant gravitational acceleration g. This model provides adequate precision to aircraft guidance and control problems and for short-range trajectory planning. It also assumes that the intensity of the wind is mild such that the airflow can be considered aligned with the vehicle fuselage, that is, that the angle-of-attack and sideslip angle are null, which are reasonable suppositions to cruise flight and coordinate maneuvers. Under these assumptions, as depicted in Figure 1, the drag force $\mathbf{D}_i(t)$, generated by the airflow, is aligned to the fuselage, pointing backward, whereas the lift force $\mathbf{L}_i(t)$ is perpendicular to the fuselage. It is assumed that the propulsive system provides a thrust force vector $\mathbf{T}_i(t)$ aligned with the fuselage/airflow and in the opposite direction to $\mathbf{D}_i(t)$ and that the net angular momentum generated by the propulsive system is null. Finally, it is assumed that the vehicle mass m_i is (approximately) constant, i.e., the propulsive system is electric or if fuel-based, that the consumption is small compared to the total vehicle mass. These simplified suppositions are commonly used in the literature [9,10,27,28]. To achieve extra precision in more aggressive maneuvers, the effect of the angle-of-attack and sideslip angle should be included in the model [14]. In general, these quantities require a dedicated sensory system—which is usually not present in small UAVs—to these angles be measured. In this work, the angle-of-attack and sideslip angle are allowed to be unmeasured but it is supposed that their effect can be incorporated in the model as a bounded disturbance $\mathbf{b}_i(t)$.

The state vector of the point-mass model of the *i*-th UAV is composed of its position vector $\mathbf{p}_i(t) = \begin{bmatrix} p_{xi}(t) & p_{yi}(t) & p_{zi}(t) \end{bmatrix}^T$ described in the inertial Cartesian reference frame NED (North-East-Down) and by its velocity, described in a spherical coordinate system composed by the ground speed $V_i(t)$, flight path angle $\gamma_i(t)$ and course angle $\chi_i(t)$. By rotating the reference frame first by $\chi_i(t)$ around the z axis and after by $\gamma_i(t)$ around the rotated y axis, the *i*-th aircraft wind frame is obtained. Since it is assumed that the aircraft velocity vector is aligned with its fuselage, $\chi_i(t)$ and $\gamma_i(t)$ are equivalent respectively to the yaw $\psi_i(t)$ and pitch $\theta_i(t)$ attitude angles. The point-mass model includes also the roll attitude angle $\phi_i(t)$, which is a rotation of the fuselage around the direction of the velocity vector. The definition of the roll, pitch and yaw angles can be seen in [29]. Figure 1 shows the *i*-th UAV and its vectors and attitude angles.

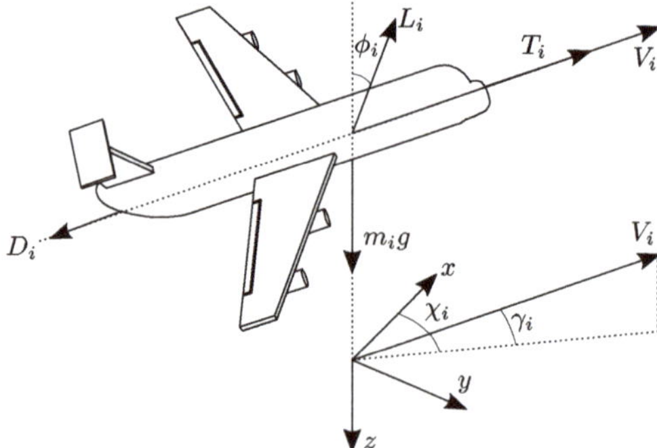

Figure 1. The *i*-th Unmanned Aerial Vehicle (UAV), its force and velocity vectors, and attitude angles.

The state change given by the derivative of $\mathbf{p}_i(t)$ is computed as

$$\dot{\mathbf{p}}_i(t) = \begin{bmatrix} \dot{p}_{xi}(t) \\ \dot{p}_{yi}(t) \\ \dot{p}_{zi}(t) \end{bmatrix} = \mathbf{R}_i(t) \begin{bmatrix} V_i(t) \\ 0 \\ 0 \end{bmatrix} = V_i(t) \begin{bmatrix} \cos \gamma_i(t) \cos \chi_i(t) \\ \cos \gamma_i(t) \sin \chi_i(t) \\ -\sin \gamma_i(t) \end{bmatrix}, \tag{1}$$

where

$$\mathbf{R}_i(t) = \begin{bmatrix} \cos \gamma_i(t) \cos \chi_i(t) & -\sin \chi_i(t) & \sin \gamma_i(t) \cos \chi_i(t) \\ \cos \gamma_i(t) \sin \chi_i(t) & \cos \chi_i(t) & \sin \gamma_i(t) \sin \chi_i(t) \\ -\sin \gamma_i(t) & 0 & \cos \gamma_i(t) \end{bmatrix} \tag{2}$$

is a rotation matrix that rotates from the wind frame to the reference frame with angular velocity

$$\boldsymbol{\omega}_i(t) = [\, -\dot{\chi}_i(t) \sin \gamma_i(t) \;\; \dot{\gamma}_i(t) \;\; \dot{\chi}_i(t) \cos \gamma_i(t) \,]^T, \tag{3}$$

and the variables $V_i(t)$, $\chi_i(t)$, and $\gamma_i(t)$ can be computed by

$$\tan \chi_i(t) = \frac{\dot{p}_{yi}(t)}{\dot{p}_{xi}(t)}, \qquad \sin \gamma_i(t) = -\frac{\dot{p}_{zi}(t)}{V_i(t)}, \qquad V_i^2(t) = \dot{p}_{xi}^2(t) + \dot{p}_{yi}^2(t) + \dot{p}_{zi}^2(t). \tag{4}$$

The UAV dynamics is described by [9]

$$\begin{aligned}
\dot{V}_i(t) &= \frac{T_i(t) - D_i(t)}{m_i} - g \sin \gamma_i(t) + b_{ti}(t), \\
\dot{\chi}_i(t) &= \frac{L_i(t) \sin \phi_i(t)}{m_i V_i(t) \cos \gamma_i(t)} + \frac{b_{\psi i}(t)}{V_i(t) \cos \gamma_i(t)}, \\
\dot{\gamma}_i(t) &= \frac{L_i(t) \cos \phi_i(t)}{m_i V_i(t)} - \frac{g \cos \gamma_i(t)}{V_i(t)} + \frac{b_{\theta i}(t)}{V_i(t)},
\end{aligned} \tag{5}$$

where the disturbance signal $\mathbf{b}_i(t) = [\, b_{ti}(t) \; b_{\theta i}(t) \; b_{\psi i}(t) \,]$ encompasses model approximations, parameter uncertainty, and disturbances in acceleration, generated by several sources, such as wind. It is supposed that $\mathbf{b}_i(t)$ and $\dot{\mathbf{b}}_i(t)$ are unknown but with known bounds. The subscripts t, θ, and ψ from the elements of $\mathbf{b}_i(t)$ means respectively thrust, pitch, and yaw. The thrust force magnitude T_i is a function of the engine throttle; $D_i(t)$ is the magnitude of the drag force; the lift force magnitude L_i is a function of several parameters, such as air density and aircraft speed, and is adjusted mainly by changing the elevator position and ϕ_i is adjusted by a combination of aileron and rudder positions. The variables $T_i(t)$, $L_i(t)$, and $\phi_i(t)$ are the control inputs at the i-th UAV. It can be seen that Equation (5) presents a singularity when $V_i(t) = 0$. Fixed-wing vehicles must maintain non-null airspeed to maintain its lift. Assuming null or mild wind speed, the ground speed $V_i(t)$ is also non-null and this singularity does not occur. It can be seen that $\cos \gamma_i(t) = 0$ also presents a singularity in Equation (5). This occurs only when the UAV is flying exactly in an up or down direction. However, this is not an achieved state, except in highly acrobatic vehicles.

By defining the load factor $n_i(t)$ as [28]

$$n_i(t) \triangleq \frac{L_i(t)}{m_i g}, \tag{6}$$

and defining the following virtual control input

$$\boldsymbol{\Gamma}_i(t) = \begin{bmatrix} a_{ti}(t) \\ a_{\psi i}(t) \\ a_{\theta i}(t) \end{bmatrix} = \begin{bmatrix} \frac{T_i(t) - D_i(t)}{m_i} \\ g n_i(t) \sin \phi_i(t) \\ g n_i(t) \cos \phi_i(t) \end{bmatrix}, \tag{7}$$

the dynamics Equation (5) can be rewritten as

$$\begin{aligned}
\dot{V}_i(t) &= a_{ti}(t) - g \sin \gamma_i(t) - b_{ti}(t), \\
\dot{\chi}_i(t) &= \frac{a_{\psi i}(t) - b_{\psi i}(t)}{V_i(t) \cos \gamma_i(t)}, \\
\dot{\gamma}_i(t) &= \frac{a_{\theta i}(t) - g \cos \gamma_i(t) - b_{\theta i}(t)}{V_i(t)}.
\end{aligned} \tag{8}$$

By deriving $\dot{\mathbf{p}}_i(t)$ from Equation (1), and applying some manipulations, is obtained that

$$\ddot{\mathbf{p}}_i(t) = \mathbf{R}_i(t)\left(\mathbf{\Gamma}_i(t) + \mathbf{b}_i(t)\right) + \mathbf{g}, \tag{9}$$

where $\mathbf{g} = [\,0\ 0\ g\,]^T$ is the gravitational acceleration vector. By defining

$$\boldsymbol{\tau}_i(t) = [\,\tau_{xi}(t)\ \tau_{yi}(t)\ \tau_{zi}(t)\,]^T \triangleq \mathbf{R}_i(t)\mathbf{\Gamma}_i(t) + \mathbf{g}, \tag{10}$$

$$\mathbf{d}_i(t) = [\,d_{xi}(t)\ d_{yi}(t)\ d_{zi}(t)\,]^T, \triangleq \mathbf{R}_i(t)\mathbf{b}_i(t), \tag{11}$$

the dynamics are finally rewritten as

$$\ddot{\mathbf{p}}_i(t) = \boldsymbol{\tau}_i(t) + \mathbf{d}_i(t), \tag{12}$$

where $\boldsymbol{\tau}_i(t) \in \mathbb{R}^3$ is a virtual controller input and $\mathbf{d}_i(t) \in \mathbb{R}^3$ is the virtual disturbance described in the reference frame.

For the controller design, the model given by Equation (12) will be used. Once the virtual control signals are known, the original variables can be obtained. Since for any rotation matrix, $\mathbf{R}_i^{-1}(t) = \mathbf{R}_i^T(t)$, the virtual input $\mathbf{\Gamma}_i(t)$ can always be obtained from $\boldsymbol{\tau}_i(t)$ by

$$\mathbf{\Gamma}_i(t) = \mathbf{R}_i^T(t)\left(\boldsymbol{\tau}_i(t) - \mathbf{g}\right), \tag{13}$$

and then $T_i(t)$, $n_i(t)$, $\phi_i(t)$ can be obtained from Equation (7), which can finally be used as the input of an inner loop controller that actuates over the engine and control surfaces [8].

2.2. UAV Formation

It is considered a formation of UAVs with a virtual leader scheme. The virtual leader is designated here as the 0-th UAV, and consists of a virtual point with a position $\mathbf{p}_0(t) = [\,p_{x0}(t)\ p_{y0}(t)\ p_{z0}(t)\,]^T$ in space, known by all UAVs, which describes a smooth trajectory as a function of time. The results of this work can also be used for a non-virtual leader configuration by assuming that the leader UAV can broadcast its position to all followers UAVs.

The fleet formation is planned by the generation of the desired position $\mathbf{p}_i^d(t) = [\,p_{xi}^d(t)\ p_{yi}^d(t)\ p_{zi}^d(t)\,]^T$ for the i-th UAV which is described as

$$\mathbf{p}_i^d(t) = \mathbf{p}_0(t) + \tilde{\mathbf{p}}_i(t), \tag{14}$$

where $\tilde{\mathbf{p}}_i(t) = [\,\tilde{p}_{xi}(t)\ \tilde{p}_{yi}(t)\ \tilde{p}_{zi}(t)\,]^T$ is the desired (time-varying) clearance, which is described in the reference frame.

To achieve a formation shape that rotates with the leader is interesting to describe the desired clearance $\tilde{\mathbf{p}}_i(t)$ in the leader's wind frame or any other frame related to the leader as

$$\tilde{\mathbf{p}}_i(t) = \mathbf{R}_r(t)\tilde{\mathbf{p}}_i^r(t), \tag{15}$$

where $\tilde{\mathbf{p}}_i^r(t)$ is the clearance vector described in a leader's frame, such as wind, and the formation rotation matrix $\mathbf{R}_r(t)$ rotates from the leader's frame to the reference frame. For example, by defining $\mathbf{R}_r(t) = \mathbf{R}_0(t)$ (see Equation (2) with $i = 0$), it is achieved formation description aligned with the (virtual) leader's trajectory as in, for example, Reference [8,9]. If, instead, $\mathbf{R}_r(t)$ is defined as

$$\mathbf{R}_r(t) = \mathbf{R}_\chi(t) \triangleq \begin{bmatrix} \cos\chi_0(t) & -\sin\chi_0(t) & 0 \\ \sin\chi_0(t) & \cos\chi_0(t) & 0 \\ 0 & 0 & 1 \end{bmatrix}, \tag{16}$$

it is achieved a formation description aligned with the horizontal projection of the (virtual) leader's trajectory, used in, for example, References [14,17].

Another option is to describe the formation using a leader's frame defined by the attitude Euler angles yaw $\psi_0(t)$, pitch $\theta_0(t)$, and roll $\phi_0(t)$. This can be useful, for example, for maneuvers involving close interaction between the leader and the followers, such as to a boom-receptacle automatic aerial refueling. In this case, $\mathbf{R}_r(t) = \mathbf{R}_b(t)$, where [29]

$$\mathbf{R}_b(t) \triangleq$$
$$\begin{bmatrix} \cos\psi_0\cos\theta_0 & \cos\psi_0\sin\theta_0\sin\phi_0 - \sin\psi_0\cos\phi_0 & \cos\psi_0\sin\theta_0\cos\phi_0 + \sin\psi_0\sin\phi_0 \\ \sin\psi_0\cos\theta_0 & \sin\psi_0\sin\theta_0\sin\phi_0 + \cos\psi_0\cos\phi_0 & \sin\psi_0\sin\theta_0\cos\phi_0 - \cos\psi_0\sin\phi_0 \\ -\sin\theta_0 & \cos\theta_0\sin\phi_0 & \cos\theta_0\cos\phi_0 \end{bmatrix} . \quad (17)$$

The derivatives of $\mathbf{p}_i^d(t)$ in Equation (14) can be computed as

$$\dot{\mathbf{p}}_i^d(t) = \dot{\mathbf{p}}_0(t) + \dot{\tilde{\mathbf{p}}}_i(t), \quad (18)$$
$$\ddot{\mathbf{p}}_i^d(t) = \ddot{\mathbf{p}}_0(t) + \ddot{\tilde{\mathbf{p}}}_i(t). \quad (19)$$

Using the Theorem of Coriolis [29], the derivatives of $\tilde{\mathbf{p}}_i(t)$ in Equation (15) can be computed as

$$\dot{\tilde{\mathbf{p}}}_i(t) = \mathbf{R}_r(t)\left[\dot{\tilde{\mathbf{p}}}_i^r(t) + \boldsymbol{\omega}_r(t) \times \tilde{\mathbf{p}}_i^r(t)\right], \quad (20)$$
$$\ddot{\tilde{\mathbf{p}}}_i(t) = \mathbf{R}_r(t)\left\{\ddot{\tilde{\mathbf{p}}}_i^r(t) + 2\boldsymbol{\omega}_r(t) \times \dot{\tilde{\mathbf{p}}}_i^r(t) + \dot{\boldsymbol{\omega}}_r(t) \times \tilde{\mathbf{p}}_i^r(t) + \boldsymbol{\omega}_r(t) \times \left[\boldsymbol{\omega}_r(t) \times \tilde{\mathbf{p}}_i^r(t)\right]\right\}, \quad (21)$$

where $\boldsymbol{\omega}_r(t)$ is the angular velocity between the rotating leader's frame and the reference frame and is given by

$$\boldsymbol{\omega}_r(t) = \begin{cases} \text{leader's gyro measurements,} & \text{if } \mathbf{R}_r(t) = \mathbf{R}_b(t), \\ \left[-\dot{\chi}_0\sin\gamma_0 \ \dot{\gamma}_0 \ \dot{\chi}_0\cos\gamma_0\right]^T, & \text{if } \mathbf{R}_r(t) = \mathbf{R}_0(t), \\ \left[0 \ 0 \ \dot{\chi}_0\right]^T, & \text{if } \mathbf{R}_r(t) = \mathbf{R}_\chi(t). \end{cases} \quad (22)$$

It is worth noting that when using the non-virtual leader's body frame, the angular velocity $\boldsymbol{\omega}_r(t)$ is the body angular velocity, which can be directly measured by a gyro sensor at the leader. By using a non-virtual leader and any of the wind frame variants, the ground velocity obtained from a GPS sensor or from a navigation algorithm must be used. When using a virtual leader approach, its trajectory is smooth, pre-known, and artificially generated, in a way that $\boldsymbol{\omega}_r(t)$ can be pre-computed analytically or numerically with arbitrary precision depending on how the trajectory is created.

2.3. Communication Graph

Each follower UAV can exchange data with their neighbors. The communication network is represented by an undirected graph, which means that, if an i-th UAV receives data from a j-th UAV, this means that the j-th UAV receives data from the i-th UAV. The set of the UAVs that are neighbors of the i-th UAV is defined as $\mathcal{N}_i$.

The Laplacian matrix $\mathbf{L}$ represents the connectivity between the UAVs

$$\mathbf{L}_{ij} = \begin{cases} -a_{ij}, & \text{if } j \neq i \text{ and } j \in \mathcal{N}_i, \\ \sum_{k\in\mathcal{N}_i} a_{ik}, & \text{if } j = i, \\ 0, & \text{otherwise,} \end{cases} \quad (23)$$

where $a_{ij} = 0$ means that there is no communication between the i-th and j-th UAVs and $a_{ij} > 0$ means that there is a communication link between the i-th and j-th UAVs, and the value of a_{ij} is used as a weight to the control algorithm that is developed in this paper. If all UAVs are reachable, that is, if someone starts from any UAV and can achieve any-other UAV via the communication links, **L** is semidefinite positive.

In decentralized controllers, the weight that is given to the information present in the own i-th UAV is also described, which is given by the diagonal matrix **Λ**. The matrix **H** includes both the own weight and the neighborhood weight. These matrices are given by

$$\mathbf{H} = \mathbf{\Lambda} + \mathbf{L}, \tag{24}$$

$$\mathbf{\Lambda} = \mathrm{diag}([\,\lambda_1\ \ldots\ \lambda_n\,]). \tag{25}$$

Note that since $\lambda_1, \ldots, \lambda_n > 0$ and **L** is semidefinite positive, the matrix **H** is invertible.

2.4. Formation Tracking and Synchronization Errors

The tracking error of each aircraft $\mathbf{e}_i(t) = [\,e_{xi}(t)\ e_{yi}(t)\ e_{zi}(t)\,]^T \in \mathbb{R}^3$, relative to a desired position in the reference frame, is defined as

$$\mathbf{e}_i(t) \triangleq \mathbf{p}_i(t) - \mathbf{p}_i^d(t). \tag{26}$$

The synchronization error $\Delta\mathbf{e}_{ij}(t) = [\,\Delta e_{xij}(t)\ \Delta e_{yij}(t)\ \Delta e_{zij}(t)\,]^T \in \mathbb{R}^3$, which can be seen as a relative position error between the UAVs, is defined as

$$\Delta\mathbf{e}_{ij}(t) \triangleq \mathbf{e}_i(t) - \mathbf{e}_j(t) = \mathbf{p}_i(t) - \tilde{\mathbf{p}}_i(t) - \left(\mathbf{p}_j(t) - \tilde{\mathbf{p}}_j(t)\right). \tag{27}$$

It can be seen that $\Delta\mathbf{e}_{ij}(t)$ can be computed without knowing the leader's position. However, since the computation of $\tilde{\mathbf{p}}_i(t)$ and $\tilde{\mathbf{p}}_j(t)$ in Equation (15) can be chosen to be dependent on the leader's flight direction or attitude angles, it is assumed here that the leader's data is available to all UAVs.

It is assumed that each i-th UAV can communicate only with a correspondent set of neighbor UAVs, $\mathcal{N}_i \subset \{1, 2, \ldots, n\}$. The communication graph is assumed to be undirected, connected, not change with time, and previously known. Each UAV receives the tracking error information of other UAVs in the fleet only through its neighbors (as, for example, in the simulation in Section 4). The virtual leader can be seen as an extra node in the graph, that connects to every other UAV in a directed way, from leader to each follower.

The coupled error at i-th UAV is defined as the weighted sum of its tracking error and the synchronization error with respect to its neighbors, that is,

$$\mathbf{e}_i^c(t) = [\,e_{xi}^c(t)\ e_{yi}^c(t)\ e_{zi}^c(t)\,]^T \triangleq \lambda_i \mathbf{e}_i(t) + \sum_{j \in \mathcal{N}_i} a_{ij}\Delta\mathbf{e}_{ij}(t) = \lambda_i \mathbf{e}_i(t) + \sum_{j=1}^{n} a_{ij}\Delta\mathbf{e}_{ij}(t), \tag{28}$$

in which $\lambda_i > 0$ weights its own tracking error and $a_{ij} > 0$ weights the error difference between the neighbor UAV j of the UAV i. In the last equality in Equation (28), if $j \notin \mathcal{N}_i$ then $a_{ij} = 0$. The synchronization control objective is to make the coupled errors approach to zero.

2.5. A Componentwise Formation Description

It is supposed that each component of $\mathbf{d}_i(t)$ is independent of each other which implies that each component of $\ddot{\mathbf{p}}_i(t)$ is independent of each other. In this way, the controller design is simplified since the description of only one axis is sufficient. A controller policy can be developed to a single axis and then it can be directly applied to the other two.

The one-dimensional dynamics from axis $l = x, y, z$, of the reference frame, is obtained from Equation (12) as

$$\ddot{p}_{li}(t) = \tau_{li}(t) + d_{li}(t). \tag{29}$$

Accordingly, the coupled tracking-synchronization error is obtained from Equation (28) as

$$e_{li}^c(t) = \lambda_i e_{li}(t) + \sum_{j \in \mathcal{N}_i} a_{ij} \left[e_{li}(t) - e_{lj}(t) \right]. \tag{30}$$

3. Proposed Controller

Here, a synchronous sliding mode controller is proposed. Figure 2 shows the proposed control structure. It achieves robustness against model uncertainty and disturbance. The chattering is attenuated by the use of a low pass filter (LPF).

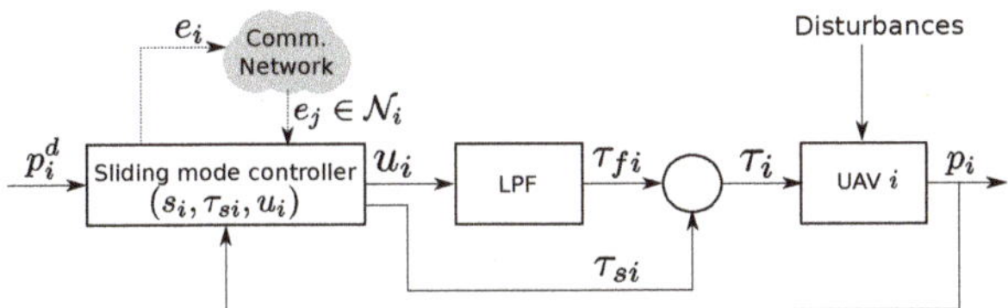

Figure 2. Block diagram of the control structure.

To achieve synchronization, each UAV uses tracking errors of its neighbors to compute a sliding surface in the coupled error space. The sliding surface at the i-th UAV for the l axis is defined as

$$s_{li}(t) = \ddot{e}_{li}^c(t) + k_d \dot{e}_{li}^c(t) + k_p e_{li}^c(t). \tag{31}$$

As usual for sliding mode controllers, it is shown in the next subsection that $s_{li}(t)$ converges to zero in finite time, and maintains equal to zero thereafter. On the sliding surface, that is, when $s_{li}(t) = 0$, the coupled error behaves according to the linear system

$$\ddot{e}_{li}^c(t) + k_d \dot{e}_{li}^c(t) + k_p e_{li}^c(t) = 0, \tag{32}$$

which has all poles in the left plane and, thereafter, is exponentially asymptotically stable for project parameters $k_d, k_p > 0$.

The proposed control law for i-th UAV is

$$\tau_{li}(t) = \tau_{li}^s(t) + \tau_{li}^f(t), \tag{33}$$

where $\tau_{li}^s(t)$ and $\tau_{li}^f(t)$ are, respectively, a smooth signal and a filtered signal of the control law, computed by

$$\tau_{li}^s(t) = \ddot{p}_{li}^d(t) - k_d \dot{e}_{li}(t) - k_p e_{li}(t), \tag{34}$$

$$\tau_{li}^f(t) + \xi_i \tau_{li}^f(t) = u_{li}(t), \tag{35}$$

$$u_{li}(t) = -\text{sign}(s_{li}(t))\eta. \tag{36}$$

Equation (35) defines a low pass filter with cutoff frequency $\xi_i > 0$ that converts a chattering signal $u_{li}(t)$ to a smooth signal $\tau_{li}^f(t)$. The parameter η must be chosen by the designer to guarantee the stability of the overall system.

The proposed control law given by Equations (33)–(36) contains only information from the virtual leader (or from a broadcasting non-virtual leader), from the own i-th UAV, and from its neighborhood $\mathcal{N}_i$. The neighborhood information is contained in $s_{li}(t)$, defined in Equation (31), which is a function of $e_{li}^c(t)$ from Equation (30), which is a function of the own local error $e_{li}(t)$ and the neighborhood errors $e_{lj}(t), j \in \mathcal{N}_i$.

Remark 1. *The variables k_p and k_d define the natural frequency and damping factor of the 2nd order local sliding surface s_{li} of the i-th UAV from Equation (31). As can be seen in [20], these gains also define a control bandwidth,* *which must be sufficiently small to account for, for example, to actuator dynamics. Since it is chosen the same gain k_p and the same gain k_d to all UAVs, it means that they have sliding surfaces that share the same control bandwidth. This is reasonable if all UAVs have similar physical, actuator, and aerodynamic characteristics. However, if there are distinct UAVs, the constants must be chosen to respect the control bandwidth of the UAV with the slowest dynamics.*

3.1. Disturbance Model

Measurement or computation errors and the effect of non-modeled dynamics are incorporated in the dynamics model, given by Equation (12), as a disturbance signal described in the reference frame, $\mathbf{d}_i = [d_{xi}\,d_{yi}\,d_{zi}]^T$. It is supposed that the controller has no access to $\mathbf{d}_i$ but there are known upper bounds Δ_{xi}, Δ_{yi} and Δ_{zi} on the magnitude of the components of $\mathbf{d}_i$ and upper bounds $\tilde{\Delta}_{xi}$, $\tilde{\Delta}_{yi}$, and $\tilde{\Delta}_{zi}$ on the derivatives of the components of $\mathbf{d}_i$, that is,

$$|d_{li}(t)| \leq \Delta_{li}, |\dot{d}_{li}(t)| \leq \tilde{\Delta}_{li}, \quad l = \{x, y, z\}. \tag{37}$$

These upper bounds are used to define the value of η in Equation (36), as explained in Section 3.2. As a contribution of this paper is shown that the upper bounds on the components in the reference frame coordinates can be computed from the upper bounds δ_{ti}, $\delta_{\theta i}$, and $\delta_{\psi i}$ on the components of the disturbance signal in the wind frame $\mathbf{b}_i(t)$,

$$|b_{ti}(t)| \leq \delta_{ti}, \quad |b_{\theta i}(t)| \leq \delta_{\theta i}, \quad |b_{\psi i}(t)| \leq \delta_{\psi i}, \tag{38}$$

and from the upper bounds $\tilde{\delta}_{ti}$, $\tilde{\delta}_{\theta i}$ and $\tilde{\delta}_{\psi i}$ for the

$$|\dot{b}_{ti}(t)| \leq \tilde{\delta}_{ti}, \quad |\dot{b}_{\theta i}(t)| \leq \tilde{\delta}_{\theta i}, \quad |\dot{b}_{\psi i}(t)| \leq \tilde{\delta}_{\psi i}. \tag{39}$$

The wind frame components of the disturbances are more naturally obtained, for example, in description of imprecision in the computation of drag or thrust forces. Assume that there is an upper bound Ω_i for the i-th UAV angular velocity ω_i and define the bounds vectors $\delta_i \triangleq [\delta_{ti}\,\delta_{\theta i}\,\delta_{\psi i}]^T$ and $\tilde{\delta}_i \triangleq [\tilde{\delta}_{ti}\,\tilde{\delta}_{\theta i}\,\tilde{\delta}_{\psi i}]^T$. From Equation (11), it can be seen that

$$|d_{li}(t)| \leq \|\mathbf{d}_i(t)\| = \|\mathbf{R}_i(t)\mathbf{b}_i(t)\| = \|\mathbf{b}_i(t)\| \leq \|\delta_i\|. \tag{40}$$

The upper bounds of each component of $\mathbf{d}_i$ are

$$\Delta_{xi} = \Delta_{yi} = \Delta_{zi} = \|\delta_i\|. \tag{41}$$

Since Equation (11) involves two frames in which one rotates related to the other, its derivative is obtained by using the Theorem of Coriolis [29]

$$\dot{\mathbf{d}}_i(t) = \mathbf{R}_i(t)\left(\dot{\mathbf{b}}_i(t) + \omega_i(t) \times \mathbf{b}_i(t)\right), \tag{42}$$

where $\dot{\mathbf{d}}_i(t)$ contains two components. The first, $\dot{\mathbf{b}}_i(t)$, is the derivative of the disturbance $\mathbf{b}_i(t)$, as seen by the wind frame. The second, $\omega_i(t) \times \mathbf{b}_i(t)$, is generated by the rotation of the wind frame related to the inertial frame. See that a constant disturbance in the wind frame is a varying disturbance in the inertial frame, because of its rotation. Finally, $\mathbf{R}_i(t)$ is used to represent the sum of these components in the inertial frame.

For the bounds δ_i, $\tilde{\delta}_i$, and Ω_i, it is obtained

$$|\dot{d}_{li}(t)| \leq \|\dot{\mathbf{d}}_i(t)\| \leq \|\dot{\mathbf{b}}_i(t)\| + \|\omega_i(t) \times \mathbf{b}_i(t)\| \leq \|\tilde{\delta}_i\| + \|\Omega_i\|\|\delta_i\|. \tag{43}$$

In this way,

$$\tilde{\Delta}_{xi} = \tilde{\Delta}_{yi} = \tilde{\Delta}_{zi} = \|\tilde{\delta}_i\| + \|\Omega_i\|\|\delta_i\|. \tag{44}$$

Equations (40) and (44) provide the upper bounds to the proposed controller.

3.2. Stability Proof

To analyze the overall fleet behavior, all local variables must be concatenated in vectors. Concatenating the positions $\mathbf{p}_i$, virtual control inputs $\tau_i(t)$, and disturbances $\mathbf{d}_i(t)$ from all UAVs of the fleet results in respectively $\mathbf{P}(t) = [\mathbf{p}_1^T(t) \ \dots \ \mathbf{p}_n^T(t)]^T$, $\tau(t) = [\tau_1^T(t) \ \dots \ \tau_n^T(t)]^T$, and $\mathbf{D}(t) = [\mathbf{d}_1^T(t) \ \dots \ \mathbf{d}_n^T(t)]^T$, all $\mathbb{R}^{3n}$ vectors. In this way, the dynamics of the fleet of UAVs is given by concatenating Equation (29) as

$$\ddot{\mathbf{P}}(t) = \tau + \mathbf{D}(t). \tag{45}$$

Similarly, the error and coupled error in x axis are $\mathbb{R}^n$ vectors given by $\mathbf{E}(t) = [\mathbf{e}_1^T(t) \ \dots \ \mathbf{e}_n^T(t)]^T$ and $\mathbf{E}^c(t) = [\mathbf{e}_1^{cT}(t) \ \dots \ \mathbf{e}_n^{cT}(t)]^T$ which are related by

$$\mathbf{E}^c(t) = (\mathbf{H} \otimes \mathbf{I}_3)\mathbf{E}(t), \tag{46}$$

where $\otimes$ denotes the Kronecker product and matrix $\mathbf{H}$ is given by Equation (24). The concatenation of the n UAVs sliding surfaces $\mathbf{S}(t) = [\mathbf{s}_1^T(t) \ \dots \ \mathbf{s}_n^T(t)]^T$ is obtained as

$$\mathbf{S}(t) = \ddot{\mathbf{E}}^c(t) + k_d \dot{\mathbf{E}}^c(t) + k_p \mathbf{E}^c(t) = (\mathbf{H} \otimes \mathbf{I}_3)\left(\ddot{\mathbf{E}}(t) + k_d \dot{\mathbf{E}}(t) + k_p \mathbf{E}(t)\right). \tag{47}$$

The proposed sliding mode control law is written as

$$\tau(t) = \tau^s(t) + \tau^f(t), \tag{48}$$

where $\tau^s(t)$ and $\tau^f(t)$ are computed by

$$\tau^s(t) = \ddot{\mathbf{P}}^d(t) - k_d \dot{\mathbf{E}}(t) - k_p \mathbf{E}(t), \tag{49}$$

$$\dot{\tau}^f(t) + \Xi \tau^f(t) = \mathbf{U}(t), \tag{50}$$

$$\mathbf{U}(t) = -\text{diag}\left\{\frac{\eta}{|s_{li}(t)|}\right\}\mathbf{S}(t), \tag{51}$$

with $\mathbf{U}(t) = [\mathbf{u}_1^T(t) \ \dots \ \mathbf{u}_n^T(t)]^T$, $\mathbf{u}_i(t) = [u_{xi}(t) \ u_{yi}(t) \ u_{zi}(t)]^T$, and $\Xi \triangleq \text{diag}([\xi_1 \ \dots \ \xi_n]) \otimes \mathbf{I}_3 \in \mathbb{R}^{3n \times 3n}$.

To analyze the fleet stability, the following Lyapunov functional candidate is proposed

$$\mathcal{V}(t) = \frac{1}{2}\mathbf{S}^T(t)(\mathbf{H} \otimes \mathbf{I}_3)^{-1}\mathbf{S}(t). \tag{52}$$

Note that, since $\mathbf{H}$ and $\mathbf{H} \otimes \mathbf{I}_3$ are a positive definite matrix, $\mathbf{H}^{-1}$ and $(\mathbf{H} \otimes \mathbf{I}_3)^{-1}$ are also a positive definite matrix, so $\mathcal{V}(t)$ is always positive for $\mathbf{S}(t) \neq 0$.

By using Equations (45), (48) and (49), the sliding surface given by Equation (47) can be rewritten as

$$\begin{aligned}
\mathbf{S}(t) &= (\mathbf{H} \otimes \mathbf{I}_3)\left(\ddot{\mathbf{P}}(t) - \ddot{\mathbf{P}}^d(t) + k_d \dot{\mathbf{E}}(t) + k_p \mathbf{E}(t)\right) \\
&= (\mathbf{H} \otimes \mathbf{I}_3)\left(\tau^s(t) + \tau^f(t) + \mathbf{D}(t) - \ddot{\mathbf{P}}^d(t) + k_d \dot{\mathbf{E}}(t) + k_p \mathbf{E}(t)\right) \\
&= (\mathbf{H} \otimes \mathbf{I}_3)\left(\mathbf{D}(t) + \tau^f(t)\right).
\end{aligned} \tag{53}$$

Since $(\mathbf{H} \otimes \mathbf{I}_3)^{-1}$ is constant, the derivative of Equation (52) is

$$\dot{\mathcal{V}}(t) = \mathbf{S}^T(t)(\mathbf{H} \otimes \mathbf{I}_3)^{-1}\dot{\mathbf{S}}(t). \tag{54}$$

By deriving Equation (53) and after using Equation (50), $\dot{\mathcal{V}}(t)$ is rewritten to

$$
\begin{aligned}
\dot{\mathcal{V}}(t) &= \mathbf{S}^T(t)\left(\dot{\mathbf{D}}(t) + \dot{\boldsymbol{\tau}}^f(t)\right)\\
&= \mathbf{S}^T(t)\dot{\mathbf{D}}(t) - \mathbf{S}^T(t)\Xi\boldsymbol{\tau}^f(t) + \mathbf{S}^T(t)\mathbf{U}(t)\\
&= \sum_{i=1}^{n}\left(\mathbf{s}_i^T(t)\dot{\mathbf{d}}_i(t) - \xi_i\mathbf{s}_i^T(t)\boldsymbol{\tau}_i^f(t) + \mathbf{s}_i^T(t)\mathbf{u}_i\right)\\
&= \sum_{i=1}^{n}\left[\sum_{l=\{x,y,z\}}(s_{li}\dot{d}_{li} - \xi_i s_{li}\tau_{li}^f - |s_{li}|\eta)\right]\\
&\leq \sum_{i=1}^{n}\left[\sum_{l=\{x,y,z\}}|s_{li}|(|\dot{d}_{li}| + \xi_i|\tau_{li}^f| - \eta)\right].
\end{aligned}
\tag{55}
$$

The upper bounds of the disturbance and its derivative are given, respectively, by $\Delta_{li} \geq |d_{li}(t)|$ and $\tilde{\Delta}_{li} \geq |\dot{d}_{li}(t)|$, which are computed by, respectively, Equations (41) and (44). It is shown in [24] that $|\tau_{li}^f(t)| \leq |d_{li}(t)| \leq \Delta_{li}$. By using these upper bounds in Equation (55), it can be seen that

$$
\dot{\mathcal{V}}(t) \leq \sum_{i=1}^{n}\left[\sum_{l=\{x,y,z\}}|s_{li}|(\tilde{\Delta}_{li} + \xi_i\Delta_{li} - \eta)\right].
\tag{56}
$$

By choosing η satisfying

$$
\eta \geq \tilde{\Delta}_{li} + \xi_i\Delta_{li} + \epsilon, \quad \forall i \in \{1,...,n\}, \quad ,\forall l \in \{x,y,z\},
\tag{57}
$$

for some arbitrarily chosen constant $\epsilon > 0$, it is obtained

$$
\dot{\mathcal{V}}(t) \leq -\sum_{i=1}^{n}\sum_{l=\{x,y,z\}}|s_{li}(t)|\epsilon = -\epsilon\sum_{i=1}^{n}\sum_{l=\{x,y,z\}}|s_{li}(t)| = -\epsilon\|\mathbf{S}(t)\|_1,
\tag{58}
$$

where $\|\mathbf{S}(t)\|_1$ is the 1-norm of $\mathbf{S}(t)$. Using the fact that the 1-norm is greater than the Euclidean norm of the same vector, then

$$
\dot{\mathcal{V}}(t) \leq -\epsilon\|\mathbf{S}(t)\|,
\tag{59}
$$

which means that $\mathcal{V}(t)$ and, therefore, $\mathbf{S}(t)$ go to zero in finite time [20]. On the sliding surface, the system behaves as a stable linear system given by Equation (32) and the error converges asymptotically to zero.

Remark 2. *Note that the sliding surface given by Equation (47), when rewritten in Equation (53), is a function only of the disturbance $\mathbf{D}(t)$ and the output of the filter $\boldsymbol{\tau}^f(t)$. This has two main implications:*

1. *Since it is shown here that $\mathbf{S}(t) \to \mathbf{0}$, it follows that $\boldsymbol{\tau}^f(t) \to -\mathbf{D}(t)$. In this way, $\boldsymbol{\tau}^f(t)$ estimates and compensates disturbances. Since the effect of airflow is not aligned to the fuselage is a disturbance, the presence of a disturbance compensation shows that the wind effect can be neglected in the initial model if this effect has known bounds.*
2. *If the disturbance is null at $t = 0$, $\mathbf{S}(0) = \mathbf{0}$ if $\boldsymbol{\tau}^f(0) = \mathbf{0}$ and the system already starts in sliding condition. Similarly, if the known disturbance upper bound is relatively small, the system starts near the sliding surface and converges fast to the sliding surface.*

4. Simulation

In this section, a simulation is made to show the effectiveness of the proposed controller. A scenario of 5 UAVs with communication links described by Figure 3 is used.

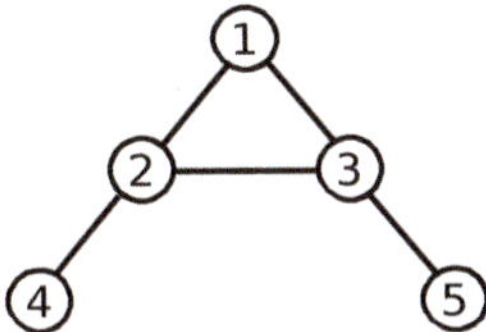

Figure 3. Five UAVs and their undirected communication links. The virtual leader is not shown here. All UAVs have access to the virtual leader's trajectory information.

The matrices

$$\mathbf{L} = \begin{bmatrix} 2 & -1 & -1 & 0 & 0 \\ -1 & 3 & -1 & -1 & 0 \\ -1 & -1 & 3 & 0 & -1 \\ 0 & -1 & 0 & 1 & 0 \\ 0 & 0 & -1 & 0 & 1 \end{bmatrix} \tag{60}$$

and $\mathbf{\Lambda} = \mathbf{I}_5$ are chosen to give the same weight for the UAV own error and for each of its relative errors. The choice $k_p = 0.5$ and $k_d = 0.0625$ provide a critically damped sliding surface with natural frequency $\omega_n = 0.25$ rad/s. These gains are chosen relatively small, as a way to limit the maximum commanded acceleration, even if the UAVs are initially far from their desired position. The low pass filters are settled such that $\mathbf{\Xi} = \mathbf{I}_5 \otimes \mathbf{I}_3$.

A fleet with a non-rectilinear 3D trajectory is described, which is defined by the virtual leader path given by

$$\begin{cases} p_{x0}(t) = 80 + 45t \quad [\text{m}], \\ p_{y0}(t) = 20\cos(0.1t) \quad [\text{m}], \\ \gamma_0(t) = \frac{\pi}{36} \text{ rad}, \quad (z_0(0) = -100 \text{ m}). \end{cases} \tag{61}$$

For easy visualization, a time-varying formation is considered, whose horizontal projection in the reference frame has a V-shape and the altitude has time-varying oscillation. Accordingly, the formation rotation matrix $\mathbf{R}_r$ is defined as $\mathbf{R}_\chi$ from Equation (16) and the clearance vectors $\tilde{\mathbf{p}}_i^r(t)$ related to the virtual leader are

$$\tilde{\mathbf{p}}_1^r(t) = \begin{bmatrix} 0 \\ 0 \\ 10\sin(0.1t) \end{bmatrix}, \quad \tilde{\mathbf{p}}_2^r(t) = \begin{bmatrix} -40 \\ -40 \\ 10\sin(0.1t + 2\pi/5) \end{bmatrix}, \quad \tilde{\mathbf{p}}_3^r(t) = \begin{bmatrix} -40 \\ 40 \\ 10\sin(0.1t + 4\pi/5) \end{bmatrix},$$

$$\tilde{\mathbf{p}}_4^r(t) = \begin{bmatrix} -80 \\ -80 \\ 10\sin(0.1t + 6\pi/5) \end{bmatrix}, \quad \tilde{\mathbf{p}}_5^r(t) = \begin{bmatrix} -80 \\ 80 \\ 10\sin(0.1t + 8\pi/5) \end{bmatrix}. \tag{62}$$

The initial position of each UAV is defined as

$$\mathbf{p}_1(0) = \begin{bmatrix} 60 \\ 0 \\ -100 \end{bmatrix}, \quad \mathbf{p}_2(0) = \begin{bmatrix} 20 \\ -30 \\ -100 \end{bmatrix}, \quad \mathbf{p}_3(0) = \begin{bmatrix} 50 \\ 20 \\ -100 \end{bmatrix}, \quad \mathbf{p}_4(0) = \begin{bmatrix} 10 \\ -50 \\ -100 \end{bmatrix}, \quad \mathbf{p}_5(0) = \begin{bmatrix} 20 \\ 80 \\ -100 \end{bmatrix}. \tag{63}$$

The initial velocity of each UAV is defined as

$$\dot{\mathbf{p}}_1(0) = \begin{bmatrix} 50 \\ 5 \\ 0 \end{bmatrix}, \quad \dot{\mathbf{p}}_2(0) = \begin{bmatrix} 40 \\ 10 \\ 0 \end{bmatrix}, \quad \dot{\mathbf{p}}_3(0) = \begin{bmatrix} 45 \\ -10 \\ 0 \end{bmatrix}, \quad \dot{\mathbf{p}}_4(0) = \begin{bmatrix} 40 \\ -5 \\ 0 \end{bmatrix}, \quad \dot{\mathbf{p}}_5(0) = \begin{bmatrix} 45 \\ 0 \\ 0 \end{bmatrix}. \tag{64}$$

The disturbance is simulated as

$$\mathbf{b}_i(t) = 0.2 \begin{bmatrix} \cos(0.5t) & \cos(0.5t) & \cos(0.5t) \end{bmatrix}^T, \qquad \forall i \in \{1,2,3,4,5\}. \tag{65}$$

From Equation (65), the magnitude of the upper bound vector δ_i of $\mathbf{b}_i(t)$ is computed as $\|\delta_i\| = 0.35$. The magnitude of the upper bound vector of $\tilde{\delta}_i$ is computed also from Equation (65) as $\|\tilde{\delta}_i\| = 0.17$.

The upper bound of each component of $\mathbf{d}_i(t)$ is computed by Equation (41) resulting in $\Delta_{xi} = \Delta_{yi} = \Delta_{zi} = 0.35$. By simulation experiments it is verified that $\Omega_i = 0.17$ rad/s is an upper bound for the angular velocity amplitude; the upper bound in $\dot{\mathbf{d}}_i(t)$ is computed by Equation (44), resulting in $\tilde{\Delta}_{xi} = \tilde{\Delta}_{yi} = \tilde{\Delta}_{zi} = 0.23$. By choosing $\epsilon = 0.42$, it is obtained from Equation (57) that $\eta = 1$.

The system is implemented using an ode4 Runge-Kutta solver, with a fixed-step size of 1 ms. Since it is impossible to perfectly simulate the effect of a chattering input signal in a continuous differential equation, the controller output is evaluated at 10 ms time steps and maintained constant between time intervals.

For comparison purposes, the unfiltered synchronous formation flight controller presented in Reference [18] is also simulated. It is configured to be as similar as possible to the proposed controller. The first order sliding surface is defined with the same natural frequency as the proposed controller, that is, $\omega_n = 0.25$ rad/s. By using the same upper bound $\Delta_{xi} = \Delta_{yi} = \Delta_{zi} = 0.35$ and by choosing the same $\epsilon = 0.42$, it is computed $\eta = 0.77$. Other parameters are exactly the same as the proposed controller.

Figure 4 shows the desired trajectory for each UAV in black, and the trajectory achieved by each UAV in distinct colors. Square and '*' markers show respectively the desired and achieved positions in specific and equally spaced time instants. When a '*' is inside the square, the UAV is in its desired position.

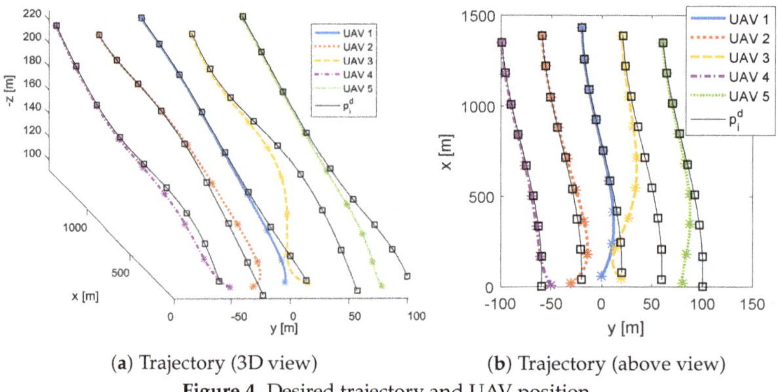

(**a**) Trajectory (3D view)   (**b**) Trajectory (above view)

Figure 4. Desired trajectory and UAV position.

Figure 5 shows the formation flight error components e_{xi}, e_{yi}, and e_{zi} for each i-th UAV for both controllers. Figure 6, shows the coupled error of each i-th UAV, which is given by Equation (46) for both controllers. It can be seen that, for both controllers, the system rapidly enters in sliding mode, the coupled errors slide in the prescribed linear sliding surface and achieve the performance described by the linear system that defines the sliding surface. It can also be seen that the error converges to zero, which shows that both controllers completely compensate for the added input disturbance.

Figure 7 shows the controller output τ_{xi}, τ_{yi}, and τ_{zi} for each i-th UAV, which is generated by adding the smooth τ_i^s control signal and τ_i^f, obtained by filtering the chattering signal $\mathbf{U}_i$ in the proposed controller, or is the unfiltered control signal in the controller from Reference [18]. As can be seen, the proposed control output is smooth, whereas the control output from the unfiltered SMC chatters.

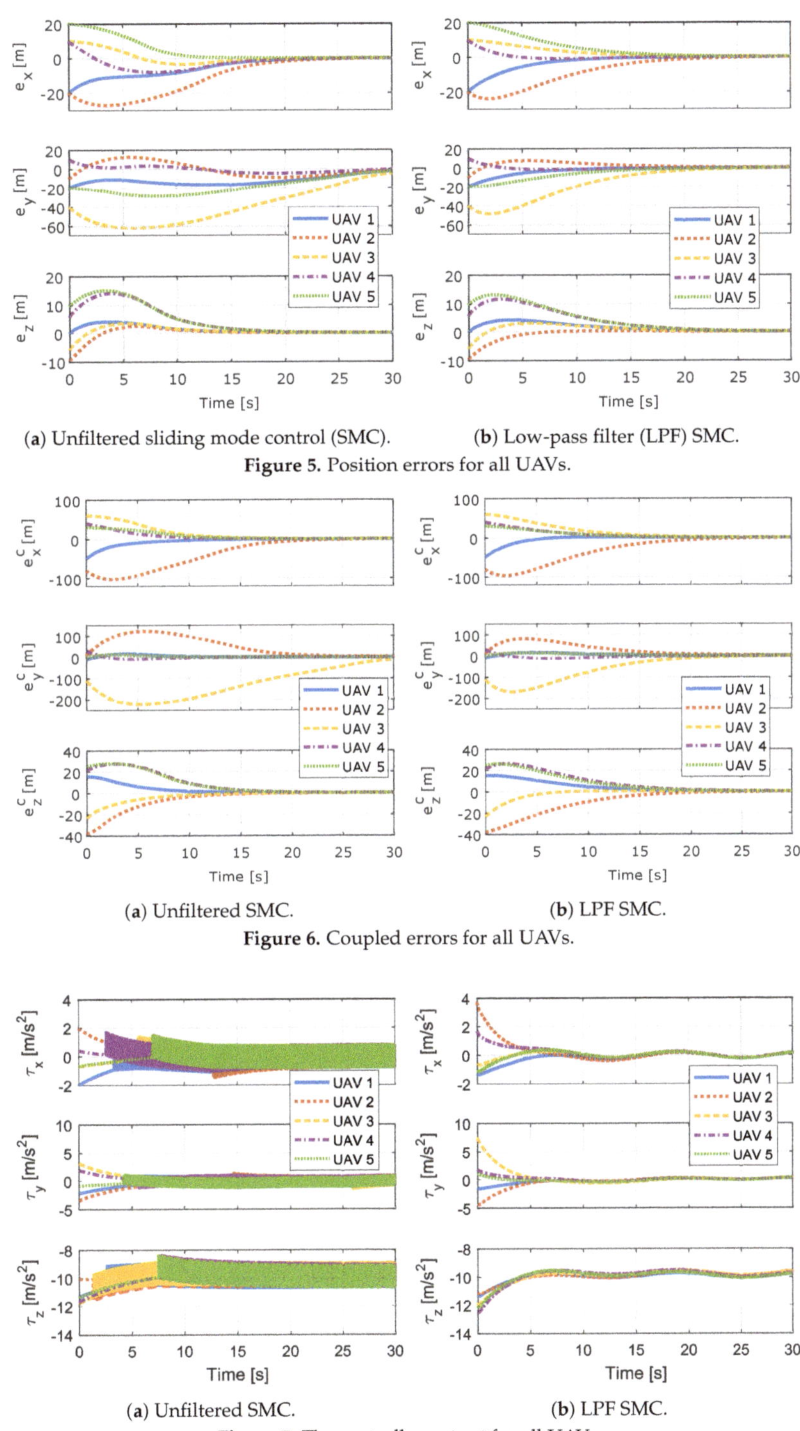

(**a**) Unfiltered sliding mode control (SMC). (**b**) Low-pass filter (LPF) SMC.

Figure 5. Position errors for all UAVs.

(**a**) Unfiltered SMC. (**b**) LPF SMC.

Figure 6. Coupled errors for all UAVs.

(**a**) Unfiltered SMC. (**b**) LPF SMC.

Figure 7. The controller output for all UAVs.

5. Conclusions

A decentralized architecture for synchronous formation flight of UAVs based on sliding mode control with a low pass filter was proposed. The use of the SMC technique provides robustness to disturbances, in a way that the system slides in the prescribed sliding surface even in the presence of disturbances. The LPF virtually removes the chattering while maintaining the convergence to a null error in steady-state. In the proposed architecture only the chattering component of the control signal is filtered. As a result, the controller has a simpler expression when compared to recent results of the literature, such as in [23]. Also, it is presented an equation that is used to compute the upper bounds in the disturbance and in its derivative to a formation described in a global frame. This equation assumes that the upper bounds are known in the wind frame of each follower UAV. It is proved that the proposed controller is stable, achieving a prescribed sliding surface in finite time.

For future work, more realistic models for UAV and wind gusts can be implemented. Also, it is desired to implement other SOSMC, such as presented in References [21,22], in the context of the synchronous formation flight.

Author Contributions: Conceptualization, J.Y.I. and H.C.F.; Formal analysis, T.F.K.C. and J.Y.I.; Investigation, T.F.K.C.; Methodology, T.F.K.C., J.Y.I. and H.C.F.; Software, T.F.K.C.; Supervision, H.C.F.; Visualization, T.F.K.C.; Writing—original draft, T.F.K.C., J.Y.I. and H.C.F. All authors have read and agree to the published version of the manuscript.

Funding: This work was supported in part by National Council for Scientific and Technological Development—CNPq, under grants 312528/2017-5/PQ and 460311/2014-0 and by District Federal Research Foundation—FAPDF, under grant 0193.001153/2015.

Conflicts of Interest: The authors declare no conflict of interest.

References

1. Zheng, Z.; Qian, M.; Li, P.; Yi, H. Distributed Adaptive Control for UAV Formation with Input Saturation and Actuator Fault. *IEEE Access* **2019**, *7*, 144638–144647. [CrossRef]
2. Cai, Z.; Wang, L.; Zhao, J.; Wu, K.; Wang, Y. Virtual target guidance-based distributed model predictive control for formation control of multiple UAVs. *Chin. J. Aeronaut.* **2019**. [CrossRef]
3. Zhao, H.; Wu, S.; Wen, Y.; Liu, W.; Wu, X. Modeling and Flight Experiments for Swarms of High Dynamic UAVs: A Stochastic Configuration Control System with Multiplicative Noises. *Sensors* **2019**, *19*, 3278. [CrossRef] [PubMed]
4. Liao, F.; Teo, R.; Wang, J.L.; Dong, X.; Lin, F.; Peng, K. Distributed Formation and Reconfiguration Control of VTOL UAVs. *IEEE Trans. Control Syst. Technol.* **2017**, *25*, 270–277. [CrossRef]
5. Park, C.; Cho, N.; Lee, K.; Kim, Y. Formation Flight of Multiple UAVs via Onboard Sensor Information Sharing. *Sensors* **2015**, *15*, 17397–17419. [CrossRef]
6. He, L.; Bai, P.; Liang, X.; Zhang, J.; Wang, W. Feedback formation control of UAV swarm with multiple implicit leaders. *Aerosp. Sci. Technol.* **2018**, *72*, 327–334. [CrossRef]
7. Brandão, A.S.; Sarcinelli-Filho, M. On the Guidance of Multiple UAV using a Centralized Formation Control Scheme and Delaunay Triangulation. *J. Intell. Robot. Syst.* **2016**, *84*, 397–413. [CrossRef]
8. Singh, S.N.; Zhang, R.; Chandler, P.; Banda, S. Decentralized nonlinear robust control of UAVs in close formation. *Int. J. Robust Nonlinear Control* **2003**, *13*, 1057–1078. [CrossRef]
9. Rezaee, H.; Abdollahi, F.; Talebi, H.A. $\mathcal{H}_\infty$ based motion synchronization in formation flight with delayed communications. *IEEE Trans. Ind. Electron.* **2014**, *61*, 6175–6182. [CrossRef]
10. Lin, W. Distributed UAV formation control using differential game approach. *Aerosp. Sci. Technol.* **2014**, *35*, 54–62. [CrossRef]
11. Li, Z.; Xing, X.; Yu, J. Decentralized output-feedback formation control of multiple 3-DOF laboratory helicopters. *J. Frankl. Inst.* **2015**, *352*, 3827–3842. [CrossRef]
12. Ren, W. Consensus strategies for cooperative control of vehicle formations. *IET Control Theory A* **2007**, *1*, 505–512. [CrossRef]
13. Oh, K.K.; Park, M.C.; Ahn, H.S. A survey of multi-agent formation control. *Automatica* **2015**, *53*, 424–440. [CrossRef]

14. Cordeiro, T.F.K.; Ferreira, H.C.; Ishihara, J.Y. Non linear controller and path planner algorithm for an autonomous variable shape formation flight. In Proceedings of the 2017 International Conference on Unmanned Aircraft Systems (ICUAS), Miami, FL, USA, 13–16 June 2017. [CrossRef]

15. Liu, W.; Gao, Z. A distributed flocking control strategy for UAV groups. *Comput. Commun.* **2020**, *153*, 95–101. [CrossRef]

16. Qiu, H.; Duan, H. A multi-objective pigeon-inspired optimization approach to UAV distributed flocking among obstacles. *Inf. Sci.* **2020**, *509*, 515–529. [CrossRef]

17. Campa, G.; Gu, Y.; Seanor, B.; Napolitano, M.R.; Pollini, L.; Fravolini, M.L. Design and flight-testing of non-linear formation control laws. *Control Eng. Pract.* **2007**, *15*, 1077–1092. [CrossRef]

18. Cordeiro, T.F.K.; Ferreira, H.C.; Ishihara, J.Y. Robust and Synchronous Nonlinear Controller for Autonomous Formation Flight of Fixed Wing UASs. In Proceedings of the 2019 International Conference on Unmanned Aircraft Systems (ICUAS), Atlanta, GA, USA, 11–14 June 2019. [CrossRef]

19. Yu, J.; Dong, X.; Li, Q.; Ren, Z. Time-varying formation tracking for high-order multi-agent systems with switching topologies and a leader of bounded unknown input. *J. Frankl. Inst.* **2018**, *355*, 2808–2825. [CrossRef]

20. Slotine, J.J.; Li, W. *Applied Nonlinear Control*; Prentice Hall: Englewood Cliffs, NJ, USA, 1990.

21. Mei, K.; Ding, S. Second-order sliding mode controller design subject to an upper-triangular structure. *IEEE Trans. Syst. Man Cybern. Syst.* **2018**. [CrossRef]

22. Ding, S.; Park, J.H.; Chen, C.C. Second-order sliding mode controller design with output constraint. *Automatica* **2020**, *112*, 108704. [CrossRef]

23. Zhao, D.; Li, C.; Zhu, Q. Low-pass-filter-based position synchronization sliding mode control for multiple robotic manipulator systems. *Proc. Inst. Mech. Eng. I J. Syst. Control Eng.* **2011**, *225*, 1136–1148. [CrossRef]

24. Chong, S.; Sheng, Y.; Xiangyuan, Z.; Liu, X. An improved chattering-free sliding mode control with finite time convergence for reentry vehicle. In Proceedings of the 2016 IEEE Chinese Guidance, Navigation and Control Conference (CGNCC), Nanjing, China, 12–14 August 2016. [CrossRef]

25. Balamurugan, S.; Venkatesh, P.; Varatharajan, M. Fuzzy sliding-mode control with low pass filter to reduce chattering effect: An experimental validation on Quanser SRIP. *Sādhanā* **2017**, *42*, 1693–1703. [CrossRef]

26. Xu, B.; Li, J.; Yang, Y.; Wu, H.; Postolache, O. A Novel Sliding Mode Control with Low-Pass Filter for Nonlinear Handling Chain System in Container Ports. *Complexity* **2020**, *2020*, 7254503. [CrossRef]

27. Han, T.; Guan, Z.H.; Wu, Y.; Zheng, D.F.; Zhang, X.H.; Xiao, J.W. Three-dimensional containment control for multiple unmanned aerial vehicles. *J. Frankl. Inst.* **2016**, *353*, 2929–2942. [CrossRef]

28. Anderson, J.D., Jr. *Introduction to Flight*, 3rd ed.; Mcgraw-Hill: New York, NY, USA, 1988.

29. Stevens, B.L.; Lewis, F.L. *Aircraft Control and Simulation*; Wiley-Interscience: New York, NY, USA, 2003.

Article

Modeling and Flight Experiments for Swarms of High Dynamic UAVs: A Stochastic Configuration Control System with Multiplicative Noises

Hongbo Zhao [1,*], Sentang Wu [1], Yongming Wen [2], Wenlei Liu [1] and Xiongjun Wu [3]

1   School of Automation Science and Electrical Engineering, Beihang University, Beijing 100191, China
2   Science and Technology on Information Systems Engineering Laboratory, Beijing Institute of Control & Electronics Technology, Beijing 100038, China
3   The 802 Institute of Shanghai Academy of Space Flight Technology, The Eighth Academy of China Aerospace Science and Technology Corporation, Shanghai 200090, China
*   Correspondence: zhaohb07@buaa.edu.cn

Received: 13 June 2019; Accepted: 23 July 2019; Published: 25 July 2019

Abstract: UAV Swarm with high dynamic configuration at a large scale requires a high-precision mathematical model to fully exploit its boundary performance. In order to instruct the engineering application with high confidence, uncertainties induced from either systematic measurement or the environment cannot be ignored. This paper investigates the *Itô* stochastic model of the UAV Swarm system with multiplicative noises. By combining the cooperative kinematic model with a simplified individual dynamic model of fixed-wing-aircraft for the first time, the configuration control model is derived. Considering the uncertainties in actual flight, multiplicative noises are introduced to complete the *Itô* stochastic model. Following that, the estimator and controller are designed to control the formation. The mean-square uniform boundedness condition of the proposed stochastic system is presented for the closed-loop system. In the simulation, the stochastic robustness analysis and design (SRAD) method is used to optimize the properties of the formation. More importantly, the effectiveness of the proposed model is also verified using real data of five unmanned aircrafts collected in outfield formation flight experiments.

Keywords: stochastic system; UAV swarm; configuration control; multiplicative noises; dynamic model; stochastic robustness analysis and design

---

1. Introduction

Swarms of UAVs, which can autonomously implement missions [1], have received significant attention in recent years. There are many application scenarios for UAV swarms, such as comprehensive combat [1], distributed reconfigurable sensor networks [2], surveillance [3], and reconnaissance systems [4]. The primary concern of the UAV swarm is the configuration control problem and related research mainly focuses on mathematical modeling [5], control strategies and methods [6,7] and collision and obstacle avoidance algorithms [8,9]. However, most of them are based on a deterministic system or a system with ideal Gaussian noises. High dynamic USCC model with multiplicative noise remains as one of the primary and practical issues for utilizing UAV swarms in engineering applications.

Generally, the stochastic differential equation refers to a stochastic process-driven system or an ordinary differential equation with a random coefficient [10–14]. Random factors are introduced into the system in the following three ways [10]: 1. the system's initial conditions or inputs are taken as random variables, 2. the system's random external disturbances, 3. the system's parameters and structures taken as random variables. The disturbances in the UAV swarm system are mainly induced from the sensor measurement, internet transmission and the task environment, and they fall under point 2 mentioned above.

Existing work on the formation configuration control problem has been extensively investigated for low dynamic deterministic systems. In [15], the formation containment problems based on linear state equations of the multiagent systems were investigated. In [6,16], some theoretical and practical problems of multiple quadrotors were studied. In [17–19], the dynamical model based formation control problems of multirobot systems were studied. Most of the studies focus on low-speed vehicles, such as multiple quadrotors or multirobot systems, with an ideal environment; however, very few studies consider the formation control problems of high dynamic fixed-wing unmanned aircraft under complex environments. The mathematical modeling of stochastic high dynamic UAV swarms remains to be solved.

In this paper, we introduce the stochastic model for the following two reasons: (1) The UAV swarm, especially for high dynamic, dense configuration and large scale swarms, requires a high-precision mathematical model to describe the dynamic relationships among formation members and fully exploit its boundary performance; (2) When the UAV swarm is carrying out missions, the influence of various uncertainties (systematic measurement random interferences, network-induced random interferences and mission environment random interferences) cannot be ignored, and the relative movements of its members are usually random. Communication topology, caused by a network change, inevitably influences the process of information sending and receiving [20]. Therefore, it is necessary to combine the mathematical model of the USCC with the stochastic system to instruct the engineering practice such as cooperative detection under complex mission environments with higher confidence.

Undoubtedly, constructing a more adaptable stochastic model for multiple vehicle systems is an urgent task. The problem of Brownian motion-driven multiagent tracking was discussed in [21] and sufficient conditions for the tracking of multi-agents were obtained by using the auxiliary function of Brownian motion and random *Itô* integral technology. A time lag multiagent system model with measurement noise was set up in [22], and the stability theory of stochastic differential equations was used. In [23], the stochastic factor has been considered in the leader-following multiagent model based on the event-triggered control strategy, *Itô* formula and stability theory. Studies [20,24–27] have considered the influence of stochastic disturbances on the multiagent system (MAS) and have established a stochastic model to control the MAS. However, existing works about the stochastic MAS are mainly based on assumed state matrices. Although the assumed formula can be applied across multiple levels, extra difficulties will occur in a certain practical system; for example, the process of combining the flight member's dynamics model and the formation's kinematics model is more complex; the modeling of process noises is more complex because the coefficients are presented in a very complex way in the practical system; the simulation and flight test are more difficult because of the complex system and high-risk environment. Therefore, numerous problems remain to be solved.

Typically, there are four main methods for formation coordination modeling reported in the literature: the leader-follower framework [28], virtual structure approach [29,30], behavior-based model [31,32] and graph theories [33]. Most of them focus on the consensus problems based on the kinematic model. In this paper, intending to improve both the formation properties and individual capabilities of UAV swarm in complex environments, we first use a simplified nonlinear dynamics model of the fixed-wing aircraft flight control system and then construct the group dynamic cooperative control system model together with the relative kinematic model. Based on Reynolds' three criteria [31], the model comprehensively considers the flight member's individual properties and the whole swarm's cooperativeness.

To the best of our knowledge, although few efforts have been devoted to the modeling of dynamic formation systems, there is almost no literature on the stochastic model of the UAV swarms. For example, a six-degree-of-freedom (DOF) unified nonlinear dynamic model of spacecraft formation was presented in [34]; In [35,36], the formation control laws for YF-22 aircraft models with six DOF dynamics plus kinematic equations were designed. Although these formation models are efficient for the cooperative of a group without stochastic disturbance, the control strategy may be invalid when control objects are moving in the noise environment.

To the best of the authors' knowledge, few papers discuss the quality of the configuration, and most of them have come up with an algorithm to control the formation, thereby achieving the desired configuration or avoiding collision [6–8]. However, they do not discuss the robustness of UAV swarms in much detail. In order to improve the robustness of the stochastic system, the parameters of the estimator and the controller are optimized by the stochastic robustness analysis and design (SRAD) method [37] in the simulation. Furthermore, few studies have carried out fixed-wing flight test experiments. A multi-UAV outfield flight experiment was carried out to verify the effectiveness of the formation collision forecast and coordination algorithm in [9]. In [36], a set of flights was performed to assess the performance of the formation control laws. To extend the previous outfield flight test results, the overall design in this paper is validated experimentally by flight testing using the leader-follower configuration.

Motivated by the discussions above, the stochastic USCC model with multiplicative noises is investigated in this paper. Compared with the existing literature, the main theoretical and experimental contributions of our work are summarized as follows:

(a) Firstly, with the aim to instruct the engineering application of UAV swarms, we construct the nonlinear formation model by combining the nonholonomic individual dynamics model of a UAV swarm with the relative cooperative kinematics model. The model comprehensively considers the personality of the individual members and the cooperation of the whole formation.
(b) Secondly, the stochastic state equation and output equation, together with the estimator and controller, finally constitute a state-feedback-based closed-loop *Itô* stochastic system, which makes full use of the platform's boundary performance and better matches the complex task environments to improve the cost-effectiveness of the UAV swarm system.
(c) Thirdly, most studies of formation configuration control focus on theoretical analysis, while the technology proposed in this paper is for engineering applications and has been verified by outfield flight tests.

The rest of this paper is structured as follows:

In Section 2, the problem's formulation and preliminary studies of the USCC stochastic system are presented. In Section 3, the mathematical model of formation control stochastic system is illustrated in detail. The estimator and controller are also designed to control the formation, and SRAD has been used to optimize the controller and estimator. The mean-square uniformly bounded condition of the proposed stochastic system is then presented. In Section 4, simulations and experiments are conducted to verify the effectiveness of the model. Finally, concluding remarks are given in Section 5.

2. Preliminary and Problem Formulation

2.1. Itô Stochastic System

Consider the linear *Itô* stochastic system as the model to be investigated as follows:

$$dx = [A(t)x + B(t)]dt + \sum_{i=1}^{m}[F_i(t)x + G_i(t)]dW_i \tag{1}$$

$$dx = A(t)xdt + \sum_{i=1}^{m}F_i(t)xdW_i \tag{2}$$

where $x, B, G_i \in R^n$, $A, F_i \in R^{n \times n}$, $W(t) = [W_1(t), W_2(t), \cdots, W_m(t)]^T, (t \geq 0)$ are m dimensional standard Wiener processes, which are defined in the complete probability space (Ω, F, P) and are independent of each other. Define the following matrices:

$$\begin{cases} M(t) = A(t) \oplus A(t) + \sum_{i=1}^{m} F_i(t) \otimes F_i(t) \\ R(t) = 2\left[I_N \otimes B^T(t)\right]K_N + 2\sum_{i=1}^{m} \left[I_N \otimes G_i^T(t)\right]K_N F_i(t) \\ K(t) = \sum_{i=1}^{m} G_i(t) \otimes G_i(t) \end{cases} \tag{3}$$

where $N = n^2$, '$\oplus$' denotes the Kronecker tensor for a matrix, and $A \oplus A = I_n \otimes A + A \otimes I_n$, '$\otimes$' is the Kronecker tensor product of a matrix.

$$K_N = \begin{bmatrix} 1 & 0\cdots & 0 & 0\cdots & 0 & 0\cdots & 0 \\ 0 & 0\cdots & 1 & 0\cdots & 0 & 0\cdots & 0 \\ & & & \cdots\cdots & & & \\ 0 & 0\cdots & 0 & 0\cdots & 1 & 0\cdots & 0 \end{bmatrix} \tag{4}$$

where the element '1' appears in the first column, $(n+1)th$ column and $[(n-1)n+1]th$ column of K_N.

2.2. Mean-Square Uniform Boundedness of the Stochastic System

Since the stochastic system is complicated by external interferences, its stability condition is relatively strict and there is no trivial solution to the equation. To solve this problem, we take advantage of the mean-square uniform boundedness of the stochastic system. The condition for boundedness is slightly less strict than that of stability. Under the condition of boundedness, the states of the system are converging to bounded areas instead of certain stable points as time tends to infinity. Therefore, for the USCC stochastic system model, stability refers to the mean-square uniform boundedness.

Definition 1. *If there is a positive number c:*

$$\limsup_{t \to \infty} E\left\{\|X_{ij}(t, t_0, X_{ij0})\|^2\right\} \leq c \tag{5}$$

Then the states $X_{ij}(t, t_0, X_{ij0})$ are mean-square bounded. The subscript '0' represents the initial value, X_{ij0} denotes the initial states of the system which is composed of two members: i and j, t_0 denotes the initial time, t denotes the current time, $\|\cdot\|$ is the Euclidean norm, $E\{\cdot\}$ is the mathematical expectation, and sup is the minimum upper bound.

Lemma 1. *[12] The necessary and sufficient condition for the mean-square boundedness of the solution for the time-varying linear stochastic system (1) is that the following time-varying linear deterministic system is bounded:*

$$\dot{y} = \begin{bmatrix} M(t) & R(t) \\ 0 & A(t) \end{bmatrix} y + \begin{bmatrix} K(t) \\ B(t) \end{bmatrix} \tag{6}$$

Lemma 2. *[12] If (2) is a time-invariant linear system (i.e., $A, F_i = const, i = 1, 2, \cdots, m$), system (2) is uniformly asymptotically stable if and only if $M = A \oplus A + \sum_{i=1}^{m} F_i \otimes F_i$ is stable, i.e., M is a Hurwitz matrix, or the real parts of the eigenvalues of matrix M are negative.*

Lemma 3. *[12] If $B(t), G_i(t), (i = 1, 2, \cdots, m)$ are bounded and system (2) is mean-square uniformly asymptotically stable, then the solution of system (1) is mean-square uniformly bounded.*

Based on Lemma 1, Lemma 2, Lemma 3, we present sufficient conditions for the stability of the *Itô* stochastic model and prove them.

Theorem 1. *The sufficient conditions for the stability of the Itô stochastic model (or the mean-square uniform boundedness of the stochastic system (1)) are:*

(1) $B(t)$ *and* $G_i(t)$ *are bounded.*
(2)   *The real part of the eigenvalues of matrix* $M(t)$ *are negative.*

Proof. For Equation (1), although we can use the conclusion of Lemma 1 to obtain the necessary and sufficient conditions for the mean-square boundedness directly, given that A and $F_k, k = 1 \sim m$ are linear time-invariant matrices, we further simplify the mean-square boundedness conditions.

According to Lemma 3, if it satisfies condition (1), the mean-square uniform boundedness of the system (1) is equivalent to system (2) and is mean-square uniformly asymptotically stable:

According to Lemma 2, system (2) is mean-square uniformly asymptotically equivalent to condition (2). Proof completed. □

2.3. Estimator of Itô Stochastic System

Lemma 4. *[38] considering the Itô stochastic system in the form as follows:*

$$dx(t) = Ax(t)dt + A_0x(t)dw(t) + dw_1(t) \tag{7}$$

$$dy(t) = Hx(t)dt + H_0x(t)dw(t) + dw_2(t) \tag{8}$$

where $x(t) \in R^n$, $y(t) \in R^m$ are system states and measured values, respectively. A, A_0, H and H_0 are constant matrices (they can also be extended to time-varying matrices if needed). $w(t)$ is a standard scalar Wiener process, as well as $w_1(t)$ and $w_2(t)$, where $w_1(t) \in R^n$ and $w_2(t) \in R^m$. The initial state $x(0)$ is a zero mean second-order stochastic process.

Assuming that $x(0)$ is independent of $w(t)$, $w_1(t)$, and $w_2(t)$, and it satisfies:

$$\begin{cases} E\{x(0)x^T(0)\} = D(0) \\ E\{dw(t)dw^T(t)\} = dt \\ E\{dw_1(t)dw_1{}^T(t)\} = Qdt \\ E\{dw_2(t)dw_2{}^T(t)\} = Rdt \end{cases} \tag{9}$$

Then the linear estimator with minimum mean square error is:

$$d\hat{x}(t) = [A - K(t)H]\hat{x}(t)dt + K(t)dy(t) \tag{10}$$

$$\hat{x}(0) = 0 \tag{11}$$

The gain of the estimator is:

$$K(t) = [P(t)H^T + A_0D(t)H_0{}^T][H_0D(t)H_0{}^T + R]^{-1} \tag{12}$$

where $P(t)$ can be obtained as follow:

$$dP(t) = \begin{aligned} &AP(t)dt + P(t)A^Tdt + A_0D(t)A_0{}^Tdt + Qdt \\ &-K(t)[H_0D(t)H_0{}^T + R]K^T(t)dt \end{aligned} \tag{13}$$

$$P(0) = D(0) \tag{14}$$

$$dD(t) = AD(t)dt + D(t)A^T dt + A_0 D(t) A_0{}^T dt + Qdt \tag{15}$$

Remark 1. *In this study, according to the above lemmas, we can construct a more applicable and effective USCC stochastic model to improve the formation properties of the UAV swarm system. Moreover, the measurement equation with Gaussian noises, the optimal estimator and controller are ingeniously involved in the closed-loop model. The mean square uniform boundedness condition of USCC stochastic system can be obtained based on the lemmas and theorem proposed above.*

3. USCC Stochastic System Modeling

The group dynamic cooperative control system model comprehensively considers the personality of the individual members and the cooperativeness of the whole formation based on Reynolds' three criteria [31]. Generally, the model is built with virtual forces: individuality and interaction forces. Individuality force describes the nodes' individual characteristics. Interaction force indicates the quality of group collaboration among nodes and describes the group dynamic cooperative characteristics, reflecting the ability to obey Reynolds' three criteria. We will use it as a general theory to guide the modeling of the USCC stochastic system.

3.1. Model of Individual Flight Control System

The individual flight control system of the UAV swarm adopts the north-up-east coordinate system.

Assumption 1. *The formation moves in a two-dimensional plane, thus the flight path inclination and pitch velocity are zero; the aircraft adopts side slip turning, thus the speed inclination angle, roll angle, roll angle velocity, angle of attack and side slip angle are all small values.*

Assumption 2. *Thrust P is independent of velocity V.*

The simplified nonlinear mathematical model of individual flight control system is:

$$\begin{cases} m & \dfrac{dV}{dt} = P - X \\ mV & \dfrac{d\varphi}{dt} = -P\beta + Z \\ mV & \dfrac{d\beta}{dt} = \omega_y + P\beta - Z \\ J_y & \dfrac{d\omega_y}{dt} = M_y \\ & \dfrac{dP}{dt} = -\dfrac{1}{T_P}P + \dfrac{K_P}{T_P}\delta_{Pc} \\ & \dfrac{d\delta_y}{dt} = -\dfrac{1}{T_{\delta y}}\delta_y + \dfrac{K_{\delta y}}{T_{\delta y}}\delta_{yc} \end{cases} \tag{16}$$

where V is the flight velocity; φ is the flight path declination; β is the lateral slip angle; ω is the rotational angle velocity of the body's coordinate system relative to the ground coordinate system. The subscript "y" denotes the y-component of ω. J_y is the y-component of inertial moments of the body's coordinate system. M_y is the y-component of moment caused by the external force (including thrust) on the mass center; P is the thrust; X is the resistance force; Z is the lateral force; δ is the rudder declination; K_δ, T_δ are gain and time constants of the control surface response, respectively (subscripts x, y, z are aileron, rudder, and elevator, respectively); K_p, T_p are gain and time constants for the thrust response, respectively; δ_c, δ_{Pc} are rudder angle command and thrust command, respectively.

By performing a small-disturbance linearization on (16) [39], we can obtain:

$$
\begin{cases}
\frac{d\Delta V}{dt} = -\frac{X^V}{m}\Delta V + \frac{1}{m}\Delta P \\
\frac{d\Delta\varphi}{dt} = \frac{P-Z^\beta}{mV}\Delta\beta - \frac{Z^{\delta y}}{mV}\Delta\delta_y \\
\frac{d\Delta\beta}{dt} = -\frac{P-Z^\beta}{mV}\Delta\beta + \Delta\omega_y + \frac{Z^{\delta y}}{mV}\Delta\delta_y \\
\frac{d\Delta\omega_y}{dt} = (\frac{M_y^\beta}{J_y} - \frac{M_y^{\dot\beta}}{J_y}\frac{P-Z^\beta}{mV})\Delta\beta + (\frac{M_y^{\omega y}}{J_y} + \frac{M_y^{\dot\beta}}{J_y})\Delta\omega_y + (\frac{M_y^{\delta y}}{J_y} + \frac{M_y^{\dot\beta}}{J_y}\frac{Z^{\delta y}}{mV})\Delta\delta_y \\
\frac{d\Delta P}{dt} = -\frac{1}{T_P}\Delta P + \frac{K_P}{T_P}\Delta\delta_{Pc} \\
\frac{d\Delta\delta_y}{dt} = -\frac{1}{T_{\delta y}}\Delta\delta_y + \frac{K_{\delta y}}{T_{\delta y}}\Delta\delta_{yc}
\end{cases}
\tag{17}
$$

where $X^V = \frac{\partial X}{\partial V}$, the same as other elements which is in the same form with X^V in (17).

3.2. Model of Formation Control System

For the convenience of outfield experiments and the safety of UAV swarm, we construct the model in a two-dimensional (2D) plane to make it more adaptive to complex task environments such as flat and dense formations, under which the aircraft would carry out missions at low altitude with almost no vertical maneuver space. Moreover, theoretical results can be covered fully and extended to the three-dimensional (3D) space. Therefore, we focus on the problem of USCC in the two-dimensional plane, as shown in Figure 1.

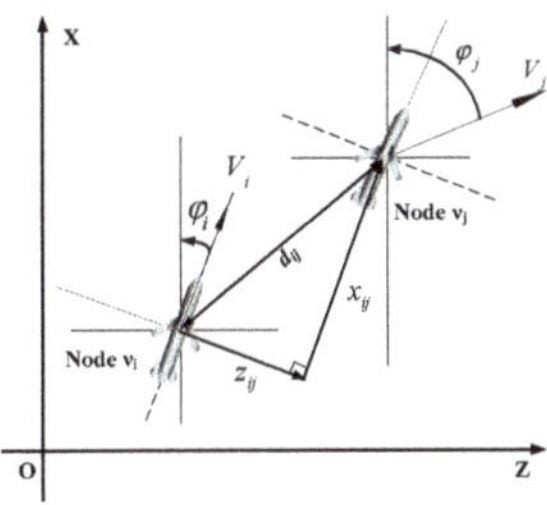

Figure 1. Relative movement between neighboring nodes.

In Figure 1, v_i and v_j are the two nodes adjacent to each other. The flight path coordinate of node v_i is set as the relative coordinate system, in which the x-axis represents the direction of the velocity and the z-axis is perpendicular to the x-axis shown as in Figure 1.

The ground coordinate system is set as the fixed coordinate system. d_{ij} is the distance between the two nodes; x_{ij} and z_{ij} are the relative distances in the forward and lateral directions of the flight path coordinate system, respectively. V_i, V_j, φ_i and φ_j represent their velocities and flight path declinations in the ground coordinate system, respectively.

With the relationship from theoretical mechanics: absolute velocity = relative velocity + convected velocity, the following kinematics equation for node v_i and node v_j can be derived:

$$
\vec{V}_j = \dot{\vec{d}}_{ij} + \vec{V}_i + \dot\varphi_i \times \vec{d}_{ij}
\tag{18}
$$

The above equation can be decomposed in the flight path coordinate system of the node v_i as:

$$
\begin{cases}
\frac{dx_{ij}}{dt} = V_j\cos(\varphi_j - \varphi_i) - V_i + \frac{d\varphi_i}{dt}z_{ij} \\
\frac{dz_{ij}}{dt} = V_j\sin(\varphi_j - \varphi_i) - \frac{d\varphi_i}{dt}x_{ij}
\end{cases}
\tag{19}
$$

Substituting the second formula in (17) into (19) and performing small perturbation linearization yields:

$$\begin{cases} \dfrac{d\Delta x_{ij}}{dt} = [-1 + \dfrac{(P_i-X_i)(-P_i\beta_i+Z_i)+m_iV_iZ_i{}^{V_i}}{m_i{}^2V_i{}^2}z_{ij}]\Delta V_i + V_j\sin(\varphi_j-\varphi_i)\Delta\varphi_i \\[2mm] \quad + \dfrac{(-P_i+Z^{\beta_i})z_{ij}}{m_iV_i}\Delta\beta_i - \dfrac{\beta_iz_{ij}}{m_iV_i}\Delta P_i - \dfrac{Z^{\delta_{iy}}z_{ij}}{m_iV_i}\Delta\delta_{iy} + \cos(\varphi_j-\varphi_i)\Delta V_j - V_j\sin(\varphi_j-\varphi_i)\Delta\varphi_j \\[2mm] \dfrac{d\Delta z_{ij}}{dt} = [\dfrac{(P_i-X_i)(-P_i\beta_i+Z_i)+m_iV_iZ_i{}^{V_i}}{m_i{}^2V_i{}^2}x_{ij}]\Delta V_i - V_j\cos(\varphi_j-\varphi_i)\Delta\varphi_i \\[2mm] \quad - \dfrac{(-P_i+Z_i{}^{\beta_i})x_{ij}}{m_iV_i}\Delta\beta_i + \dfrac{\beta_ix_{ij}}{m_iV_i}\Delta P_i + \dfrac{Z_i{}^{\delta_{iy}}x_{ij}}{m_iV_i}\Delta\delta_{iy} + \sin(\varphi_j-\varphi_i)\Delta V_j + V_j\cos(\varphi_j-\varphi_i)\Delta\varphi_j \end{cases} \tag{20}$$

Note that the aircraft flight momentum m_iV_i is relatively large. Moreover, β_i has small values according to Assumption 1, and $\varphi_j-\varphi_i \approx 0$, thus $\dfrac{(P_i-X_i)(-P_i\beta_i+Z_i)}{m_i{}^2V_i{}^2}\Delta V_i$, $\dfrac{\beta_i}{m_iV_i}\Delta P$, and $\sin(\varphi_j-\varphi_i)\Delta\varphi_i$ are second-order small quantities. Meanwhile, $\cos(\varphi_j-\varphi_i) \approx 1$, $\sin(\varphi_j-\varphi_i) \approx \varphi_j-\varphi_i$. Ignoring these second-order small quantities and simplifying (20), we can get:

$$\begin{cases} \dfrac{d\Delta x_{ij}}{dt} = (\dfrac{Z_i{}^{V_i}z_{ij}}{m_iV_i} - 1)\Delta V_i + \dfrac{(-P_i+Z_i{}^{\beta_i})z_{ij}}{m_iV_i}\Delta\beta_i - \dfrac{Z_i{}^{\delta_{iy}}z_{ij}}{m_iV_i}\Delta\delta_{iy} + \Delta V_j \\[2mm] \dfrac{d\Delta z_{ij}}{dt} = \dfrac{Z_i{}^{V_i}x_{ij}}{m_iV_i}\Delta V_i - V_j\Delta\varphi_i - \dfrac{(-P_i+Z_i{}^{\beta_i})x_{ij}}{m_iV_i}\Delta\beta_i + \dfrac{Z_i{}^{\delta_{iy}}x_{ij}}{m_iV_i}\Delta\delta_{iy} + V_j\Delta\varphi_j \end{cases} \tag{21}$$

Combining Equations (17) with (19) and (21) yields the formation control system model:

$$\begin{cases} \dfrac{d\Delta x_{ij}}{dt} = a_1\Delta V_i + a_2\Delta\beta_i - a_3\Delta\delta_{iy} + \Delta V_j \\[2mm] \dfrac{d\Delta z_{ij}}{dt} = a_4\Delta V_i - V_j\Delta\varphi_i - a_5\Delta\beta_i + a_6\Delta\delta_{iy} + V_j\Delta\varphi_j \\[2mm] \dfrac{d\Delta V_i}{dt} = -a_7\Delta V_i + \dfrac{1}{m_i}\Delta P_i \\[2mm] \dfrac{d\Delta\varphi_i}{dt} = a_8\Delta\beta_i - a_9\Delta\delta_{iy} \\[2mm] \dfrac{d\Delta\beta_i}{dt} = -a_8\Delta\beta_i + \Delta\omega_{iy} + a_9\Delta\delta_{iy} \\[2mm] \dfrac{d\Delta\omega_{iy}}{dt} = a_{10}\Delta\beta_i + a_{11}\Delta\omega_{iy} + a_{12}\Delta\delta_{iy} \\[2mm] \dfrac{d\Delta P_i}{dt} = -\dfrac{1}{T_{iP}}\Delta P_i + \dfrac{K_{iP}}{T_{iP}}\Delta\delta_{iPc} \\[2mm] \dfrac{d\Delta\delta_{iy}}{dt} = -\dfrac{1}{T_{i\delta y}}\Delta\delta_{iy} + \dfrac{K_{i\delta y}}{T_{i\delta y}}\Delta\delta_{iyc} \end{cases} \tag{22}$$

where $a_1 = \dfrac{Z_i{}^{V_i}z_{ij}}{m_iV_i} - 1$, $a_2 = \dfrac{(-P_i+Z_i{}^{\beta_i})z_{ij}}{m_iV_i}$, $a_3 = \dfrac{Z_i{}^{\delta_{iy}}z_{ij}}{m_iV_i}$, $a_4 = \dfrac{Z_i{}^{V_i}x_{ij}}{m_iV_i}$, $a_5 = \dfrac{(-P_i+Z_i{}^{\beta_i})x_{ij}}{m_iV_i}$, $a_6 = \dfrac{Z_i{}^{\delta_{iy}}x_{ij}}{m_iV_i}$,

$a_7 = \dfrac{X_i{}^{V_i}}{m_i}$, $a_8 = \dfrac{P_i-Z_i{}^{\beta_i}}{m_iV_i}$, $a_9 = \dfrac{Z_i{}^{\delta_{iy}}}{m_iV_i}$, $a_{10} = \dfrac{M_{iy}^{\beta_i}}{J_{iy}} - \dfrac{M_{iy}^{\dot\beta_i}}{J_{iy}}\dfrac{P_i-Z_i{}^{\beta_i}}{m_iV_i}$, $a_{11} = \dfrac{M_{iy}^{\omega_{iy}}}{J_{iy}} + \dfrac{M_{iy}^{\dot\beta_i}}{J_{iy}}$, $a_{12} = \dfrac{M_{iy}^{\delta_{iy}}}{J_{iy}} + \dfrac{M_{iy}^{\dot\beta_i}}{J_{iy}}\dfrac{Z_i{}^{\delta_{iy}}}{m_iV_i}$.

Note that: the state coefficients V_i, P_i, x_{ij}, z_{ij} and $Z_i{}^{V_i}$ in (22) are obtained at the balanced point.

3.3. Random Noises Analysis and Its Modeling

3.3.1. Process Noises

The formation could be easily affected by various forces in the atmosphere that cannot be accurately measured in advance. Therefore, the process noises cannot be ignored.

For the node v_i, assuming that the mass m_i, velocity V_j and flight path declination angle φ_j of the adjacent node v_j, which are obtained from the supporting network, are given values (i.e., consider that V_j and φ_j are already estimated in v_j, and ignore the random transmission interference), but x_{ij}, z_{ij}, V_i, P_i, $X_i{}^{V_i}$, $Z_i{}^{V_i}$, $Z_i{}^{\beta_i}$, $Z_i{}^{\delta_{iy}}$, $M_{iy}^{\beta_i}$, $M_{iy}^{\dot\beta_i}$, $M_{iy}^{\omega_{iy}}$ and $M_{iy}^{\delta_{iy}}$ are determined by the aircraft's instantaneous state (such as velocity, altitude, attack angle, yaw angle, etc.); These states and their influences are random in the real flight environment. Therefore, based on the central limit theorem, we assume that the above parameters approximately obey the Gaussian distribution, that is:

$$
(1)\begin{cases} x_{ij} = x_{ijb} + w_{x_{ij}} \\ w_{x_{ij}} \sim N(0,\sigma_{x_{ij}}^2) \end{cases}, \quad
(2)\begin{cases} z_{ij} = z_{ijb} + w_{z_{ij}} \\ w_{z_{ij}} \sim N(0,\sigma_{z_{ij}}^2) \end{cases}, \quad
(3)\begin{cases} V_i = V_{ib} + w_{V_i} \\ w_{V_i} \sim N(0,\sigma_{V_i}^2) \end{cases}, \quad
(4)\begin{cases} P_i = P_{ib} + w_{P_i} \\ w_{P_i} \sim N(0,\sigma_{P_i}^2) \end{cases},
$$

$$
(5)\begin{cases} X_i^{V_i} = X_{ib}^{V_i} + w_{X_i^{V_i}} \\ w_{X_i^{V_i}} \sim N(0,\sigma_{X_i^{V_i}}^2) \end{cases}, \quad
(6)\begin{cases} Z_i^{V_i} = Z_{ib}^{V_i} + w_{Z_i^{V_i}} \\ w_{Z_i^{V_i}} \sim N(0,\sigma_{Z_i^{V_i}}^2) \end{cases}, \quad
(7)\begin{cases} Z_i^{\beta_i} = Z_{ib}^{\beta_i} + w_{Z_i^{\beta_i}} \\ w_{Z_i^{\beta_i}} \sim N(0,\sigma_{Z_i^{\beta_i}}^2) \end{cases}, \quad
(8)\begin{cases} Z_i^{\delta_{iy}} = Z_{ib}^{\delta_{iy}} + w_{Z_i^{\delta_{iy}}} \\ w_{Z_i^{\delta_{iy}}} \sim N(0,\sigma_{Z_i^{\delta_{iy}}}^2) \end{cases}, \qquad (23)
$$

$$
(9)\begin{cases} M_{iy}^{\beta_i} = M_{iyb}^{\beta_i} + w_{M_{iy}^{\beta_i}} \\ w_{M_{iy}^{\beta_i}} \sim N(0,\sigma_{M_{iy}^{\beta_i}}^2) \end{cases}, \quad
(10)\begin{cases} M_{iy}^{\dot{\beta}_i} = M_{iyb}^{\dot{\beta}_i} + w_{M_{iy}^{\dot{\beta}_i}} \\ w_{M_{iy}^{\dot{\beta}_i}} \sim N(0,\sigma_{M_{iy}^{\dot{\beta}_i}}^2) \end{cases}, \quad
(11)\begin{cases} M_{iy}^{\omega_{iy}} = M_{iyb}^{\omega_{iy}} + w_{M_{iy}^{\omega_{iy}}} \\ w_{M_{iy}^{\omega_{iy}}} \sim N(0,\sigma_{M_{iy}^{\omega_{iy}}}^2) \end{cases}, \quad
(12)\begin{cases} M_{iy}^{\delta_{iy}} = M_{iyb}^{\delta_{iy}} + w_{M_{iy}^{\delta_{iy}}} \\ w_{M_{iy}^{\delta_{iy}}} \sim N(0,\sigma_{M_{iy}^{\delta_{iy}}}^2) \end{cases}.
$$

The subscript "*b*" denotes that the values are determined and they are obtained at the balanced point. The formation states do not change much when they fly around the balanced point; thus, the variances of the random variables are approximately constant, and it can be assumed that the above parameters are independent of each other. In the following, we use $n_1, n_2, \cdots, n_{12}$ to represent the random variables in (23).

For $a_1 = \dfrac{Z_i^{V_i} z_{ij}}{m_i V_i} - 1$, substituting (2) (3) (6) in (23) into a_1 yields:

$$
a_1 = \frac{Z_i^{V_i} z_{ij}}{m_i V_i} - 1 = \frac{(Z_{ib}^{V_i} + w_{Z_i^{V_i}})(z_{ijb} + w_{z_{ij}})}{m_i(V_{ib} + w_{V_i})} - 1 = \frac{Z_{ib}^{V_i} z_{ijb}}{m_i(V_{ib} + w_{V_i})} - 1 + \frac{Z_{ib}^{V_i} w_{z_{ij}}}{m_i(V_{ib} + w_{V_i})} + \frac{z_{ijb} w_{Z_i^{V_i}}}{m_i(V_{ib} + w_{V_i})} + \frac{w_{Z_i^{V_i}} w_{z_{ij}}}{m_i(V_{ib} + w_{V_i})} \tag{24}
$$

Assume that w_{V_i} is relatively small compared to the aircraft speed V_{ib}. Since $V_{ib} + w_{V_i}$ is in the denominator, the impact of w_{V_i} on a_1 is small and can be ignored; assuming that both $w_{Z_i^{V_i}}$ and $w_{z_{ij}}$ are small, $w_{Z_i^{V_i}} w_{z_{ij}}$ is a second-order small quantity and can be ignored. Then we get:

$$
a_1 = \left(\frac{Z_{ib}^{V_i} z_{ijb}}{m_i V_{ib}} - 1\right) + \frac{z_{ijb} w_{Z_i^{V_i}}}{m_i V_{ib}} + \frac{Z_{ib}^{V_i} w_{z_{ij}}}{m_i V_{ib}} = \left(\frac{Z_{ib}^{V_i} z_{ijb}}{m_i V_{ib}} - 1\right) + \frac{Z_{ib}^{V_i} \sigma_{z_{ij}}}{m_i V_{ib}} n_2 + \frac{z_{ijb} \sigma_{Z_i^{V_i}}}{m_i V_{ib}} n_6 \triangleq a_{1b} + a_{1b2} n_2 + a_{1b6} n_6 \tag{25}
$$

Assuming that all the random parts of (23) are small, and ignoring the second-order small quantity. For the same reason as a_1, the expression of the coefficients $a_2 \sim a_{15}$ could be derived. The results are given as follows:

$$
a_2 = \frac{(-P_{ib} + Z_{ib}^{\beta_i}) z_{ijb}}{m_i V_{ib}} + \frac{(-P_{ib} + Z_{ib}^{\beta_i})\sigma_{z_{ij}}}{m_i V_{ib}} n_2 + \frac{-z_{ijb}\sigma_{P_i}}{m_i V_{ib}} n_4 + \frac{z_{ijb}\sigma_{Z_i^{\beta_i}}}{m_i V_{ib}} n_7 \triangleq a_{2b} + a_{2b2} n_2 + a_{2b4} n_4 + a_{2b7} n_7 \tag{26}
$$

$$
a_3 = \frac{Z_{ib}^{\delta_{iy}} z_{ijb}}{m_i V_{ib}} + \frac{Z_{ib}^{\delta_{iy}}\sigma_{z_{ij}}}{m_i V_{ib}} n_2 + \frac{z_{ijb}\sigma_{Z_i^{\delta_{iy}}}}{m_i V_{ib}} n_8 \triangleq a_{3b} + a_{3b2} n_2 + a_{3b8} n_8 \tag{27}
$$

$$
a_4 = \frac{Z_{ib}^{V_i} x_{ijb}}{m_i V_{ib}} + \frac{Z_{ib}^{V_i}\sigma_{x_{ij}}}{m_i V_{ib}} n_1 + \frac{x_{ijb}\sigma_{Z_i^{V_i}}}{m_i V_{ib}} n_6 \triangleq a_{4b} + a_{4b1} n_1 + a_{4b6} n_6 \tag{28}
$$

$$
a_5 = \frac{(-P_{ib} + Z_{ib}^{\beta_i}) x_{ijb}}{m_i V_{ib}} + \frac{(-P_{ib} + Z_{ib}^{\beta_i})\sigma_{x_{ij}}}{m_i V_{ib}} n_1 + \frac{-x_{ijb}\sigma_{P_i}}{m_i V_{ib}} n_4 + \frac{x_{ijb}\sigma_{Z_i^{\beta_i}}}{m_i V_{ib}} n_7 \triangleq a_{5b} + a_{5b1} n_1 + a_{5b4} n_4 + a_{5b7} n_7 \tag{29}
$$

$$
a_6 = \frac{Z_{ib}^{\delta_{iy}} x_{ijb}}{m_i V_{ib}} + \frac{Z_{ib}^{\delta_{iy}}\sigma_{x_{ij}}}{m_i V_{ib}} n_1 + \frac{x_{ijb}\sigma_{Z_i^{\delta_{iy}}}}{m_i V_{ib}} n_8 \triangleq a_{6b} + a_{6b1} n_1 + a_{6b8} n_8 \tag{30}
$$

$$
a_7 = \frac{X_{ib}^{V_i}}{m_i} + \frac{\sigma_{X_i^{V_i}}}{m_i} n_5 \triangleq a_{7b} + a_{7b5} n_5 \tag{31}
$$

$$
a_8 = \frac{P_{ib} - Z_{ib}^{\beta_i}}{m_i V_{ib}} + \frac{\sigma_{P_i}}{m_i V_{ib}} n_4 + \frac{-\sigma_{Z_i^{\beta_i}}}{m_i V_{ib}} n_7 \triangleq a_{8b} + a_{8b4} n_4 + a_{8b7} n_7 \tag{32}
$$

$$
a_9 = \frac{Z_{ib}^{\delta_{iy}}}{m_i V_{ib}} + \frac{\sigma_{Z_i^{\delta_{iy}}}}{m_i V_{ib}} n_8 \triangleq a_{9b} + a_{9b8} n_8 \tag{33}
$$

$$
\begin{aligned}
a_{10} &= \left(\frac{M_{iyb}^{\beta_i}}{J_{iy}} - \frac{M_{iyb}^{\dot{\beta}_i}}{J_{iy}}\frac{P_{ib} - Z_{ib}^{\beta_i}}{m_i V_{ib}}\right) + \frac{-M_{iyb}^{\dot{\beta}_i}\sigma_{P_i}}{J_{iy} m_i V_{ib}} n_4 + \frac{M_{iyb}^{\dot{\beta}_i}\sigma_{Z_i^{\beta_i}}}{J_{iy} m_i V_{ib}} n_7 + \frac{\sigma_{M_{iy}^{\beta_i}}}{J_{iy}} n_9 + \frac{-(P_{ib} - Z_{ib}^{\beta_i})\sigma_{M_{iy}^{\dot{\beta}_i}}}{J_{iy} m_i V_{ib}} n_{10} \\
&\triangleq a_{10b} + a_{10b4} n_4 + a_{10b7} n_7 + a_{10b9} n_9 + a_{10b10} n_{10}
\end{aligned} \tag{34}
$$

$$a_{11} = \frac{M_{iyb}^{\omega_{iy}} + M_{iyb}^{\dot{\beta}_i}}{J_{iy}} + \frac{\sigma_{M_{iy}^{\dot{\beta}_i}}}{J_{iy}}n_{10} + \frac{\sigma_{M_{iy}^{\omega_{iy}}}}{J_{iy}}n_{11} \triangleq a_{11b} + a_{11b10}n_{10} + a_{11b11}n_{11} \tag{35}$$

$$a_{12} = \left(\frac{M_{iyb}^{\delta_{iy}}}{J_{iy}} + \frac{M_{iyb}^{\dot{\beta}_i}}{J_{iy}}\frac{Z_{ib}^{\delta_{iy}}}{m_i V_{ib}}\right) + \frac{M_{iyb}^{\dot{\beta}_i}\sigma_{Z_i^{\delta_{iy}}}}{J_{iy}m_i V_{ib}}n_8 + \frac{Z_{ib}^{\delta_{iy}}\sigma_{M_{iy}^{\dot{\beta}_i}}}{J_{iy}m_i V_{ib}}n_{10} + \frac{\sigma_{M_{iy}^{\delta_{iy}}}}{J_{iy}}n_{12}$$
$$\triangleq a_{12b} + a_{12b8}n_8 + a_{12b10}n_{10} + a_{12b12}n_{12} \tag{36}$$

The states of the system are $X_{ij} = \begin{bmatrix} \Delta x_{ij} & \Delta z_{ij} & \Delta V_i & \Delta\varphi_i & \Delta\beta_i & \Delta\omega_{iy} & \Delta P_i & \Delta\delta_{iy} \end{bmatrix}^T$; the inputs are $U_{ij} = \begin{bmatrix} \Delta\delta_{iPc} & \Delta\delta_{iyc} & \Delta V_j & \Delta\varphi_j \end{bmatrix}^T$; ΔV_j and $\Delta\varphi_j$ are determined random inputs. Substituting (25) to (36) into (22), then the open-loop state equation of the formation stochastic system could be get:

$$\dot{X}_{ij} = A_{ij}X_{ij} + B_{ij}U_{ij} + \sum_{k=1}^{12} F_{ijk}X_{ij}n_k \tag{37}$$

where $A_{ij} = \begin{bmatrix} 0 & 0 & a_{1b} & 0 & a_{2b} & 0 & 0 & -a_{3b} \\ 0 & 0 & a_{4b} & -V_j & -a_{5b} & 0 & 0 & a_{6b} \\ 0 & 0 & -a_{7b} & 0 & 0 & 0 & 1/m_i & 0 \\ 0 & 0 & 0 & 0 & a_{8b} & 0 & 0 & -a_{9b} \\ 0 & 0 & 0 & 0 & -a_{8b} & 1 & 0 & a_{9b} \\ 0 & 0 & 0 & 0 & a_{10b} & a_{11b} & 0 & a_{12b} \\ 0 & 0 & 0 & 0 & 0 & 0 & -1/T_{iP} & 0 \\ 0 & 0 & 0 & 0 & 0 & 0 & 0 & -1/T_{i\delta_y} \end{bmatrix}, B_{ij} = \begin{bmatrix} 0 & 0 & 1 & 0 \\ 0 & 0 & 0 & V_j \\ 0 & 0 & 0 & 0 \\ 0 & 0 & 0 & 0 \\ 0 & 0 & 0 & 0 \\ 0 & 0 & 0 & 0 \\ K_{iP}/T_{iP} & 0 & 0 & 0 \\ 0 & K_{i\delta_y}/T_{i\delta_y} & 0 & 0 \end{bmatrix}.$$

For convenience of description, we define the square matrix T_{mn} whose order is eight and has only one nonzero element; the value of the nonzero element is '1' and it lies in column m and row n. Then matrices $F_{ijk}(k = 1, 2, \ldots, 12)$ can be described as follows:

$$F_{ij1} = a_{4b1}T_{23} - a_{6b1}T_{25} + a_{7b1}T_{28},$$
$$F_{ij2} = a_{2b2}T_{15} - a_{3b2}T_{18},$$
$$F_{ij3} = a_{1b3}T_{13},$$
$$F_{ij4} = a_{2b4}T_{15} - a_{5b4}T_{25} + a_{8b4}T_{45} - a_{89b4}T_{55} + a_{10b4}T_{65},$$
$$F_{ij5} = -a_{7b5}T_{33},$$
$$F_{ij6} = a_{1b6}T_{13} + a_{4b6}T_{23},$$
$$F_{ij7} = a_{2b7}T_{15} - a_{5b7}T_{25} + a_{8b7}T_{45} - a_{8b7}T_{55} + a_{10b7}T_{65},$$
$$F_{ij8} = -a_{3b8}T_{18} + a_{6b8}T_{28} - a_{9b8}T_{48} + a_{9b8}T_{58} + a_{12b8}T_{68},$$
$$F_{ij9} = a_{10b9}T_{65},$$
$$F_{ij10} = a_{10b10}T_{65} + a_{11b10}T_{66} + a_{12b10}T_{68},$$
$$F_{ij11} = a_{11b11}T_{66},$$
$$F_{ij12} = a_{12b12}T_{68}.$$

3.3.2. Measurement Noises

The states of the stochastic system (37) are measured by the support network and relative navigation in the UAV swarm's information acquisition system. The measuring vector is defined as:
$Y_{ij} = \begin{bmatrix} \Delta x_{ijm} & \Delta z_{ijm} & \Delta V_{im} & \Delta\varphi_{im} & \Delta\beta_{im} & \Delta\omega_{iym} & \Delta\delta_{iym} \end{bmatrix}^T$, assuming that ΔP_i cannot be measured. These measured values are mainly affected by sensor measurement, network transmission and random disturbances in the flight environment. Assuming that the measured noises of the system approximately obey the Gaussian distribution, whose mathematical expectation is 0 and variance is $\sigma_m{}^2$, the measurement equation is:

$$\begin{cases} \Delta x_{ijm} = \Delta x_{ij} + \sigma_{\Delta x_{ijm}} n_{13} \\ \Delta z_{ijm} = \Delta z_{ij} + \sigma_{\Delta z_{ijm}} n_{14} \\ \Delta V_{im} = \Delta V_i + \sigma_{\Delta v_{im}} n_{15} \\ \Delta \varphi_{im} = \Delta \varphi_i + \sigma_{\Delta \varphi_{im}} n_{16} \\ \Delta \beta_{im} = \Delta \beta_i + \sigma_{\Delta \beta_{im}} n_{17} \\ \Delta \omega_{iym} = \Delta \omega_{iy} + \sigma_{\Delta \omega_{iym}} n_{18} \\ \Delta \delta_{iym} = \Delta \delta_{iy} + \sigma_{\Delta \delta_{iym}} n_{19} \end{cases} \tag{38}$$

The subscript "m" means the measured value of the system, $n_{13}, n_{15}, \cdots, n_{19}$ are standard Gaussian white noises independent of each other. They are also independent of $n_1, n_2, \cdots, n_{12}$.

In summary, the measurement equation is:

$$Y_{ij} = H_{ij} X_{ij} + \sum_{k=13}^{19} E_{ijk} n_k \tag{39}$$

$$\text{where, } H_{ij} = \begin{bmatrix} 1 & 0 & & & & & & \\ & 1 & & & & & & \\ & & 1 & & & & & \\ & & & 1 & \ddots & & & \\ & & & & 1 & & & \\ & & & & & 1 & 0 \\ & & & & & 0 & 1 \end{bmatrix}, \; E_{ij13} = \begin{bmatrix} \sigma_{\Delta x_{ijm}} \\ 0 \\ 0 \\ 0 \\ 0 \\ 0 \\ 0 \end{bmatrix}, \; E_{ij14} = \begin{bmatrix} 0 \\ \sigma_{\Delta z_{ijm}} \\ 0 \\ 0 \\ 0 \\ 0 \\ 0 \end{bmatrix}, \; E_{ij15} = \begin{bmatrix} 0 \\ 0 \\ \sigma_{\Delta v_{im}} \\ 0 \\ 0 \\ 0 \\ 0 \end{bmatrix},$$

$$E_{ij16} = \begin{bmatrix} 0 \\ 0 \\ 0 \\ \sigma_{\Delta \varphi_{im}} \\ 0 \\ 0 \\ 0 \end{bmatrix}, \; E_{ij17} = \begin{bmatrix} 0 \\ 0 \\ 0 \\ 0 \\ \sigma_{\Delta \beta_{im}} \\ 0 \\ 0 \end{bmatrix}, \; E_{ij18} = \begin{bmatrix} 0 \\ 0 \\ 0 \\ 0 \\ 0 \\ \sigma_{\Delta \omega_{iym}} \\ 0 \end{bmatrix}, \; E_{ij19} = \begin{bmatrix} 0 \\ 0 \\ 0 \\ 0 \\ 0 \\ 0 \\ \sigma_{\Delta \delta_{iym}} \end{bmatrix}.$$

3.4. Formation Estimator Design

For the state estimating problem of the formation control stochastic system (37) and (39), we use a novel *Itô* stochastic system estimator which has a fixed gain according to [38].

In Lemma 4, the gain of the estimator is time-varying, but in this paper, the USCC problem is investigated during cruising and the formation maintains a certain configuration with certain speed and height in that period, the formation does not change very much. Therefore, we use an estimator with fixed gain in (10) to estimate states, which can significantly simplify the computation and improve the real-time performance of the system.

The formation control stochastic system (37) and (39) can be described as the following *Itô* stochastic system:

$$dX_{ij}(t) = A_{ij} X_{ij}(t) dt + B_{ij} U_{ij}(t) dt + \sum_{k=1}^{12} F_{ijk} X_{ij}(t) dW_k(t) \tag{40}$$

$$dY_{ij}(t) = H_{ij} dX_{ij}(t) + \sum_{k=13}^{19} E_{ijk} dW_k(t) \tag{41}$$

where $W_k(t)$ ($k = 1 \sim 12$) in (40) are standard scalar Wiener processes. $\sum_{k=13}^{19} E_{ijk} dW_k(t)$ is a 7-dimension Wiener process.

The initial states $X_{ij}(0)$ satisfy:

$$\begin{cases} E\{X_{ij}(0)X_{ij}{}^T(0)\} = D_{ij}(0) \\ E\{dW_k(t)dW_k{}^T(t)\} = dt \\ E\{[\sum_{k=13}^{19} E_{ijk}dW_k(t)][\sum_{k=13}^{19} E_{ijk}dW_k(t)]^T\} = R_{ij}dt \end{cases} \tag{42}$$

Then, substituting (40) into (41) and the fixed gain estimator of the formation control stochastic system can be obtained by Lemma 4:

$$d\hat{X}_{ij}(t) = (A_{ij} - K_f H_{ij} A_{ij})\hat{X}_{ij}(t)dt + B_{ij}U_{ij}(t)dt + K_f dY_{ij}(t) \tag{43}$$

$$\hat{X}_{ij}(0) = 0 \tag{44}$$

where $\hat{X}_{ij} = [\ \Delta\hat{x}_{ij}\ \ \Delta\hat{z}_{ij}\ \ \Delta\hat{V}_i\ \ \Delta\hat{\varphi}_i\ \ \Delta\hat{\beta}_i\ \ \Delta\hat{\omega}_{iy}\ \ \Delta\hat{P}_i\ \ \Delta\hat{\delta}_{iy}\ ]^T$ is the state estimate vector; K_f is fixed gain of the estimator. U_{ij} is the control input.

3.5. Formation Controller Design

It can be seen from (22) that the system states have a high degree of coupling between each other (such as the forward distance Δx_{ij} and the sideslip angle $\Delta\beta_i$, the lateral distance Δz_{ij} and ΔV_i, they all have high degree of coupling between each other), so the forward and lateral distance will be controlled with a couple in this paper. The PID-based formation control system structure is a commonly used design method in the engineering application at present [40]. According to the clustering algorithm proposed in [41], the PID formation controller we adopted is:

$$U_{ij} = K_c K_{\omega ij}(\hat{X}_{ij} - X_{ij}^*) + U_{jd} \tag{45}$$

$$X_{ij}^* = [\Delta x_{ij}^*, \Delta z_{ij}^*, \Delta V_i^*, \Delta\varphi_i^*, \Delta\beta_i^*, 0, 0, 0]^T \tag{46}$$

$$K_{\omega ij} = diag(\omega_{ij}, \omega_{ij}, 1, 1, 1, 1, 1, 1) \tag{47}$$

$$U_{jd} = [0, 0, \Delta V_j, \Delta\varphi_j]^T \tag{48}$$

where $\hat{X}_{ij}$ is the output of the estimator; X_{ij}^* is the system command; superscript "*" indicates the command, the same as below; U_{jd} is the determined interference input vector of the adjacent node; $K_c \in R^{4\times8}$ is the control law, in which the last two rows in K_c are zero vectors because ΔV_j and $\Delta\varphi_j$ are the determined interference inputs in U_{ij}; $K_{\omega ij}$ is the adjacency adjustment matrix, $0 \le \omega_{ij} \le 1$ are adjacency coefficients. The larger it is, the stronger the adjacency relationship between node v_i and v_j.

The key to designing a better PID controller is to contrive the proper PID gain parameters. In order to get a better parameter of the feedback coefficients, we use the SRAD method which combines the genetic algorithm and Monte Carlo simulation to improve the robustness of the stochastic system.

We can see from (45) that $K_{c\Delta v_i}(\Delta\hat{V}_i - \Delta V_i^*)$ and $K_{c\Delta\varphi_i}(\Delta\hat{\varphi}_i - \Delta\varphi_i^*)$ reflect the individuality forces of the individual aircraft, $K_{c\Delta x_{ij}}\omega_{ij}(\Delta\hat{x}_{ij} - \Delta x_{ij}^*)$ and $K_{c\Delta z_{ij}}\omega_{ij}(\Delta\hat{z}_{ij} - \Delta z_{ij}^*)$ reflecting the interaction forces which represent the quality of the formation cooperation. Both of them can contribute to maintaining the configuration during formation maneuvering.

3.6. Closed-Loop USCC Stochastic System

In summary, the *Itô* stochastic system of the USCC is as follows:

$$
\begin{cases}
dX_{ij} = A_{ij}X_{ij}dt + B_{ij}U_{ij}dt + \sum_{k=1}^{12} F_{ijk}X_{ij}dW_k \\
dY_{ij} = H_{ij}dX_{ij} + \sum_{k=13}^{19} E_{ijk}dW_k \\
U_{ij} = K_c K_{\omega ij}(\hat{X}_{ij} - X_{ij}^*) + U_{jd} \\
d\hat{X}_{ij} = (A_{ij} - K_f H_{ij}A_{ij})\hat{X}_{ij}dt + B_{ij}U_{ij}dt + K_f dY_{ij}
\end{cases}
\tag{49}
$$

where the first equation is the state equation, the second is the measurement equation, the third is the control input, and the fourth is the state estimation equation. The above four equations together construct the expansion closed-loop equation for the stochastic system of USCC (as shown in Figure 2):

$$
d\overline{X}_{ij} = \left[\overline{A}_{ij}\overline{X}_{ij} + \overline{B}_{ij}(t)\right]dt + \sum_{k=1}^{19}\left[\overline{F}_{ijk}\overline{X}_{ij} + \overline{G}_{ijk}\right]dW_i
\tag{50}
$$

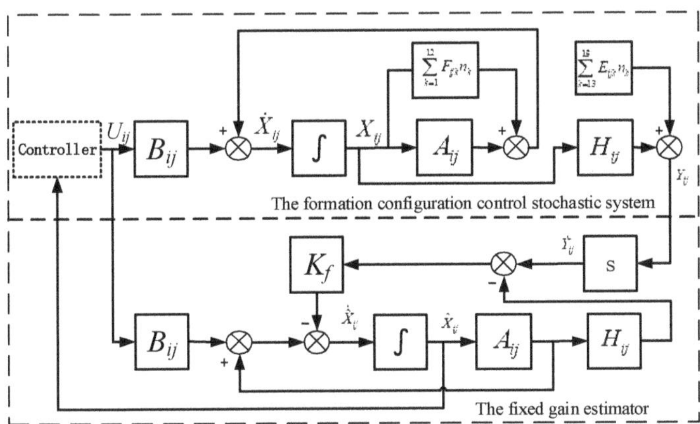

Figure 2. The framework of the closed-loop system.

The states of the expansion system are:

$$
\overline{X}_{ij} = \begin{bmatrix} X_{ij} \\ \hat{X}_{ij} \end{bmatrix}
\tag{51}
$$

where $\overline{X}_{ij} \in R^{16\times1}$; $X_{ij} \in R^{8\times1}$ are original states; $\hat{X}_{ij} \in R^{8\times1}$ are estimated states.

The state transfer matrix of the expansion system is:

$$
\overline{A}_{ij} = \begin{bmatrix} A_{ij} & B_{ij}K_c K_{\omega ij} \\ K_f H_{ij}A_{ij} & [(I_{8\times8} - K_f H_{ij})A_{ij} + (I_{8\times8} + K_f H_{ij})B_{ij}K_c K_{\omega ij}] \end{bmatrix}
\tag{52}
$$

where $\overline{A}_{ij} \in R^{16\times16}$; $A_{ij} \in R^{8\times8}$ is the state transfer matrix of the original system; $B_{ij} \in R^{8\times4}$ is the input matrix of the original system; $H_{ij} \in R^{7\times8}$ is the estimate matrix of the original system; $K_{\omega ij} \in R^{8\times8}$ is the adjacent adjustment matrix in U_{ij}; $K_c \in R^{4\times8}$ is the control law; $K_f \in R^{8\times7}$ is the gain of the estimator.

The input matrix of the expansion system is:

$$\overline{B}_{ij}(t) = \left[\begin{array}{c} -B_{ij}(K_cK_{\omega ij}X^*_{ij} - U_{jd}) \\ -(I_{8\times 8} + K_fH_{ij})B_{ij}(K_cK_{\omega ij}X^*_{ij} - U_{jd}) \end{array} \right] \tag{53}$$

where $\overline{B}_{ij}(t) \in R^{16\times 1}$; $X^*_{ij} \in R^{8\times 1}$ is the system command; $U_{jd} \in R^{4\times 1}$ is the determined input of the adjacent node in U_{ij}. Note that $\overline{D}_{ij}(t)$ is a time-varying matrix because X^*_{ij} is time-varying.

The expansion stochastic state transfer matrix is:

$$\overline{F}_{ijk} = \left[\begin{array}{cc} F_{ijk} & 0 \\ K_fH_{ij}F_{ijk} & 0 \end{array} \right] \tag{54}$$

where $\overline{F}_{ijk} \in R^{16\times 16}$; $F_{ijk} = 0^{8\times 8}(k = 13 \sim 19)$.

The stochastic input matrix of the expansion system is:

$$\overline{G}_{ijk} = \left[\begin{array}{c} 0 \\ K_fE_{ijk} \end{array} \right] \tag{55}$$

where $\overline{G}_{ijk} \in R^{16\times 1}$; $E_{ijk} = 0^{7\times 1}(k = 1 \sim 12)$.

The standard Wiener process is:

$$W = [W_1, W_2, \cdots, W_{19}]^T \tag{56}$$

where $W_1 \sim W_{19}$ are the Wiener processes in (49); W is an independent 19-dimension standard Wiener process defined in the complete probability space: (Ω, F, P).

3.7. Main Results

Define the following matrix:

$$M_{ij} = \overline{A}_{ij} \oplus \overline{A}_{ij} + \sum_{k=1}^{19} \overline{F}_{ijk} \otimes \overline{F}_{ijk} \tag{57}$$

According to Theorem 1, the sufficient conditions of the stability of USCC stochastic model are:

(1) $\overline{B}_{ij}(t), \overline{G}_{ijk}, k = 1 \sim 19$ are bounded.
(2) The real part of the eigenvalues of matrix M_{ij} are negative.

It can be seen from (53) and (55) that all the elements in the matrices $\overline{B}_{ij}(t), \overline{G}_{ijk}$ are bounded according to their definition, because in B_{ij}, K_{ip}, T_{ip} are constant parameters of the thrust response and $K_{i\delta_y}, T_{i\delta_y}$ are constant parameters of the elevator response; $K_c \in R^{4\times 8}$ is the control law which can be obtained after simulation; $K_{\omega ij}$ is an adjacency adjustment matrix whose elements are in the range of $(0,1)$; X^*_{ij} are bounded command values; U_{jd} is the determined interference input vector of the adjacent node; $K_f \in R^{8\times 7}$ is the fixed gain of the estimator which can be determined from simulation; the nonzero elements in the matrix $H_{ij} \in R^{7\times 8}$ are '1'; and in $\overline{G}_{ijk}, E_{ijk}(k = 13 \sim 19)$ is the bounded variance vector of measuring noises. Therefore, we can further deduce the following:

Proposition 1. *Under normal circumstances, a sufficient condition for the stability (or the mean square uniform boundedness of the stochastic system (50)) of the USCC system is that the real parts of the eigenvalues are negative, that is:*

$$max\left\{Re\lambda(M_{ij})\right\} < 0 \tag{58}$$

where the $max\left\{Re\lambda(M_{ij})\right\}$ is the maximum real part of the eigenvalue of M_{ij}.

4. Simulation and Experiments

With the aim of cooperative detection under complex environments, we ran simulations with two aircraft to verify the effectiveness of the proposed stochastic model. Moreover, the equivalent outfield flight test was carried out to complete the mission of cooperative detection in a certain area. The results demonstrate that the formation could be achieved effectively and accurately.

4.1. Simulation Results

4.1.1. Initial State

To make it convenient for analysis, we set the number of UAV swarm $n = 2$. The configuration of the formation during cruising is $x_{12b} = 100\,\mathrm{m}$, $z_{12b} = -173.2\,\mathrm{m}$ (that is, v_2 located 100 meters forward and 173.2 meters left of v_1), and the cruising speeds are $V_{1b} = V_{2b} = 100\mathrm{m/s}$. The cruising trajectory declination is $\varphi_{1b} = \varphi_{2b} = 0\,\mathrm{rad}$. The current configuration of the formation is the cruising formation, the current speed is $V_1 = V_2 = 100\,\mathrm{m/s}$, and the current flight path declination is $\varphi_1 = \varphi_2 = 0\,\mathrm{rad}$. The original mass is $m_1 = m_2 = 1400\,\mathrm{Kg}$, the y-components of inertial moments are $I_{y1} = I_{y2} = 3980\,\mathrm{Kg \cdot m^2}$.

4.1.2. Formation Parameters

Assume that the supporting network is strongly connected and that the adjacency coefficient ω_{ij} in (47) is 1.

4.1.3. Standard deviation of random interference

Assume that the standard deviations of random interference in Equations (23) and (38) are:

$$\sigma_{x_{ij}} = 1\,\mathrm{m},\ \sigma_{z_{ij}} = 1\,\mathrm{m},\ \sigma_{V_i} = 0.1\,\mathrm{m/s},\ \sigma_{P_i} = 5\,\mathrm{N},\ \sigma_{X_i V_i} = 0.25\,\mathrm{kg/s},\ \sigma_{Z_i V_i} = 0.5\,\mathrm{kg/s},$$

$$\sigma_{Z_i \beta_i} = 0.5\,\mathrm{N/rad},\ \sigma_{Z_i \delta_{iy}} = 1.0\,\mathrm{N/rad},\ \sigma_{M_{iy}^{\beta_i}} = 1.0\,\mathrm{Nm/rad},\ \sigma_{\dot{M}_{iy}^{\beta_i}} = 0.5\,\mathrm{Nms/rad},$$

$$\sigma_{\dot{M}_{iy}^{\omega_{iy}}} = 1.0\,\mathrm{Nms/rad},\ \sigma_{M_{iy}^{\delta_{iy}}} = 1.0\,\mathrm{Nm/rad},\ \sigma_{\Delta x_{ijm}} = 4.8\,\mathrm{m},\ \sigma_{\Delta z_{ijm}} = 4.8\,\mathrm{m},$$

$$\sigma_{\Delta v_{im}} = 6.9\,\mathrm{m/s},\ \sigma_{\Delta \varphi_{im}} = 0.005\,\mathrm{rad},\ \sigma_{\Delta \beta_{im}} = 0.01\,\mathrm{rad},\ \sigma_{\Delta \omega_{iym}} = 0.01\,\mathrm{rad/s},$$

$$\sigma_{\Delta \delta_{iym}} = 0.001\,\mathrm{rad}.$$

4.1.4. Formation Commands

The expected configuration of the system are $x_{ij}^* = 50\,\mathrm{m}$, $z_{ij}^* = -73.2\,\mathrm{m}$, the expected formation speed is $V_f = 100\,\mathrm{m/s}$, and the expected formation declination is $\varphi_f = 0\,\mathrm{rad}$.

4.1.5. Optimal Design of Estimator Gain and Control Law

In order to improve the robustness of the stochastic system, the estimator gain in (43) and the control law in (45) are optimized by the SRAD [37].

The SRAD design flow is shown as in Figure 3, which is composed of two parts: a modern optimization algorithm and a control structure design. MCE denotes Monte Carlo evaluation, SRA denotes stochastic robustness analysis.

The cost function we designed is $J_{ij} = \sum_{i=1}^{12} w_i I_i^2(q_i)$, where q_i represents the 12 indicators which are shown in Table 1, w_i is the weight of each indicator, and $I_i(\cdot)$ is the membership function of each indicator which obeys the rising-ridge distribution (59) or 0–1 distribution (60).

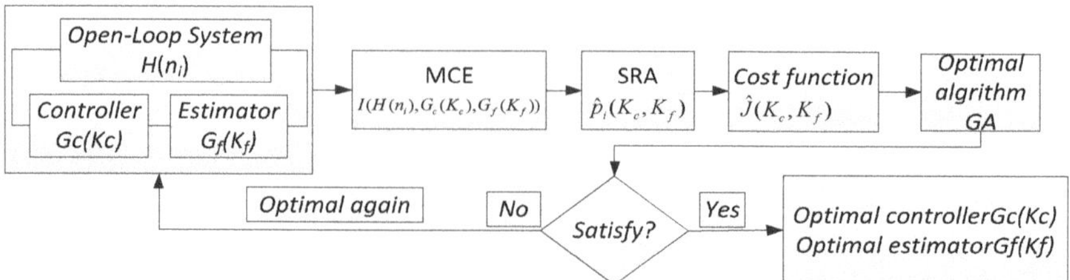

Figure 3. The design flow of SRAD.

Table 1. The stability and indicators.

Indicators	Weight	Membership (a,b)	Results
1.Stability (the real part of the maximum eigenvalue is negative)	10	$(-0.0001, 0)$	-0.00001
2. Forward distance adjust time	1	$(0, 75\ \text{s})$	54.7 s
3. Lateral distance adjust time	1	$(0, 75\ \text{s})$	49.8 s
4. Forward distance overshoot	5	$(0, 10\%)$	1.42%
5. Lateral distance overshoot	5	$(0, 10\%)$	2.51%
6. Forward distance steady error	1	$(0, 1\ \text{m})$	0.0332 m
7. Lateral distance steady error	1	$(0, 1\ \text{m})$	0.0944 m
8. Forward distance fluctuation	1	$(0, 2\ \text{m})$	0.2586 m
9. Lateral distance fluctuation	1	$(0, 2\ \text{m})$	0.9057 m
10. Average velocity instruct	3	$(90\ \text{m/s}, 110\ \text{m/s})$	100.07 m/s
11. Average flight path declination instruct	3	$(-1.57\ \text{rad}, 1.57\ \text{rad})$	0.0644 rad
12. The weighted variance of the estimation error	2	$(0, 10.0)$	3.86

For indicators 1–9 and 12, the membership function obeys the rising-ridge distribution, i.e.,

$$
I(x) = \begin{cases} 0 & x \le a \\ \frac{1}{2} + \frac{1}{2}\sin\frac{\pi}{b-a}\left(x - \frac{a+b}{2}\right) & a < x \le b \\ 1 & x > b \end{cases}
\tag{59}
$$

where a, b are the best and allowable value of the indicators respectively. x is the simulation result.

For indicators 10 and 11, the membership function obeys the 0–1 distribution, i.e.,

$$
I(x) = \begin{cases} 1 & x < a \\ 0 & a \le x \le b \\ 1 & x > b \end{cases}
\tag{60}
$$

where (a, b) is the allowable range of the indicators, x is the simulation result.

$\hat{p}_i(K_c, K_f)$ is the probability of indicator i that cannot satisfy the stability and requirements whose probability distribution function is $I_i(\cdot)$.

The minimum cost function J reflects the minimum probability that any indicator cannot satisfy the stability and requirements. It also indicates the minimum errors of the selected properties. Therefore, the obtained controller has high-quality robustness, and the probability that the control system does not meet the requirements is significantly reduced after multiple simulations.

The design steps in Figure 3 are as follows:

(1) Design the controller $G_c(K_c)$ and estimator $G_f(K_f)$ for the controlled object $H(n_i)$;

(2) Define the indicators for SRAD: $I(H(n_i), G_c(K_c), G_f(K_f))$;

(3) Carry out Monte Carlo simulation on the closed-loop system to obtain the probability $\hat{p}_i(K_c, K_f)$ that cannot satisfy the stability and performance;

(4) Constitute a random cost function $\hat{J}(K_c, K_f)$ to satisfy both of robust stability and performance;

(5) Apply a modern optimization algorithm to get an optimal value. After we get the minimum value of $\hat{J}(K_c, K_f)$, then we obtain a stochastic robust optimal controller and an optimal estimator.

The optimization process of the designed parameters K_f and K_c is shown in Figure 4. After iterating 15 times with the genetic algorithm and running the Monte Carlo simulation 100 times per iteration, we got the optimal cost value: $J = 4.56$ and the optimal parameters:

$$K_f = \begin{bmatrix} 0.172 & & & & & & & \\ 0 & 0.172 & & & & & & \\ & 0 & 0.014 & & & & & \\ & & 0 & 0.619 & & & & \\ & & & 0 & 0.523 & & & \\ & & & & 0 & 0.521 & & \\ & & & & & 0 & 0 & \\ & & & & & & 0.796 \end{bmatrix} \tag{61}$$

$$K_c = \begin{bmatrix} 0.0408 & -0.0009 & -0.1157 & -0.0039 & 0.0164 & 0.0176 & 0 & 0.1894 \\ 0.0089 & 0.0048 & -0.0274 & -0.8921 & 0.3046 & 0.3350 & 0 & -0.0012 \\ 0 & & \cdots & \cdots & & & & 0 \\ 0 & & \cdots & \cdots & & & & 0 \end{bmatrix} \tag{62}$$

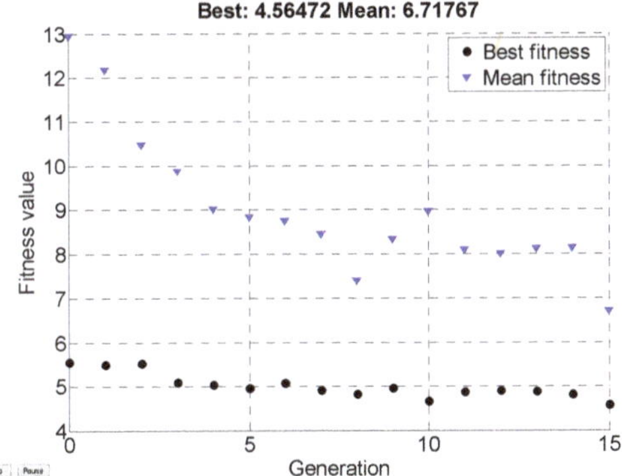

Figure 4. The iterative process of the genetic algorithm. (Note that the best fitness is the minimum value of the cost function in each iteration and the mean fitness is the mean value of the 100 Monte Carlo simulations in each iteration).

4.1.6. Simulation Framework

The framework of the simulation model is shown in Figure 5.

In the framework shown in Figure 5, the flight members in the formation are referred to nodes in the supporting network. The node obtains information through the formation support network and the sensor system, including neighbor nodes' information and environment information. The decision management system allocates missions to flight members and plans the flight route for the formation. Finally, formation control system and member flight control system carry out missions using the *Itô* stochastic system model (49) and the parameters presented above.

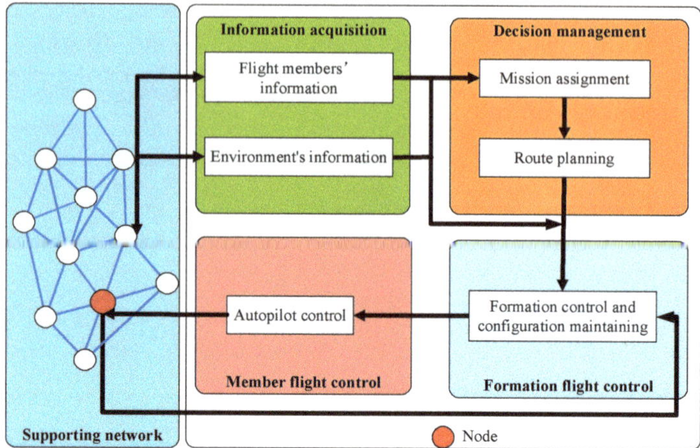

Figure 5. Simulation framework.

4.1.7. Simulation Results

In order to better reflect the performance of the USCC stochastic system, we define the following indicator.

Definition 2. *The weighted variance of the estimate error* $\widetilde{X}_{ij} = X_{ij} - \hat{X}_{ij}$ *is:*

$$V(\widetilde{X}_{ij}) = \frac{1}{t_f - t_0} \int_{t_0}^{t_f} (\widetilde{X}_{ij} - E\{\widetilde{X}_{ij}\})^T W_v (\widetilde{X}_{ij} - E\{\widetilde{X}_{ij}\}) dt \tag{63}$$

where the diagonal matrix $W_v \in R^{8\times 8}$ is the weighted matrix of weighted variances, and the diagonal elements correspond to the states $\hat{X}_{ij}$ of the estimator. The larger the weight is, the more accurate the estimation of the corresponding state becomes. The trace of W_v is $tr(W_v) = 1$. Assuming that the estimation accuracies of $\Delta\hat{x}_{ij}$, $\Delta\hat{z}_{ij}$, $\Delta\hat{V}_i$ and $\Delta\hat{\varphi}_i$ are required to be higher, the weighting matrix we designed is: $W_v = diag(0.2, 0.2, 0.2, 0.2, 0.1, 0.05, 0, 0.05)$.

The simulation results correspond with the optimal cost value and the optimal parameters are shown as in Table 1 and Figure 6. Note that the fluctuation in the table refers to the standard deviation of the difference between the real-time configuration and the expected configuration.

It can be observed from Table 1 that all the indicators are in the appropriate range. The real part of the maximum eigenvalue is negative and the weighted variance of the estimation error calculated from simulation results is 3.86. From Figure 6 we can observe that the two aircraft achieved the desired configuration after 54.7 s, and the forward distance steady error is 0.0332 m, the lateral distance steady error is 0.0944 m. The formation speed and the formation declination meet the designed requirements well. Thus the system is mean-square uniform bounded according to Proposition 1.

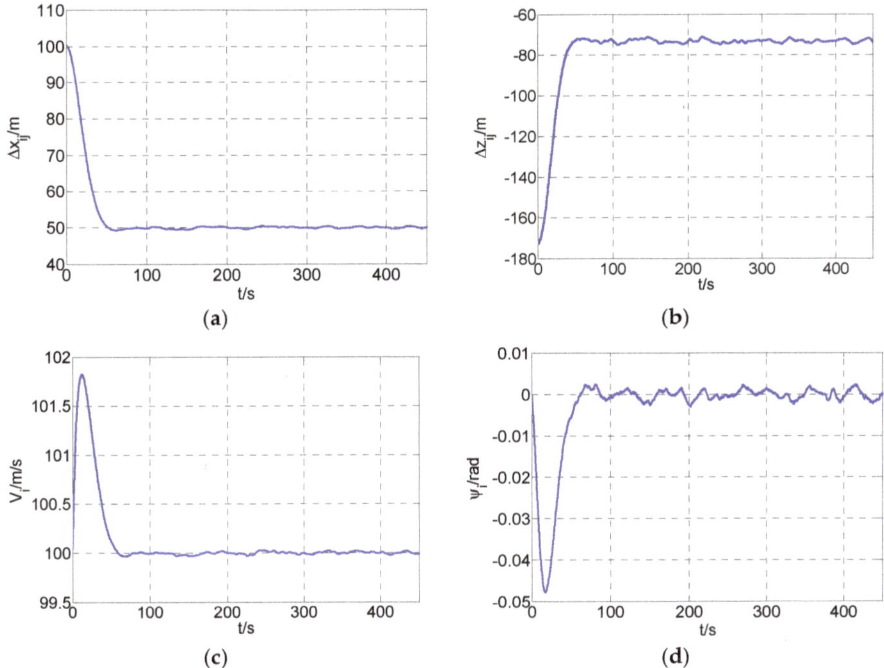

Figure 6. Simulation results of node v_i. (**a**) forward distance; (**b**) lateral distance; (**c**) speed; (**d**) flight path declination.

4.2. Autonomous Flight Experiments

In order to observe the performance of the model and verify the effectiveness of the USCC stochastic system, we conducted an equivalent outfield autonomous formation flight test by using multiple UAVs. As shown in Figure 7, we carried out experiments with seven nonholonomic UAVs. The UAV swarm can cooperatively search the certain area with different configurations.

Figure 7. Seven UAVs.

The loads in each cabin of the UAV are shown in Figure 8, including the power module, formation cooperative guidance module, autopilot module, detection module, formation communication module and flight data transmission module. In the experiment, we adopted the proposed USCC stochastic model into the formation cooperative guidance module to instruct the flight members to reach the desired position and maintain a steady configuration. The framework of the whole system is shown in Figure 9.

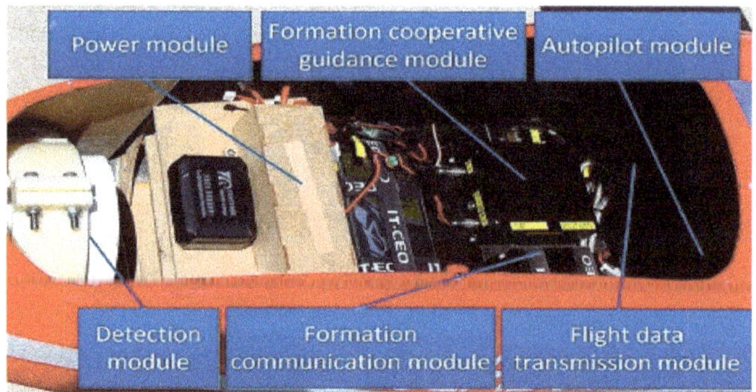

Figure 8. The loads in the cabin of the UAV.

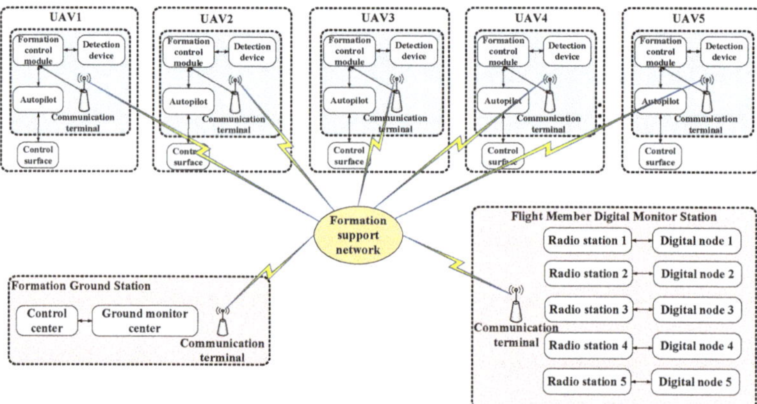

Figure 9. The framework of the USCC outfield flight system.

The formation ground station is set to observe the real-time formation flight process, and upload commands to instruct the formation to change its configuration or adjust relative distances. It can also control the pod to capture the configuration. The flight member's digital monitor station is set to observe the real-time flight member's flight statuses and to ensure the safety of the whole flight process.

Two configurations of five drones we designed in the experiments are shown in Figure 10a,b. The lateral and forward distance between neighbor UAVs was set to 50 m. Moreover, the configurations in Figure 10c,d were captured by the UAV that was flying higher with a pod.

It can be observed from Figure 10 that the UAVs achieved the desired configuration smoothly and maintained the formation effectively under the proposed model.

For the convenience of analysis, the actual flight data could be saved through the formation monitoring station. In this paper, we take the flight data of the wedge configuration with and without the USCC stochastic system model in the experiment to invert the flight process and evaluate the effectiveness of the USCC stochastic system. The results are shown in Figure 11.

The curves shown as in Figure 11 demonstrate that the five UAVs achieved the desired configuration and steadily maintained the formation. The flight paths in the rectangle that the arrows "1" and "2" point to in Figure 11a are amplified in Figure 11c,d. The data are summarized in Table 2, from which we can observe that in the flight test not using the model, the relative height of the UAVs should be maintained at the very least at 50 m to avoid the risk of collision, while the heights of the five UAVs with USCC stochastic system model converged to 200 m and the relative distance in height was zero. The average 3D distance between flight members in the flight test with the proposed model was shortened by 32.14% compared to the test without the model.

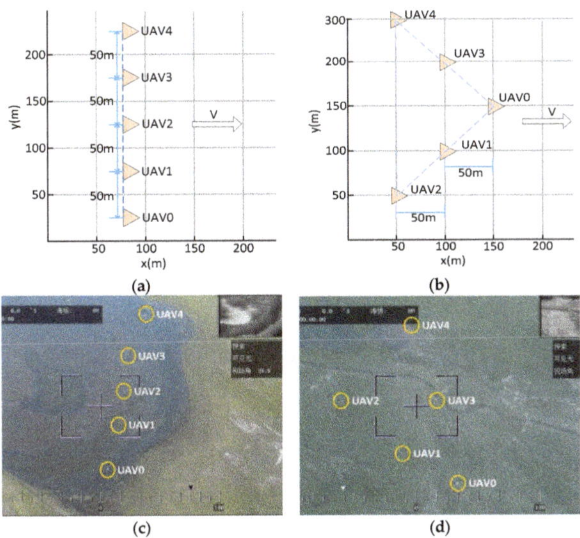

Figure 10. Configurations of five drones. (**a**) Designed lateral configuration; (**b**) Designed wedge configuration; (**c**) Real time lateral configuration; (**d**) Real time wedge configuration.

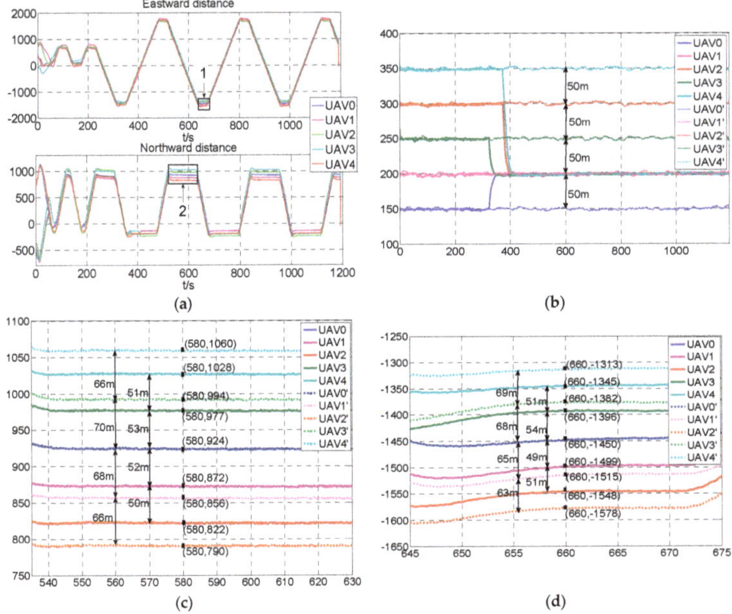

Figure 11. Experimental results of the wedge configuration. (**a**). The flight path of the five UAVs; (**b**). The height of five UAVs. The flight paths in full line belong to the UAV0 to UAV4 with USCC stochastic system model while the dotted line belong to the UAV0′ to UAV4′ are without the model; (**c**). Northward distances between the five UAVs which are indicated by "2" in (**a**). The flight paths in full line of UAV0 to UAV4 utilize the USCC stochastic system model while the dotted line of UAV0′ to UAV4′ do not use the model; (**d**). Eastward distances between the five UAVs which are indicated by "1" in (**a**). The flight paths in full line of UAV0 to UAV4 use USCC stochastic system model while the dotted line of UAV0′ to UAV4′ do not use the model.

Table 2. The data of configurations with and without USCC stochastic system model.

Indicators	Configuration Data of Proposed Model	Original Configuration Data
Eastward average distance	51.25 m	66.25 m
Northward average distance	51.5 m	67.5 m
Average relative height	0 m	50 m
Average 3D relative distance	72.6 m	107.0 m

The results show that the introduction of multiplicative noises improves the formation maintenance performance and extends the boundary properties of the formation flight, such as the safety distance is shorter and the relative height can be eliminated. Therefore, the proposed stochastic model could provide the UAV swarm with a larger maneuvering space, and then improve the efficiency and quality of mission execution, enhance the operational capability in high-risk environments and improve the adaptation of the system to the complex environment.

5. Conclusions

In this paper, the problem of the state estimation and control of the UAV swarm system with the consideration of multiplicative noises is studied. The closed-loop *Itô* stochastic system we constructed is the combination of a state equation introduced from group kinematic model and individual dynamic model considering the multiplicative noises, an observation equation considers the measurement noises, an estimator and a controller. Following that, the proof to verify the mean-square uniform boundedness of the system is presented. The optimal estimator and controller are obtained with the use of SRAD in the simulation. Finally, simulation results show that the system is stable and the selected indicators meet the requirements. The outfield experiment results demonstrate that the configuration with the proposed model could be significantly condensed by 32.14% compared to the test with the traditional model. Therefore, the stochastic system of USCC with multiplicative noises proposed in this paper could contribute to effectively exploiting the boundary performances of the system and constructing a high dynamic formation in practical application, thus better matching the actual environment.

However, there is much to be researched further in this area. For example, in the practical application, the time delay cannot be ignored, especially for large scale formations. The modeling of the USCC stochastic system considering time delay is currently under investigation.

Author Contributions: H.Z., S.W. and Y.W. proposed the research ideas and conducted modeling design. H.Z. developed the software and wrote the manuscript, S.W. reviewed the model and instructed the simulation and experiments, Y.W. improved the software and implemented simulation, W.L. carried out flight experiments, and X.W. reviewed the modeling process and polished the paper.

Funding: This work was funded by the Industrial Technology Development Program under Grant B1120131046.

Acknowledgments: The authors appreciate editors and reviewers for their suggestions which are of great value to the paper.

Conflicts of Interest: The authors declare no conflict of interest.

References

1. Wu, S.T. Flight Control System of MAF. In *Cooperative Guidance & Control of Missile Autonomous Formation*; National Defense Industry Press: Beijing, China, 2015; Chapter 5, Section 3; pp. 183–198.
2. Cai, D.; Wu, S.; Deng, J. Distributed Global Connectivity Maintenance and Control of Multi-Robot Networks. *IEEE Access* **2017**, *5*, 1. [CrossRef]
3. Casbeer, D.W.; Kingston, D.B.; Beard, R.W.; McLain, T.W. Cooperative forest fire surveillance using a team of small unmanned air vehicles. *Int. J. Syst. Sci.* **2006**, *37*, 351–360. [CrossRef]
4. Altshuler, Y.; Pentland, A.; Bruckstein, A.M. The cooperative hunters–Efficient and scalable drones swarm for multiple targets detection. In *Swarms and Network Intelligence in Search*; Springer: Cham, Switzerland, 2018; pp. 187–205.

5. Pachter, M.; D'Azzo, J.; Dargan, J.L. Automatic formation flight control. *J. Guid. Control Dyn.* **1994**, *17*, 1380–1383. [CrossRef]

6. Du, H.; Zhu, W.; Wen, G.; Duan, Z.; Lü, J. Distributed Formation Control of Multiple Quadrotor Aircraft Based on Nonsmooth Consensus Algorithms. *IEEE Trans. Cybern.* **2017**, *99*, 1–12. [CrossRef] [PubMed]

7. Cao, Y.; Ren, W.; Li, Y. Distributed discrete-time coordinated tracking with a time-varying reference state and limited communication. *Autom* **2009**, *45*, 1299–1305. [CrossRef]

8. Jin, J.; Gans, N. Collision-free formation and heading consensus of nonholonomic robots as a pose regulation problem. *Robot. Auton. Syst.* **2017**, *95*, 25–36. [CrossRef]

9. Wen, Y.M.; Wu, S.T.; Liu, W.L.; Deng, J.; Wu, X. A Collision Forecast and Coordination Algorithm in Configuration Control of Missile Autonomous Formation. *IEEE Access* **2017**, *5*, 1188–1199. [CrossRef]

10. Gao, J.G. *Stability, Stabilization and Optimization of Nonlinear Stochastic Systems*; South China University of Technology: Guangzhou, China, 2010.

11. Shen, Z.M.; Hu, L.J. Asymptotic mean square boundedness of numerical solutions to stochastic delay differential equations. *Commun. Appl. Math. Comput.* **2016**, *30*, 60–70.

12. Deng, F.Q.; Feng, Z.S.; Liu, Y.Q. Necessary and Sufficient Conditions for Mean-Square Boundedness and Stability of Linear Time-Varying Stochastic Systems. *J. South China Univ. Technol.* **1997**, *25*, 17–21.

13. Arnold, L. The Stochastic Integral as a Stochastic Process, Stochastic Differentials. In *Stochastic Differential Equations*; John Wiley & Sons, Inc.: Hoboken, NJ, USA, 1974; Chapter 5, Section 3–5; pp. 88–96.

14. Schuss, Z. Stochastic Differential Equations. In *Theory and Applications of Stochastic Differential Equations*; John Wiley & Sons, Inc.: Hoboken, NJ, USA, 1980; Chapter 4, Section 1; pp. 85–96.

15. Zuo, S.; Song, Y.; Lewis, F.L.; Davoudi, A. Time-Varying Output Formation-Containment of General Linear Homogeneous and Heterogeneous Multiagent Systems. *IEEE Trans. Control Netw. Syst.* **2018**, *6*, 537–548. [CrossRef]

16. Liu, Y.; Montenbruck, J.M.; Zelazo, D.; Odelga, M.; Rajappa, S.; Bülthoff, H.H.; Allgöwer, F.; Zell, A. A Distributed Control Approach to Formation Balancing and Maneuvering of Multiple Multirotor UAVs. *IEEE Trans. Robot.* **2018**, *34*, 870–882. [CrossRef]

17. De La Cruz, C.; Carelli, R. Dynamic model based formation control and obstacle avoidance of multi-robot systems. *Robot* **2008**, *26*, 345–356. [CrossRef]

18. Ailon, A.; Zohar, I. Controllers for trajectory tracking and string-like formation in Wheeled Mobile Robots with bounded inputs. In Proceedings of the Melecon 2010–2010 15th IEEE Mediterranean Electrotechnical Conference, Valletta, Malta, 26–28 April 2010; pp. 1563–1568.

19. Ren, W.; Sorensen, N. Distributed coordination architecture for multi-robot formation control. *Robot. Auton. Syst.* **2008**, *56*, 324–333. [CrossRef]

20. Zhao, B.R.; Peng, Y.J.; Deng, F.Q. Leader-Following Almost Surely Exponential Consensus for General Linear Multiagent Systems with Stochastic Disturbances. In Proceedings of the Chinese Control and Decision Conference, Chongqing, China, 28–30 May 2017; pp. 7196–7201.

21. Liu, W.H.; Yang, C.J.; Deng, F.Q. Synchronization of General Linear Multiagent Systems with Measurement Noises. *Asian J. Control* **2017**, *19*, 510–520. [CrossRef]

22. Ren, H.W.; Deng, F.Q. Mean Square Consensus of Leader-Following Multiagent Systems with Measurement Noises and Time Delays. *ISA Trans.* **2017**, *71*, 76–83. [CrossRef] [PubMed]

23. Tan, X.; Cao, J.; Li, X.; Alsaedi, A. Leader-following mean square consensus of stochastic multiagent systems with input delay via event-triggered control. *IET Control Theory Appl.* **2018**, *12*, 299–309. [CrossRef]

24. Ren, H.W.; Peng, Y.J.; Deng, F.Q.; Zhang, C. Impulsive Pinning Control Algorithm of Stochastic Multiagent Systems with Unbounded Distributed Delays. *Nonlinear Dyn.* **2018**, *92*, 1453–1467. [CrossRef]

25. Hu, A.-H.; Hu, M.-F.; Guo, L.-X.; Cao, J.-D. Consensus of a leader-following multi-agent system with negative weights and noises. *IET Control. Theory Appl.* **2014**, *8*, 114–119. [CrossRef]

26. Zhao, B.R.; Peng, Y.J.; Deng, F.Q. Consensus Tracking for General Linear Stochastic Multiagent Systems: A Sliding Mode Variable Structure Approach. *IET Control Theory Appl.* **2017**, *11*, 2910–2915. [CrossRef]

27. Ren, H.; Deng, F.; Peng, Y.; Zhan, B.; Zhang, C. Exponential Consensus of Non-Linear Stochastic Multiagent Systems with ROUs and RONs via Impulsive Pinning Control. *IET Control Theory Appl.* **2017**, *11*, 225–236. [CrossRef]

28. Shen, D.B.; Sun, W.J.; Sun, Z.D. Adaptive PID Formation Control of Nonholonomic Robots without Leader's Velocity Information. *ISA Trans.* **2014**, *53*, 474–480. [CrossRef] [PubMed]

29. Lewis, M.A.; Tan, K.-H. High Precision Formation Control of Mobile Robots Using Virtual Structures. *Auton. Robot.* **1997**, *4*, 387–403. [CrossRef]
30. Ren, W.; Beard, R. Formation feedback control for multiple spacecraft via virtual structures. *IEE Proc. Control. Theory Appl.* **2004**, *151*, 357–368. [CrossRef]
31. Reynolds, C.W. Flocks, herds and schools: A distributed behavioral model. *ACM SIGGRAPH Comput. Graph.* **1987**, *21*, 25–34. [CrossRef]
32. Balch, T.; Arkin, R.C. Behavior-based formation control for multi-robot teams. *IEEE Trans. Robot. Autom.* **1998**, *14*, 926–939. [CrossRef]
33. Wu, Z.G.; Xu, Y.; Pan, Y.-J.; Su, H.; Tang, Y. Event-Triggered Control for Consensus Problem in Multiagent Systems with Quantized Relative State Measurements and External Disturbance. *IEEE Trans. Circuits Syst.* **2018**, *65*, 2232–2242. [CrossRef]
34. Kristiansen, R.; Nicklasson, P.J.; Gravdahl, J.T. Spacecraft coordination control in 6DOF: Integrator backstepping vs passivity-based control. *Autom* **2008**, *44*, 2896–2901. [CrossRef]
35. Wan, S.; Campa, G.; Napolitano, M.; Seanor, B.; Gu, Y. Design of Formation Control Laws for Research Aircraft Models. In Proceedings of the AIAA Guidance, Navigation, and Control Conference, Austin, TX, USA, 11–14 August 2003.
36. Campa, G.; Gu, Y.; Seanor, B.; Napolitano, M.R.; Pollini, L.; Fravolini, M.L. Design and flight-testing of non-linear formation control laws. *Control Eng. Pract.* **2007**, *15*, 1077–1092. [CrossRef]
37. Wu, S.T. Basis of SRAD for Guidance and Control System. In *Stochastic Robustness Analysis and Design for Guidance and Control System of Winged Missile*; National Defense Industry Press: Beijing, China, 2010; Chapter 2, Section 1; pp. 8–20.
38. Song, X.M. *Estimation and Quadratic Optimal Control for Linear Systems with Multiplicative Noise*; Shandong University: Jinan, China, 2010.
39. Qian, X.F.; Lin, R.X.; Zhao, Y.N. Missile Kinematic Equations. In *Missile Flight Aerodynamics*; Beijing Institute of Technology Press: Beijing, China, 2013; Chapter 2, Section 3; pp. 36–48.
40. Montijano, E.; Cristofalo, E.; Zhou, D.; Schwager, M.; Sagues, C. Vision-Based Distributed Formation Control without an External Positioning System. *IEEE Trans. Robot.* **2016**, *32*, 339–351. [CrossRef]
41. Olfati-Saber, R. Flocking for Multi-Agent Dynamic Systems: Algorithms and Theory. *IEEE Trans. Autom. Control* **2006**, *51*, 401–420. [CrossRef]

Article

Data Offloading in UAV-Assisted Multi-Access Edge Computing Systems: A Resource-Based Pricing and User Risk-Awareness Approach

Giorgos Mitsis [1], Eirini Eleni Tsiropoulou [2],* and Symeon Papavassiliou [1]

[1] School of Electrical and Computer Engineering, National Technical University of Athens, 15780 Athina, Greece; gmitsis@netmode.ntua.gr (G.M.); papavass@mail.ntua.gr (S.P.)
[2] Department of Electrical and Computer Engineering, University of New Mexico, Albuquerque, NM 87131, USA
* Correspondence: eirini@unm.edu

Received: 21 March 2020; Accepted: 15 April 2020; Published: 24 April 2020

Abstract: Unmanned Aerial Vehicle (UAV)-assisted Multi-access Edge Computing (MEC) systems have emerged recently as a flexible and dynamic computing environment, providing task offloading service to the users. In order for such a paradigm to be viable, the operator of a UAV-mounted MEC server should enjoy some form of profit by offering its computing capabilities to the end users. To deal with this issue in this paper, we apply a usage-based pricing policy for allowing the exploitation of the servers' computing resources. The proposed pricing mechanism implicitly introduces a more social behavior to the users with respect to competing for the UAV-mounted MEC servers' computation resources. In order to properly model the users' risk-aware behavior within the overall data offloading decision-making process the principles of Prospect Theory are adopted, while the exploitation of the available computation resources is considered based on the theory of the Tragedy of the Commons. Initially, the user's prospect-theoretic utility function is formulated by quantifying the user's risk seeking and loss aversion behavior, while taking into account the pricing mechanism. Accordingly, the users' pricing and risk-aware data offloading problem is formulated as a distributed maximization problem of each user's expected prospect-theoretic utility function and addressed as a non-cooperative game among the users. The existence of a Pure Nash Equilibrium (PNE) for the formulated non-cooperative game is shown based on the theory of submodular games. An iterative and distributed algorithm is introduced which converges to the PNE, following the learning rule of the best response dynamics. The performance evaluation of the proposed approach is achieved via modeling and simulation, and detailed numerical results are presented highlighting its key operation features and benefits.

Keywords: data offloading; UAV-enabled computing; resource-based pricing; risk-awareness; multi-access edge computing systems

---

1. Introduction

Towards realizing the emerging applications supported by the fifth generation (5G) wireless networks and the Internet of Things (IoT), while demanding ultra-reliable and low latency communication (URLLC), ubiquitous, distributed, and intelligent computing is one of the key enabling technologies. IoT is foreseen to reach 500 billion devices connected to the Internet by 2030 [1], while the global mobile traffic is expected to increase seven-fold by 2021 [2]. Thus, it is evident that traditional cloud computing architectures cannot support the latency constraints of the next generation networking environments, such as Tactile Internet [3]. The reasons are that the powerful cloud centers are often deployed far away from the end users; thus, huge amounts of traffic are usually transmitted through intermediate nodes resulting in heavy load, congestion, delay uncertainties, high energy consumption, and multiple security threats [4]. Thus, multi-access edge computing (MEC), which brings computing resources from the core network to the edge network, becomes a natural and promising solution to support these applications.

Combined with the MEC concept, Unmanned Aerial Vehicles (UAVs), equipped with communication and computing facilities, become a core component of next generation networks due to their salient attributes, such as hovering ability, flexibility and effortless deployment, maneuverability, mobility, low cost, strong line-of-sight (LoS) connection links, adjustable usage, and adaptive altitude [5]. The MEC servers are embedded in the UAVs that fly in closer proximity to the users compared to the conventional MEC servers typically residing at the Macro Base Stations (MBSs) of the macrocells or at the Access Points (APs) of the small cells [6]. Thus, the UAV-mounted MEC servers more efficiently support the end users applications' data offloading and processing at the flying edge servers, by creating a flexible and dynamic computing environment paradigm [7].

1.1. Related Work

Cheng et al. [8] discuss the benefits introduced by the UAV-mounted MEC servers with respect to caching and computing, in a hybrid architecture consisting of UAV-mounted and ground MEC servers. Luo et al. [9] introduce a cloud-based UAV-assisted system and study its stability with respect to the sensors big data offloading rate. Valentino et al. [10] consider a fleet of UAV-mounted MEC servers and the optimization problem of increasing the UAVs fleet lifetime, while decreasing the overall computation time of the users' offloaded tasks is formulated and solved. In particular, the authors exploit neighboring UAV clusters with sufficient computing resources to offload the users' computation tasks.

Xiong et al. [11] formulate a joint optimization problem to optimize the users' data offloading to the UAV-mounted MEC servers, the UAVs' trajectory, and the data allocation during transmission to the different UAVs. An end-to-end solution is introduced by Jeong et al. [12], where the authors jointly optimize the users' data offloading to the UAV-mounted MEC servers (i.e., uplink) and the output processed data returned to the users (i.e., downlink), while considering the computation tasks' latency constraints. Jeong et al. [13] focus on the UAV-mounted MEC servers' energy constraints to jointly optimize the users' data offloading by considering orthogonal and non-orthogonal communication multiple access techniques, and the UAVs' trajectory. Furthermore, Zhou et al. consider a wireless powered communication environment in [14,15], where the UAVs except from acting as UAV-mounted MEC servers providing computing services to the end-users, they also provide energy to them. Accordingly, the users can exploit the harvested energy to perform local computing and/or transmit their data to the UAV-mounted MEC servers.

1.2. Motivation and Contributions

All the aforementioned research works, though having demonstrated significant benefits and potential, have made two fundamental assumptions regarding the examined UAV-assisted multi-access edge computing system: (i) the UAV-mounted MEC servers offer their communication and computing services to the users for free; and (ii) the users act as neutral utility maximizers aiming at simply maximizing their perceived satisfaction from offloading and processing their data to the UAV-mounted MEC servers, thus exhibiting a risk-neutral behavior. However, in a realistic communication and computing environment, both assumptions may not always hold true.

The operator of the UAV-mounted MEC server should enjoy some form of profit by offering its computing services and capabilities to the end-users. Depending on the operation mode, business model and use case under consideration, the profit could be—either explicit or implicit—expressed in different forms (e.g., monetary cost, etc.) [16,17]. In our work, we consider that the profit for the UAV-Mounted MEC server originates directly from applying a usage-based charging policy for allowing the exploitation of the server's computing resources.

Moreover, it has been argued recently that in real life the end-users are characterized by loss averse and risk seeking behavior in terms of exploiting the system's available resources, especially in resource-constrained environments (e.g., [18–20]). In our case, the resources that the users may opt for and compete for refer to the UAV-mounted MEC server's computing resources. Specifically, based on the users' behavioral characteristics, some users may act aggressively and opportunistically in terms of offloading their data to the UAV-mounted MEC servers in order to avoid consuming their personal resources to process their data. Those users exhibit risk seeking behavior, as the UAV-mounted MEC server may not be able to serve all the users' data offloading and processing requests. On the other hand, there are more conservative users, who exhibit loss averse behavior, thus being more willing to process their data locally at their devices, instead of taking the risk of finally not being served by the UAV-mounted MEC server due to the potential overexploitation of the latter.

Towards jointly addressing the aforementioned assumptions and filling the respective research gaps, in this paper, we exploit the power and principles of Prospect Theory [18] to capture the users risk-based behavior in their data offloading decision-making process, under the operation framework of a usage-based pricing mechanism of the UAV-mounted MEC servers' computing resources. To the best of our knowledge, this is the first research work in the existing literature that jointly combines a pricing-aware and risk-aware framework to deal with the data offloading problem in UAV-mounted multi-access edge computing systems.

In particular, we assume that the users have two available options for executing their tasks, namely the local computation and the remote computation, the latter achieved through data offloading. The local computation resources of the user's device act as safe resources, since the users do not compete with each other for consuming those resources. On the other hand, the computation resources of the UAV-mounted MEC server are treated as a Common Pool of Resources (CPR), as they are non-excludable, i.e., all the users have the right to exploit them, while they are rivalrous and subtractable, i.e., their exploitation by one user reduces the ability to be exploited by another user. In principle, the UAV-mounted MEC server resources have the potential to provide significantly higher satisfaction to the user (compared to the lower satisfaction that could be obtained through the limited user local computation resources), if properly utilized and allocated. However, if the users selfishly offload their data to the UAV-mounted MEC server, then the computing capabilities of the latter will be overexploited resulting in suboptimal outcomes for the entire set of users, possibly leading to the complete "failure" of the CPR UAV-mounted MEC server. The failure of the CPR UAV-mounted MEC server refers to its inability to concurrently handle the large amount of offloaded data and corresponding computation tasks by the users, due to its limited computation capability.

To treat this situation and differentiate the performance and usage of the available computation resources, we capitalize on the theory of Tragedy of the Commons [21,22], while also introducing a usage-based pricing mechanism to capture the users' cost for exploiting the server's computation resources. The proposed pricing mechanism implicitly introduces a more social behavior to the users and supports the fairness among them, in terms of competing for the UAV-mounted MEC server's computation resources. The users' behavioral characteristics in the data offloading decision-making process, is captured through the principles of Prospect Theory that models users' decisions under the uncertainty of the available computation resources at the UAV-mounted MEC server. Prospect Theory is a well known behavioral economic theory studying the autonomous decision-making of the individuals under risk and uncertainty of the associated payoff of their choices, which is estimated with some probability [23]. The performance evaluation and validation of the proposed pricing and risk-aware data offloading framework in UAV-assisted MEC systems is achieved via modeling and simulation, in terms of efficient exploitation of all the available computation resources, realistic capturing of users' interaction with the computing environment, and its scalability.

1.3. Outline

The rest of this paper is organized as follows. Section 2 presents the considered system model, while Section 3 initially discusses the main principles of Prospect Theory and the theory of the Tragedy of the Commons, and then accordingly designs and formulates the users' prospect-theoretic utility function. Section 4 introduces the pricing and risk-aware data offloading framework in a UAV-assisted multi-access edge computing system by formulating and solving the corresponding optimization data offloading problem, via adopting the theory of S-modular games. In Section 5, a low complexity and iterative algorithm is introduced to determine the Pure Nash Equilibrium (PNE) of the users' data offloading optimization problem. Section 6 contains the performance evaluation of the proposed framework, while Section 7 concludes the paper.

2. System Model

A UAV-assisted multi-access edge computing system is considered consisting of a set of mobile users $\mathcal{N} = \{1, \ldots, n, \ldots, N\}$ and a UAV-mounted MEC server attached to the UAV. Each user n has a computation task J_n that needs to execute. Each task is accordingly defined as $J_n = (b_n, d_n)$, where b_n [bits] is the user's n size of the input data needed for the computation task and d_n [CPU-cycles] is the number of CPU cycles required in order to accomplish the computation task. The UAV-mounted MEC server is available to the users to offload and process their data remotely instead of processing them locally on their device and consuming their own local resources. Each user decides to offload b_n^{MEC} [bits] data to the UAV-mounted MEC server, while the rest $(b_n - b_n^{MEC})$ [bits] data are processed locally on the user's device. An indicative topology of the considered UAV-assisted MEC system is presented in Figure 1. In this work we mainly focus on the modeling and provisioning of the computing resources, rather than on the user to UAV wireless communication aspects. The UAV flexibility and adaptability capabilities can ensure strong communication channels and links with the users.

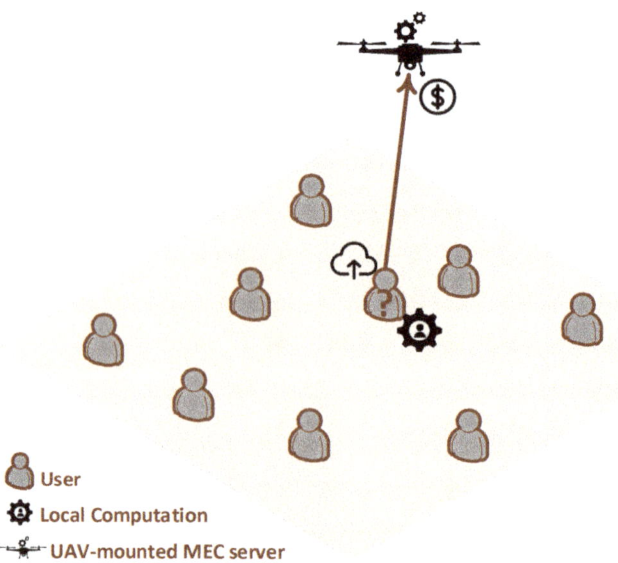

Figure 1. UAV-assisted multi-access edge computing system.

For each user n, the time $\hat{t}_n$ [s] to process the whole amount of data b_n locally is defined as:

$$\hat{t}_n = \frac{d_n}{f_n} \tag{1}$$

where f_n [CPU-cycles/s] is the computation capability of each user's n device. Apart from the processing time needed, each computation task has some energy requirements as well. The energy $\hat{e}_n$ [J] needed to process the whole amount of data b_n locally for each user n is defined as:

$$\hat{e}_n = \gamma_n d_n \tag{2}$$

where γ_n [J/CPU-cycle] is the coefficient denoting the consumed energy per CPU cycle locally at each user's n device.

We assume that the UAV-mounted MEC server applies a fair usage-based pricing policy to the users, while charging them proportionally to their offloaded data and to their demand of consuming computation resources, as they are indicated by the nature of their computation task. Thus, the cost imposed by the UAV-mounted MEC server to the user n in order to process the user's offloaded data b_n^{MEC} is defined as:

$$c_n(b_n^{MEC}) = cd_n \frac{b_n^{MEC}}{b_n} \tag{3}$$

where c [1/CPU-cycles] represents a constant pricing factor imposed by the UAV-mounted MEC server to every user. Intuitively, the cost imposed to each user (Equation (3)) is proportional to the percentage of the number of CPU cycles d_n of the user's computation task that is actually offloaded, i.e., the greater the part of the computation task offloaded to the UAV-mounted MEC server is, the greater is the cost that the user experiences by the UAV-mounted MEC server to process remotely its data. It is noted that, without loss of generality, the cost $c_n(b_n^{MEC})$ imposed by the UAV-mounted MEC server to the user n in order to process the offloaded data of the latter is assumed to be a unitless metric in this research work, and can represent any type of usage-based cost or monetary cost in a realistic implementation. Based on the above proposed model, we can therefore formulate the problem of determining the optimal b_n^{MEC*} that each user should offload considering each user's risk-aware behavioral characteristics and the pricing imposed by the UAV-mounted MEC server.

3. Users Prospect-Theoretic Utility Function in UAV-Assisted MEC Environment

In the dynamic computation environment considered in this research work, consisting of the UAV-mounted MEC server's and the users' local computing capabilities, the users exhibit a risk-aware behavior in terms of deciding where to process the data of their computation tasks. Therefore, the users do not act as risk-neutral utility maximizers following the conventional Expected Utility Theory (EUT) [23], but instead they rather exhibit a loss averse or gain seeking behavior when utilizing the UAV-mounted MEC server's computation resources. To capture the exploitation and usage characteristics and principles of the available computation resources in the considered UAV-assisted MEC system, we adopt the theory of the Tragedy of the Commons [22]. Specifically, the UAV-mounted MEC server's computation resources are considered as a Common Pool of Resources (CPR), as all the users have access to them and can offload their data to the UAV-mounted MEC server in order to be processed. If the users overexploit the computation resources of the UAV-mounted MEC server, the latter will fail to serve their computation demands and none of the users will be satisfied. On the other hand, the user's device's local computation resources are considered as safe resources, as each user exclusively exploits them for its own benefit. It is noted that the safe resources provide a guaranteed satisfaction to the user; however, the user can potentially experience lower satisfaction compared to exploiting the CPR, as the user has to spend its own resources, e.g., energy to process locally its data.

As mentioned before, towards capturing the users' loss averse and gain seeking behavior in terms of exploiting the CPR and safe computation resources, the principles of Prospect Theory are adopted [24]. Prospect Theory is a behavioral economic theory that quantifies individuals' behavioral patterns, which demonstrate systematic deviations from the Expected Utility Theory. Under the prospect theoretic model, the users experience greater dissatisfaction from a potential outcome of losses compared to their satisfaction from gains of the same amount. In addition, the level of the users' satisfaction and dissatisfaction is evaluated with respect to a reference point, which is considered as the ground truth of the examined system. Recently, several efforts have appeared in the literature, where Prospect Theory has been adopted in various environments and application domains, including dynamic resource management in 5G wireless networks [18,25], public safety networks [19], anti-jamming communications in cognitive radio networks [20], users' transmission power management and anti-jamming techniques in UAV-assisted networks [5], and Quality of Experience in cyber-physical social systems [21].

Following the principles of Prospect Theory, the user's prospect theoretic utility is defined as [24,26]:

$$P_n(U_n) = \begin{cases} (U_n - U_{n,0})^{\alpha_n}, & \text{if } U_n \geq U_{n,0} \\ -k_n(U_{n,0} - U_n)^{\beta_n}, & \text{otherwise} \end{cases} \tag{4}$$

where $U_{n,0} = \frac{1}{\bar{t}_n \bar{e}_n} b_n$ denotes the reference point expressing the user's n perceived satisfaction by processing all of its data locally at its device, which is the safe choice in terms of receiving a guaranteed satisfaction. Similarly, U_n denotes the user's actual perceived satisfaction from offloading part of its data to the UAV-mounted MEC server, and is given by Equation (5) below.

The parameters α_n, β_n where $\alpha_n, \beta_n \in (0,1]$ express the sensitivity of users to the gains and losses of their actual perceived satisfaction U_n, respectively. In particular, the user's risk averse behavior in gains and risk seeking behavior in losses is captured by small values of the parameter $\alpha_n \in (0,1]$. Similarly, a small value of the parameter $\beta_n \in (0,1]$ captures a higher decrease in the user's prospect theoretic utility, when its actual perceived satisfaction is close to the reference point. It is noted that the values of the parameters α_n, β_n can be determined and quantified based on statistical analysis of existing open datasets stemming from qualitative results of users' behavioral models (e.g., [27]). Furthermore, the loss aversion parameter $k_n \in \mathbb{R}^+$ quantifies the impact of losses compared to the gains in user's prospect theoretic utility. Specifically, for $k_n > 1$, the user weighs the losses more than the gains, while the exact opposite holds true for $0 \leq k_n \leq 1$. For simplicity and without loss of generality, in this work, we assume $\alpha_n = \beta_n$.

Specifically, the user's actual perceived satisfaction from offloading part of its data (denoted by b_n^{MEC}) to the UAV-mounted MEC server is denoted as $U_n(b_n^{MEC})$ and is formally defined as follows:

$$U_n(\mathbf{b^{MEC}}) = \begin{cases} \frac{1}{\bar{t}_n \bar{e}_n} b_n, & \text{if } b_n^{MEC} = 0 \\ \frac{1}{\bar{t}_n \bar{e}_n}(b_n - b_n^{MEC}) + b_n^{MEC} RoR(d_\tau) - c_n(b_n^{MEC}), & \text{if } b_n^{MEC} \neq 0 \text{ and MEC survives} \\ \frac{1}{\bar{t}_n \bar{e}_n}(b_n - b_n^{MEC}) - c_n(b_n^{MEC}), & \text{if } b_n^{MEC} \neq 0 \text{ and MEC fails} \end{cases} \tag{5}$$

The first branch of Equation (5) expresses the user's actual perceived satisfaction from processing all of its data locally to its mobile device. The second branch of Equation (5) captures the user's actual perceived satisfaction by processing part of its data locally (first term) and part of them to the UAV-mounted MEC server (second term), while experiencing the corresponding usage-based cost (third term) for exploiting the UAV-mounted MEC server's computation resources in the case that the MEC server can process all the users' requests. The third branch of Equation (5) represents the user's utility in the case that the MEC server fails to process the users' data due to its overexploitation. The user's actual perceived satisfaction from processing part of its data to the UAV-mounted MEC server depends on the server's rate of return function $RoR(d_\tau)$, where $d_\tau(\mathbf{b^{MEC}})$, $\mathbf{b^{MEC}} = (b_1^{MEC}, \dots, b_N^{MEC})$ is a normalized increasing function with respect to the users' total demand of computation resources by the UAV-mounted MEC server. The vector $\mathbf{b^{MEC}} = (b_1^{MEC}, \ldots, b_N^{MEC})$ denotes the data offloading strategies of all the users in the examined system to the UAV-mounted MEC server. For demonstration purposes and without loss of generality, the users' total demand function $d_\tau(\mathbf{b^{MEC}}) \in [0,1]$ of computation resources by the UAV-mounted MEC server is defined as follows:

$$d_\tau(\mathbf{b^{MEC}}) = -1 + \frac{2}{1 + e^{-\theta \sum\limits_{n=1}^{N} d_n \frac{b_n^{MEC}}{b_n}}} \tag{6}$$

where $\theta > 0$ is a positive constant calibrating the sigmoidal curve of Equation (6) based on the computing capabilities of the UAV-mounted MEC server. The users' total computation demand function $d_\tau(\mathbf{b^{MEC}})$ is a continuous and strictly increasing function with respect to the users' total amount of offloaded data. Equation (6) is a representative example of the users' total computation demand function, while any other function that follows the above described properties can be adopted for the following analysis without loss of generality. In a nutshell, the UAV-mounted MEC server's rate of return function $RoR(d_\tau)$ provides positive experience, i.e., $RoR(d_\tau) > 0$, if the server has sufficient computation resources to serve the users' total computation demand $d_\tau(\mathbf{b^{MEC}})$. The UAV-mounted MEC server's rate of return function $RoR(d_\tau)$ is a continuous, monotonically decreasing, and concave function with respect to the users' total demand of computation resources, since the server's computation resources assigned to each user and correspondingly the users' perceived actual satisfaction decrease for increasing values of the users' total computation demand [28]. For demonstration purposes, in this paper, we adopt an indicative rate of return function that respects all aforementioned properties and is defined as follows:

$$RoR(d_\tau) = 2 - e^{d_\tau - 1} \tag{7}$$

Following the above discussion and focusing on the user's prospect theoretic utility function, as defined in Equation (4), it is noted that the first branch of Equation (4) expresses the user's n risk-aware satisfaction in the case that the UAV-mounted MEC server survives and can support the users' total computation demand. In that case, each user targets at the maximization of its gains, while, in the opposite case, i.e., the second branch of Equation (4), the user targets at the minimization of its losses, as the UAV-mounted MEC server has failed due to overexploitation.

If the UAV-mounted MEC-server survives, then the user's actual utility is determined by the second branch of Equation (5), given that the user offloaded part of its data to the MEC server. Thus, in combination with the first branch of Equation (4), the user's prospect theoretic utility is given as follows:

$$\begin{aligned} P_n^{surv.}(U_n) &= (U_n - U_{n,0})^{\alpha_n} \\ &= (b_n^{MEC})^{\alpha_n} [(2 - e^{d_\tau - 1}) - \frac{1}{\hat{t}_n \hat{e}_n} - c\frac{d_n}{b_n}]^{\alpha_n} \end{aligned} \tag{8}$$

If the opposite holds true, that is, the UAV-mounted MEC server's computation resources are overexploited by the users and the server fails to serve them, then by combining the second branch of Equation (4) and the third branch of Equation (5), the user's prospect theoretic utility can be written as follows:

$$\begin{aligned} P_n^{fail}(U_n) &= -k_n(U_{n,0} - U_n)^{\alpha_n} \\ &= -k_n(b_n^{MEC})^{\alpha_n} (\frac{1}{\hat{t}_n \hat{e}_n} + c\frac{d_n}{b_n})^{\alpha_n} \end{aligned} \tag{9}$$

Furthermore, the probability of failure of the UAV-mounted MEC server, which is the server's probability to fail serving the users' total computation demand d_τ (Equation (6)), is denoted by $Pr(d_\tau)$. The UAV-mounted MEC server's probability of failure function $Pr(d_\tau), 0 \leq Pr(d_\tau) \leq 1$ is assumed to be continuous, strictly increasing, convex, and twice differentiable function with respect to the users' total computation demand d_τ. In the following, we adopt the square function to present the UAV-mounted MEC server's probability of failure, as shown below:

$$Pr(d_\tau) = d_\tau^2 \tag{10}$$

It is noted that the rest of the paper's analysis still holds true for any probability of failure function that is characterized by the properties described above and the selection of the square function for the probability of failure is mainly made for presentation purposes. Accordingly, the UAV-mounted MEC server's probability to survive and process the users' total amount of offloaded data are $(1 - Pr(d_\tau))$. Moreover, due to the nature of the user's total computation demand (Equation (6)), the UAV-mounted MEC server's probability of failure (Equation (10)) is convex on low to medium users' computation demand and concave on high demand, while it asymptotically converges to one, as shown in Figure 2.

Combining Equations (8)–(10), the user's expected prospect theoretic utility by offloading b_n^{MEC} data to the UAV-mounted MEC server is defined as follows, jointly capturing the uncertainty of the UAV-mounted MEC server's computation resources, the pricing of the UAV-mounted MEC server, as well as the user's risk-aware characteristics in its data offloading decision:

$$\mathbb{E}(U_n) = P_n^{surv.}(U_n)(1 - Pr(d_\tau)) + P_n^{fail}(U_n)Pr(d_\tau). \tag{11}$$

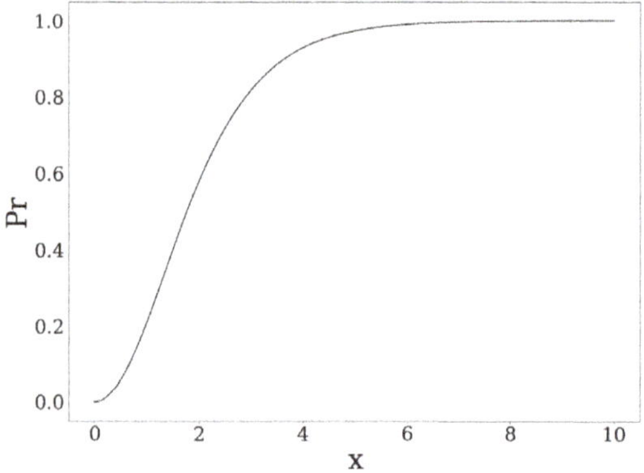

Figure 2. Probability of failure vs x when $Pr(x) = (-1 + \frac{2}{1+e^{-x}})^2$.

4. Pricing and Risk-Aware Data Offloading in UAV-Assisted MEC Systems

In this section, the distributed pricing and risk-aware data offloading problem in UAV-assisted multi-access edge computing systems is formulated by adopting the principles of non-cooperative game theory and solved based on the theory of S-modular games.

4.1. Problem Formulation

Each user aims at maximizing its expected prospect theoretic utility function (Equation (11)) by distribution and autonomously deciding its optimal data offloading strategy b_n^{MEC*} to the UAV-mounted MEC server, while considering the imposed pricing policy and its personal risk-aware characteristics. Accordingly, the users' pricing and risk-aware data offloading problem is formulated as a distributed optimization problem as follows:

$$\max_{b_n^{MEC} \in [0, b_n]} \mathbb{E}(U_n(b_n^{MEC}, \mathbf{b_{-n}^{MEC}})) \tag{12a}$$

$$s.t. \quad 0 \leq b_n^{MEC} \leq b_n \tag{12b}$$

where $\mathbf{b_{-n}^{MEC}}$ denotes the amount of the offloaded data by the rest of the users except for user n.

The distributed optimization problem of users' data offloading can be formulated as a non-cooperative game among the users $G = [\mathcal{N}, A_n, \mathbb{E}(U_n(b_n^{MEC}, \mathbf{b_{-n}^{MEC}}))]$, where $\mathcal{N}$ is the set of users, $A_n = [0, b_n]$ is the user's n data offloading strategy space, and $\mathbb{E}(U_n(b_n^{MEC}, \mathbf{b_{-n}^{MEC}}))$ denotes the user's n expected prospect theoretic utility function, as defined in the previous section. The solution of the non-cooperative game G should determine each user's optimal data offloading strategy b_n^{MEC*} in order to maximize its expected prospect theoretic utility. The Pure Nash Equilibrium (PNE) approach is adopted and described below, towards analytically seeking the solution of the pricing and risk-aware data offloading problem (Equation (12a) and (12b)).

Definition 1. *(Pure Nash Equilibrium Point): A data offloading vector $\mathbf{b_n^{MEC*}} = (b_1^{MEC*}, \ldots, b_N^{MEC*})$ in the strategy space $b_n^{MEC*} \in A_n = [0, b_n]$ is a Pure Nash Equilibrium point if for every user n the following condition holds true:*

$$\mathbb{E}(U_n(b_n^{MEC*}, \mathbf{b_{-n}^{MEC*}})) \geq \mathbb{E}(U_n(b_n^{MEC}, \mathbf{b_{-n}^{MEC*}})) \tag{13}$$

for all $b_n^{MEC} \in A_n$.

The physical interpretation of the above definition is that, at the Pure Nash Equilibrium point, no user has the incentive to unilaterally change its data offloading strategy to the UAV-mounted MEC server given the data offloading strategies of the rest of the users, as its achieved expected prospect theoretic utility cannot be improved.

4.2. Problem Solution

In order to prove the existence of at least one PNE of the non-cooperative game G, as a solution of the maximization problem (Equation (12a) and (12b)), the theory of submodular games is adopted [29]. The submodular games are characterized by strategic substitutes, i.e., when a user offloads more data to the UAV-mounted MEC server, the rest of the users tend to avoid following similar behavior, as the UAV-mounted MEC server's computation resources can become overexploited and none of the users be satisfied. The submodular games are of great interest and practical importance as an optimization tool, due to the fact that they guarantee the existence of at least one PNE, while learning and adjustment tools (such as the best response dynamics) can be used in order to determine such a point.

Definition 2. *(Submodular Games): The non-cooperative game $G = [\mathcal{N}, A_n, \mathbb{E}(U_n(b_n^{MEC}, \mathbf{b_{-n}^{MEC}}))]$ is submodular, if, for all the users, the following conditions hold true [30]:*

1. *A_n is a compact subset of an Euclidean space.*
2. *$\mathbb{E}(U_n(b_n^{MEC}, \mathbf{b_{-n}^{MEC}}))$ is smooth, submodular in b_n^{MEC}, and has non-increasing differences in $(b_n^{MEC}, \mathbf{b_{-n}^{MEC}})$, i.e., $\frac{\partial^2 E_n(\vec{b}^{MEC})}{\partial b_j^{MEC} \partial b_n^{MEC}} \leq 0$.*

Additionally, in a submodular game, there always exist external equilibria [31]: a largest best response strategy $\overline{b_n^{MEC}} = sup\{b_n^{MEC} \in A_n : BR(b_n^{MEC}, \mathbf{b_{-n}^{MEC}}) \geq b_n^{MEC}\}$ and a smallest best response strategy: $\underline{b_n^{MEC}} = inf\{b_n^{MEC} \in A_n : BR(b_n^{MEC}, \mathbf{b_{-n}^{MEC}}) \leq b_n^{MEC}\}$ of the non-empty set of Pure Nash Equilibria, where $BR(b_n^{MEC}, \mathbf{b_{-n}^{MEC}})$ denotes the user's n best response strategy to the other users' strategies.

Theorem 1. *The non-cooperative game $G = [\mathcal{N}, A_n, \mathbb{E}(U_n(b_n^{MEC}, \mathbf{b_{-n}^{MEC}}))]$ is submodular for all $d_\tau \in (0, \mu)$, where $\mu \in (0, 1)$, and $c < \frac{b_n}{d_n}(1 - \frac{1}{\bar{t}_n \bar{e}_n})$, and has at least one Pure Nash Equilibrium point.*

Proof. The strategy space $A_n = [0, b_n]$ is a compact subset of a Euclidean space. The user's expected prospect theoretic utility function $\mathbb{E}(U_n(b_n^{MEC}, \mathbf{b_{-n}^{MEC}}))$, as defined in Equation (11), is smooth, as it has derivatives of all orders everywhere in its domain A_n. Towards showing that the user's expected prospect theoretic utility function is submodular in b_n and has non-increasing differences in $(b_n^{MEC}, \mathbf{b_{-n}^{MEC}})$, we examine the properties of the second order partial derivative of the user's expected prospect theoretic utility function, i.e., $\frac{\partial^2 \mathbf{E}_n(\vec{b}^{MEC})}{\partial b_j^{MEC} \partial b_n^{MEC}} \leq 0$.

We can rewrite Equation (11) using Equations (8) and (9), as follows:

$$\mathbb{E}(U_n(b_n^{MEC}, \mathbf{b_{-n}^{MEC}})) = (b_n^{MEC})^{\alpha_n} \{ [(2 - e^{d_\tau - 1}) - \tfrac{1}{\hat{t}_n \hat{e}_n} - c\tfrac{d_n}{b_n}]^{\alpha_n} (1 - Pr(d_\tau)) - k_n (\tfrac{1}{\hat{t}_n \hat{e}_n} + c\tfrac{d_n}{b_n})^{\alpha_n} Pr(d_\tau) \} \tag{14}$$

We define $\overline{RoR}(d_\tau) = [(2 - e^{d_\tau - 1}) - \tfrac{1}{\hat{t}_n \hat{e}_n} - c\tfrac{d_n}{b_n}]^{\alpha_n}$ as the user's specific rate of return, which should be positive in order for the user to have an incentive to offload part of its data to the UAV-mounted MEC server. From Equation (7), the UAV-mounted MEC server's rate of return function $RoR(d_\tau)$ is decreasing. Thus, the minimum value of $RoR(d_\tau)$, and correspondingly of the function $\overline{RoR}(d_\tau)$, is determined at $d_\tau = 1$. The physical notion of $d_\tau = 1$ is that all the users offload their total amount of data to the UAV-mounted MEC server for further processing. Following this observation, we can determine the boundaries of the constant pricing factor c that the UAV-mounted MEC server imposes on the users, in order for the latter to still have an incentive to offload part of their data to the MEC server without the imposed pricing to become a prohibitive factor. Therefore, the feasible boundaries of the constant pricing factor are determined as follows:

$$\overline{RoR}(d_\tau = 1) > 0 \Rightarrow c < \frac{b_n}{d_n}\left(1 - \frac{1}{\hat{t}_n \hat{e}_n}\right) \tag{15}$$

In addition, the following conditions hold true by performing the corresponding derivations:
$\frac{\partial d_\tau}{\partial b_n^{MEC}} > 0, \frac{\partial d_\tau}{\partial b_j^{MEC}} > 0, \frac{\partial \overline{RoR}(d_\tau)}{\partial b_n^{MEC}} < 0, \frac{\partial \overline{RoR}(d_\tau)}{\partial b_j^{MEC}} < 0, \frac{\partial^2 \overline{RoR}(d_\tau)}{\partial b_j^{MEC} \partial b_n^{MEC}} < 0, \frac{\partial Pr(d_\tau)}{\partial b_n^{MEC}} > 0, \frac{\partial Pr(d_\tau)}{\partial b_j^{MEC}} > 0, \frac{\partial^2 Pr(d_\tau)}{\partial b_n^{MEC} \partial b_j^{MEC}} = 0.$
For notational convenience, we set $A = k_n(\tfrac{1}{\hat{t}_n \hat{e}_n} + c\tfrac{d_n}{a_n})^{\alpha_n} > 0$, and we calculate the second order partial derivative of the user's expected prospect theoretic utility function, as follows:

$$
\begin{aligned}
\frac{\partial^2 \mathbb{E}(U_n(b_n^{MEC}, \mathbf{b_{-n}^{MEC}}))}{\partial b_j^{MEC} \partial b_n^{MEC}} =\ & \alpha_n (b_n^{MEC})^{\alpha_n - 1} \Big\{ \frac{\partial \overline{RoR}(d_\tau)}{\partial b_j^{MEC}} [1 - Pr(d_\tau)] - \overline{RoR}(d_\tau) \frac{\partial Pr(d_\tau)}{\partial b_j^{MEC}} - A \frac{\partial Pr(d_\tau)}{\partial b_j^{MEC}} \Big\} + \\
& (b_n^{MEC})^{\alpha_n} \Big\{ \frac{\partial^2 \overline{RoR}(d_\tau)}{\partial b_j^{MEC} \partial b_n^{MEC}} [1 - Pr(d_\tau)] - \frac{\partial \overline{RoR}(d_\tau)}{\partial b_n^{MEC}} \frac{\partial Pr(d_\tau)}{\partial b_j^{MEC}} - \frac{\partial \overline{RoR}(d_\tau)}{\partial b_j^{MEC}} \frac{\partial Pr(d_\tau)}{\partial b_n^{MEC}} \Big\} \\
=\ & (b_n^{MEC})^{\alpha_n - 1} \Big\{ \alpha_n \frac{\partial \overline{RoR}(d_\tau)}{\partial b_j^{MEC}} [1 - Pr(d_\tau)] - \alpha_n \overline{RoR}(d_\tau) \frac{\partial Pr(d_\tau)}{\partial b_j^{MEC}} - A\alpha_n \frac{\partial Pr(d_\tau)}{\partial b_j^{MEC}} + \\
& b_n^{MEC} \frac{\partial^2 \overline{RoR}(d_\tau)}{\partial b_j^{MEC} \partial b_n^{MEC}} [1 - Pr(d_\tau)] - b_n^{MEC} \frac{\partial \overline{RoR}(d_\tau)}{\partial b_n^{MEC}} \frac{\partial Pr(d_\tau)}{\partial b_j^{MEC}} - b_n^{MEC} \frac{\partial \overline{RoR}(d_\tau)}{\partial b_j^{MEC}} \frac{\partial Pr(d_\tau)}{\partial b_n^{MEC}} \Big\}
\end{aligned}
\tag{16}
$$

Let $\psi(d_\tau) = \frac{\partial \overline{RoR}(d_\tau)}{\partial b_j^{MEC}} [\alpha_n - \alpha_n Pr(d_\tau) - b_n^{MEC} \frac{\partial Pr(d_\tau)}{\partial b_j^{MEC}}] - b_n^{MEC} \frac{\partial \overline{RoR}(d_\tau)}{\partial b_n^{MEC}} \frac{\partial Pr(d_\tau)}{\partial b_j^{MEC}}$. We can rewrite Equation (16), as follows:

$$\frac{\partial^2 \mathbb{E}(U_n(b_n^{MEC}, \mathbf{b_{-n}^{MEC}}))}{\partial b_j^{MEC} \partial b_n^{MEC}} = (b_n^{MEC})^{\alpha_n - 1} \{ \psi(d_\tau) - \alpha_n \overline{RoR}(d_\tau) \frac{\partial Pr(d_\tau)}{\partial b_j^{MEC}} - A\alpha_n \frac{\partial Pr(d_\tau)}{\partial b_j^{MEC}} + b_n^{MEC} \frac{\partial^2 \overline{RoR}(d_\tau)}{\partial b_j^{MEC} \partial b_n^{MEC}} [1 - Pr(d_\tau)] \} \tag{17}$$

It is observed that the last three terms of Equation (17) are negative; thus, we study the properties of the function $\psi(d_\tau)$, $\forall n \in \mathcal{N}$. For $d_\tau = 0$, we have $b_n^{MEC} = 0$. Thus, we calculate:

$$\psi(d_\tau = 0) = \frac{\partial \overline{RoR}(0)}{\partial b_j^{MEC}} \alpha_n < 0 \tag{18}$$

For $d_\tau \approx 1$, we have $b_n^{MEC} = b_n$, $\forall n \in \mathcal{N}$. Thus, we calculate:

$$\psi(d_\tau \approx 1) - b_n\left[\frac{\partial \overline{RoR}(1)}{\partial b_j^{MEC}}\frac{\partial Pr(1)}{\partial b_n^{MEC}} + \frac{\partial \overline{RoR}(1)}{\partial b_n^{MEC}}\frac{\partial Pr(1)}{\partial b_j^{MEC}}\right] > 0 \tag{19}$$

Since $\psi(d_\tau)$ is continuous, using the Bolzano Theorem [32], we conclude that there exists at least one $\mu \in (0,1)$ such that $\psi(d_\tau = \mu) = 0$. Given that $\psi(d_\tau = 0) < 0$ (Equation (18)), then, if μ is the smallest possible value in $(0,1)$ such that $\psi(d_\tau = \mu) = 0$, then $\psi(d_\tau) < 0, \forall d_\tau \in (0,\mu)$. Thus, we conclude that

$$\frac{\partial^2 \mathbb{E}(U_n(b_n^{MEC}, \mathbf{b_{-n}^{MEC}}))}{\partial b_j^{MEC} \partial b_n^{MEC}} < 0, \forall d_\tau \in (0,\mu), \mu \in (0,1). \tag{20}$$

Thus, the non-cooperative game G is submodular $\forall d_\tau \in (0,\mu), \mu \in (0,1)$ and $c < \frac{b_n}{d_n}(1 - \frac{1}{\hat{t}_n \hat{e}_n})$. Therefore, the non-cooperative game $G = [\mathcal{N}, A_n, \mathbb{E}(U_n(b_n^{MEC}, \mathbf{b_{-n}^{MEC}}))]$ has at least one Pure Nash Equilibrium point $\mathbf{b_n^{MEC*}} = (b_1^{MEC*}, \ldots, b_N^{MEC*})$ [33]. $\square$

5. Pricing and Risk-Aware Distributed Data Offloading Algorithm

Towards enabling the users to determine their optimal data offloading strategy b_n^{MEC*} in a distributed manner, the Best Response Dynamics (BRD) approach is adopted. The best response strategy of each user subject to the selected data offloading strategies of the rest of the users is formally determined as follows:

$$BR(b_n^{MEC}, \mathbf{b_{-n}^{MEC}}) = b_n^{MEC*} = \underset{b_n^{MEC} \in [0, b_n]}{\arg\max} \mathbb{E}(U_n(b_n^{MEC}, \mathbf{b_{-n}^{MEC}})). \tag{21}$$

Given that we have already proven that the non-cooperative game $G = [\mathcal{N}, A_n, \mathbb{E}(U_n(b_n^{MEC}, \mathbf{b_{-n}^{MEC}}))]$ belongs to the class of submodular games as stated above, and therefore possesses at least one PNE point, it also readily follows that the iterated best-response dynamics always converges to a Pure Nash Equilibrium point [34,35].

Subsequently, capitalizing on the above argumentation, a distributed iterative and low-complexity algorithm is introduced in order to determine the users' optimal data offloading strategies to the UAV-mounted MEC server (see Algorithm 1). The proposed algorithm follows the philosophy and principles of the best response dynamics learning mechanism, and, at each iteration, each user aims at maximizing its expected prospect theoretic utility given the data offloading strategies of the rest if the users. The complexity of the pricing and risk-aware data offloading algorithm is $O(N * Ite * A)$, where Ite is the total number of iterations in order for the algorithm to converge to the PNE, and A is the complexity of solving Equation (21). Detailed numerical results regarding the operation performance and scalability of our approach and algorithm, in terms of iterations, are presented in the following section as well.

Algorithm 1 Pricing and risk-aware data offloading algorithm

Input: $N, c, b_n, d_n, f_n, \gamma_n, \forall n \in \mathcal{N}$
Output: $\mathbf{b^{MEC*}}$
Initialization: $ite = 0$, $Convergence =$ **false**, $b_n^{MEC^{(ite=0)}}, \forall n \in \mathcal{N}$
while $Convergence ==$ **false do**

  $ite = ite + 1$;
  for $n = 1$ to N **do**

    user n determines $b_n^{MEC*^{(ite)}}$ w.r.t. $b_n^{MEC*^{(ite-1)}}$ (Equation (21)) and receives $\mathbb{E}(U_n)^{(ite)}$
  end for
  if $b_n^{MEC*^{(ite)}} = b_n^{MEC*^{(ite-1)}}$ **then**

    $Convergence =$ **true**
  end if
end while

---

6. Numerical Results

In this section, we provide a series of numerical results, obtained via modeling and simulation, evaluating the performance and the inherent attributes of the proposed pricing and risk-aware data offloading framework. Initially, the pure operational characteristics of the proposed framework are presented (Section 6.1), while the impact of the introduced usage-based pricing scheme is quantified and studied (Section 6.2). Moreover, a scalability analysis of the proposed framework is performed in Section 6.3, while the impact of the prospect theoretic parameters reflecting the user behavioral pattern in terms of loss aversion and sensitivity, on the overall system performance is evaluated in Section 6.4. The performed simulations were executed on an Intel Core i5-4300U CPU @ 1.90 GHz $\times$ 4 with 8 GB RAM (New York, NY, USA). The main parameters used in our simulation, along with their typical values, are presented in Table 1. In the rest of the analysis, and in particular in Sections 6.1 and 6.2, we have considered $N = 25$ users, and sensitivity (k_n) and loss aversion (α_n) parameter values as indicated in Table 1. However, in Sections 6.3 and 6.4, a wider range of the number of users and the loss aversion and sensitivity parameters are considered.

Table 1. Typical values for simulation parameters [36,37].

Parameter	Value	Description
b_n	$10^7 \pm 10^6$ [bytes]	User's n computation task's input data
d_n	$8 * 10^9 \pm 10^9$ [CPU cycles]	CPU cycles required to accomplish user's n computation task
f_n	$6 * 10^9 \pm 10^9$ [CPU cycles/sec]	User's n device's computational capability
γ_n	$4 * 10^{-9} \pm 10^{-9}$ [Joules/CPU cycles]	Coefficient of the locally consumed energy per CPU cycle
a_n	0.2	User's n sensitivity on gains and losses
k_n	1.2	User's n loss aversion parameter
c_n	$0.5 * \frac{b_n}{d_n}\left(1 - \frac{1}{\hat{t}_n \hat{e}_n}\right)$ [1/CPU cycles]	Pricing factor (satisfies condition of Theorem 1)
θ	$2 * 10^{-11}$	Parameter denoting the processing capability of the server, centers the sigmoid on to realistic values of offloading data

6.1. Pure Operation of the Framework

Figure 3 presents the amount of offloaded data by each user to the UAV-mounted MEC server, as well as the average amount of offloaded data as a function of the pricing and risk-aware data offloading algorithm's iterations. The results reveal that the introduced best response dynamics-based algorithm converges to the PNE quite fast and in small iterations (less than 10 iterations are required for all users). Moreover, Figures 4 and 5 illustrate each user's expected prospect-theoretic utility and the corresponding usage-based pricing imposed by the UAV-mounted MEC server as a function of the algorithm's iterations. The corresponding results reveal that initially the users tend to offload a great portion of their data to the MEC server, as observed in Figure 3, and therefore their expected prospect-theoretic utility increases (Figure 4). Specifically, at the first iteration of the algorithm, the users present an aggressive behavior in terms of offloading a large amount of data to the UAV-mounted MEC server (Figure 3) towards enjoying a high expected utility. (Figure 4). However, at the same time, this behavior is expected to lead to the increase of the probability of failure of the UAV-mounted MEC server (as it is confirmed below in Figure 7), and accordingly to the users being penalized with a high price. This is demonstrated in Figure 5, where, due to the fact that the users exploit more the computing capabilities of the MEC server, the latter imposes on them a higher usage-based pricing. Consequently, in combination with the impact of probability of failure and rate of return, as the iterations evolve, the users decrease the amount of data that they offload to the MEC server (Figure 3) following the learning mechanism of the best response dynamics, in order to experience a lower pricing (Figure 5) and finally they converge to the PNE.

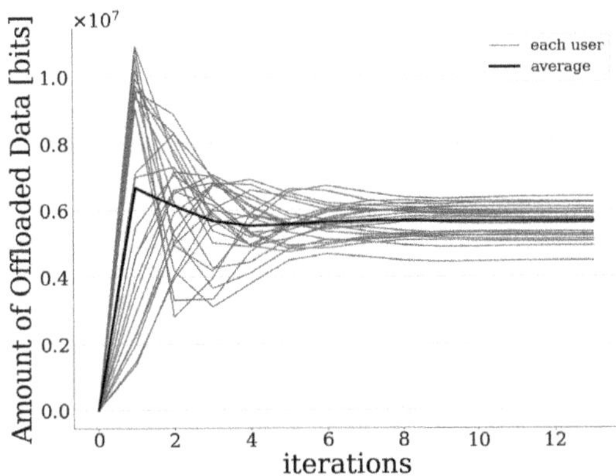

Figure 3. Amount of data offloaded by each user vs. iterations.

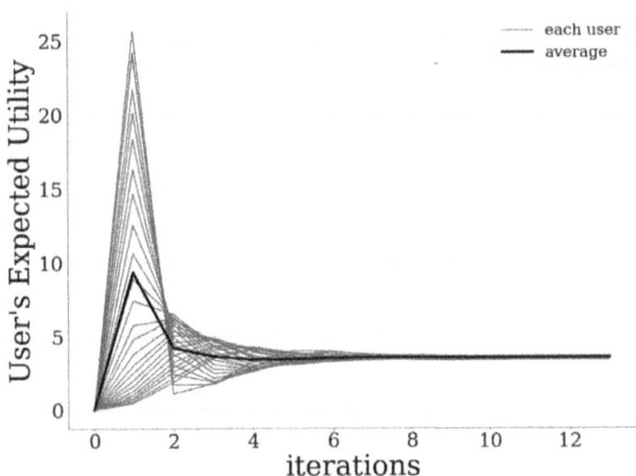

Figure 4. Expected utility of each user vs. iterations.

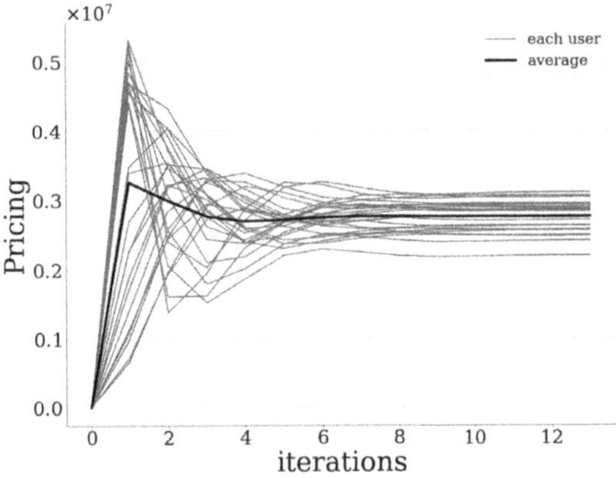

Figure 5. Pricing imposed by the server on each user vs. iterations.

Figure 6 depicts the users' average expected prospect-theoretic utility and the users' average experienced usage-based pricing for exploiting the UAV-mounted MEC server's computing capabilities, as a function of the algorithm's iterations. In addition, Figure 7 presents the UAV-mounted MEC server's probability of failure as a function of the algorithm's iterations. The above described trend in users' data offloading strategies is observed from the system's point of view. Specifically, all the users tend initially to aggressively offload a large amount of data to the MEC server in order to achieve a greater utility (Figure 6). However, the probability of failure of the UAV-mounted MEC server increases due to the over-exploitation of its computing capabilities (Figure 7). Thus, the MEC server imposes a higher pricing on the users (Figure 6) to control their greedy and selfish data offloading behavior.

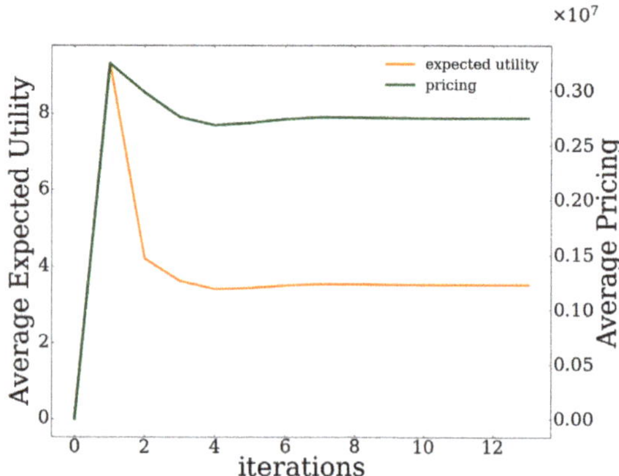

Figure 6. Average users' expected utility and average users' pricing vs. iterations.

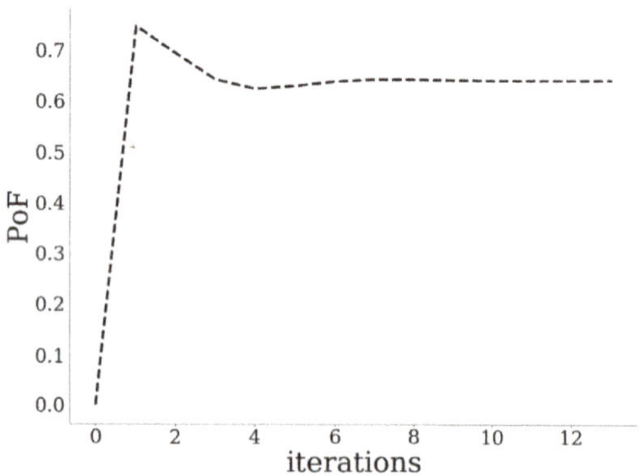

Figure 7. Probability of failure of MEC server vs. iterations.

6.2. Impact of Usage-Based Pricing

In this section, we study the impact of the usage-based pricing imposed by the UAV-mounted MEC server, on the users' data offloading strategies, as well as on the overall operation of the system. Specifically, Figure 8 presents the probability of failure of the MEC server as a function of the pricing factor c (Equation (3)). Moreover, the users' average expected utility, the users' average amount of offloaded data, and the pricing imposed by the MEC server are presented in Figure 9, as a function of the pricing factor c as well. The results reveal that, as the pricing policy becomes stricter (i.e., increasing values of the pricing factor), the usage-based pricing experienced by the users increases (Figure 9) and the exploitation of the MEC server's computing capabilities becomes cost inefficient after some point (with respect to the total offloaded data). Consequently, the users tend to offload a smaller amount of data to the MEC server (Figure 9), and the MEC server becomes less congested in terms of processing the users' computation tasks, and its probability of failure decreases (Figure 8).

Based on the results presented in Figure 9, it is observed that the users' average expected utility is concave with respect to the pricing factor. Specifically, small values of the pricing factor correspond to less-strict pricing policies; thus, the users over-exploit the MEC server's computing capabilities (i.e., high values of MEC server's probability of failure are observed), resulting in low values of expected utility. On the other hand, high values of the pricing factor result in discouraging the users to exploit the UAV-mounted MEC server's computing capabilities, thus concluding again to low levels of users' average expected utility. Therefore, a balanced pricing policy is required to keep the quality of experience of the users at high levels.

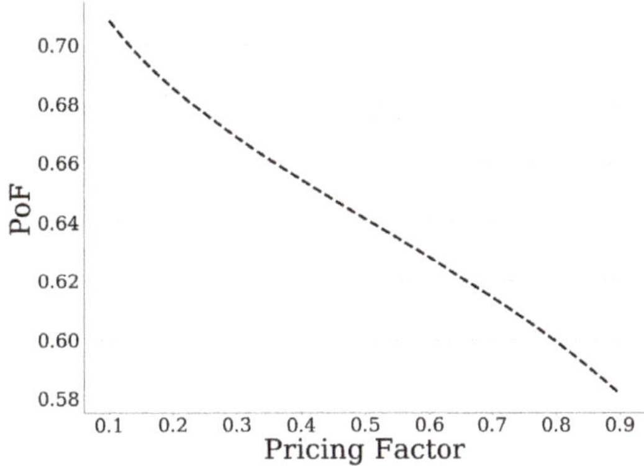

Figure 8. Probability of failure of MEC server vs. the pricing factor.

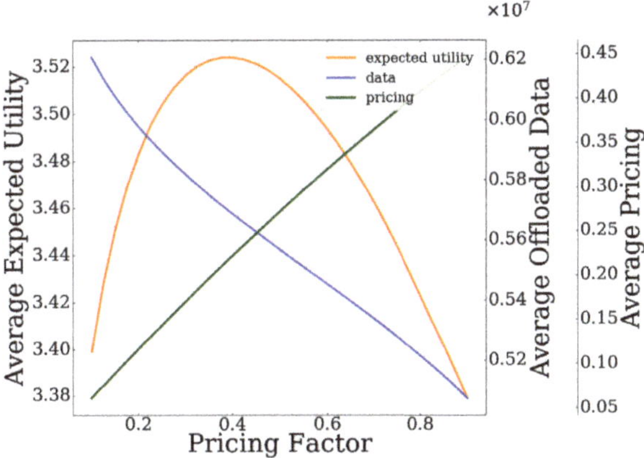

Figure 9. Average expected utility, offloaded data, and pricing vs. the pricing factor.

6.3. Scalability Evaluation

In this section, a scalability evaluation of the proposed pricing and risk-aware data offloading framework is provided considering an increasing number of users in the system. Table 2 presents the iterations and the overall corresponding execution time of the proposed algorithm in order to converge to the PNE point. Given the distributed nature of the best response dynamics approach, we observe that its execution time scales quite well for increasing number of users, achieving a close to real-time implementation in realistic scenarios. Respectively, the users' average expected utility, the users' average amount of offloaded data, and the imposed pricing by the UAV-mounted MEC server are presented in Figure 10, as a function of the number of users. The scalability evaluation is complemented by the results presented in Figure 11 that depict the convergence of the users' average amount of offloaded data as a function of the required number of iterations, for different numbers of users. In particular, we observe that, as the number of users in the system increases, they tend to offload a lower average amount of data to the MEC server (Figures 10 and 11), as the latter becomes over-congested. Thus, they experience both lower pricing (Figure 10) and lower expected utility (Figure 10), as they drive themselves in processing more data locally on their local devices and accordingly consume their own resources, i.e., battery. It is also observed that the user's experienced pricing $c_n(b_n^{MEC})$ and the user's offloaded data b_n^{MEC} (Figure 10) has the same trend, due to their one-to-one relationship stemming from Equation (3), while the corresponding curves also appear to be overlapping. However, it should be noted here that the actual values for the two curves are different, since there are two different right vertical axes in Figure 10 (each one reflecting the values of each curve respectively).

Table 2. Algorithm's execution time per user for a different number of users.

N	Iterations	Time Per User [s]
1	3	0.0036
2	3	0.0042
5	3	0.0049
10	6	0.0095
25	14	0.0122
50	31	0.0282
75	54	0.0640
100	83	0.0979

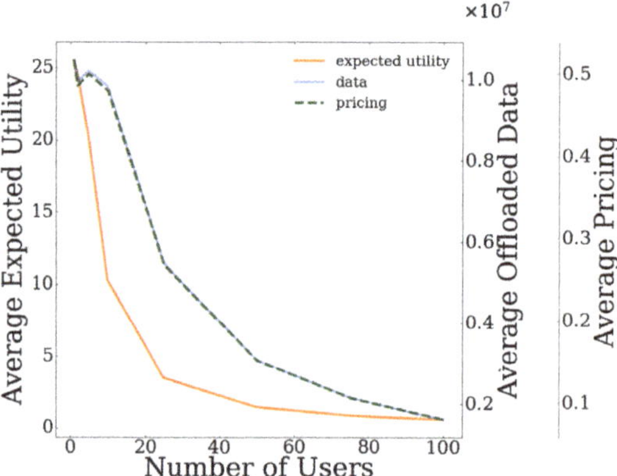

Figure 10. Users' average expected utility, users' average offloaded data and pricing at the PNE vs. number of users on the system.

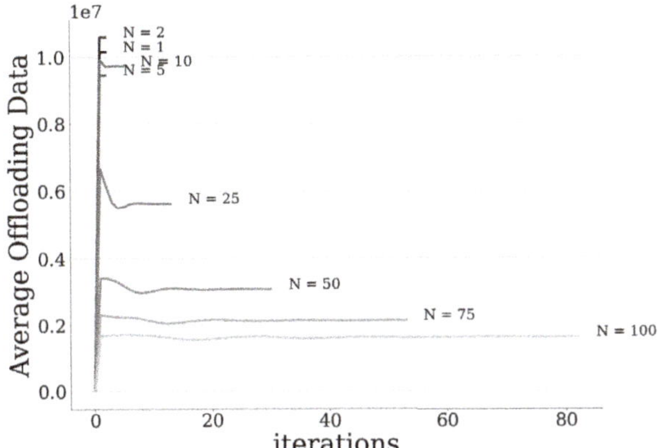

Figure 11. Users' average data offloading vs. iterations for different numbers of users.

6.4. Impact of Prospect Theoretic Parameters and User Competition

In the following, the impact of the prospect theoretic parameters, reflecting the user behavioral pattern in terms of loss aversion and sensitivity, on the overall system performance is evaluated.

Specifically, in Figures 12 and 13, initially we present the average user offloaded data and corresponding Probability of failure, as functions of the sensitivity parameter α_n and the loss aversion index k_n, respectively. As can be seen from Figure 12, by increasing the sensitivity parameter α_n, the users tend to offload more data to the MEC server since they opt to value more the larger gains, compared to those of smaller magnitude. The increased volume of data offloaded results in an increase in the corresponding Probability of Failure of the server as well. In Figure 13, on the other hand, we can see that, as the loss aversion index k_n increases, less data are offloaded to the server, since higher value signifies more loss aversion for the users, resulting in smaller Probabilities of Failure of the server.

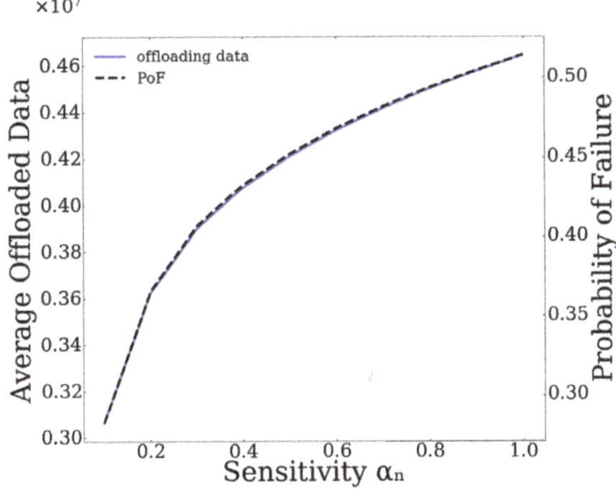

Figure 12. Average offloading data and PoF vs. sensitivity parameter α_n.

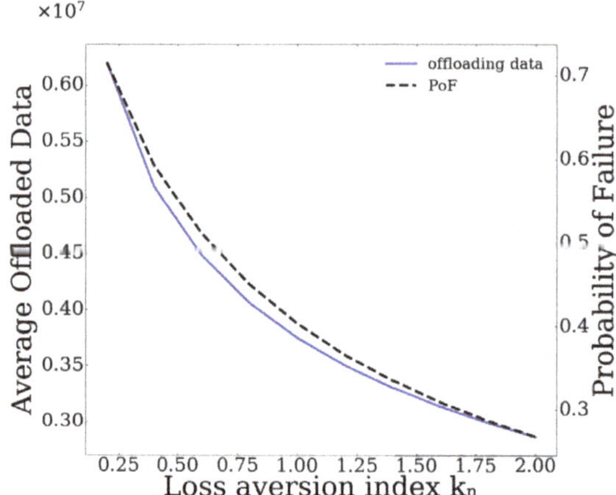

Figure 13. Average offloading data and PoF vs. loss aversion index k_n.

In order to further study the effect of competition of users for the CPR (i.e., UAV-mounted MEC server), we use the Fragility under Competition (FuC) metric [38]. This metric is expressed as the ratio between the Probability of Failure of the MEC server when N users are competing for the MEC server's resources at the equilibrium state, versus the Probability of Failure of the MEC server when there is only one user offloading data. Formally, the Fragility under Competition is defined as: $FuC = \frac{Pr(b_N^{MEC^*})}{Pr(b_1^{MEC^*})}$, where $b_N^{MEC^*}$ denotes the equilibrium point when N users are present and $b_1^{MEC^*}$ denotes the corresponding equilibrium point if only one user was present, with the same risk preferences as the group of N users.

In Figures 14 and 15, we present the FuC metric as a function of the number of users in the system, for different values of the sensitivity parameter α_n and the loss aversion index k_n, respectively. In both figures, we observe that, as the number of users increases, the FuC increases as well, since more users are competing for the CPR and consequently more data are offloaded to the server, until it eventually plateaus. Concerning the effect that the prospect theoretic parameters have on the FuC metric, in Figure 14, we can see that the higher the value of the sensitivity parameter α_n, the higher the FuC as well. This is justified by the fact that, the higher the values of α_n, the greater the sensitivity of the users towards gains and losses of higher magnitude compared to those of smaller magnitude (Figure 12). As a result, users tend to offload more data to the MEC server and the server is more prone to failure, and accordingly an increase in FuC is expected. With respect now to the loss aversion index k_n, we can see in Figure 15 that, as k_n increases, the FuC decreases. This is due to the fact that, as k_n increases, users become more loss averse and thus they tend to offload less data to the MEC server in order to avoid potential failure as already shown in Figure 13. The less data are offloaded to the server, the less the probability that the server will fail, thus resulting in lower FuC. Furthermore, based on the results of Figures 14 and 15, the FuC appears initially more sensitive to the number of users in the case α_n compared to k_n, our setting and experiments. It is clarified that the overall observed increasing trend of the FuC w.r.t. to the increasing number of users in these figures is well aligned with the fact that the failure probability is an increasing function of the total offloaded data of all users. However, the actual slope of the corresponding curves mainly depends on the used values for α_n and k_n for the generation of these curves, which are selected here only for demonstration purposes, and are not correlated with each other in any way.

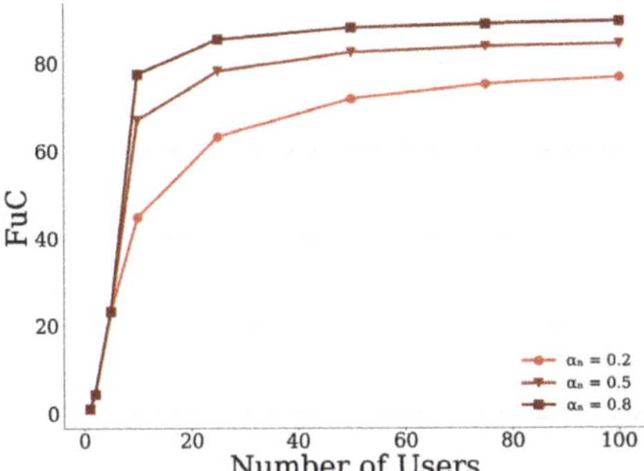

Figure 14. Fragility under Competition vs. no. of users for different sensitivity parameters α_n.

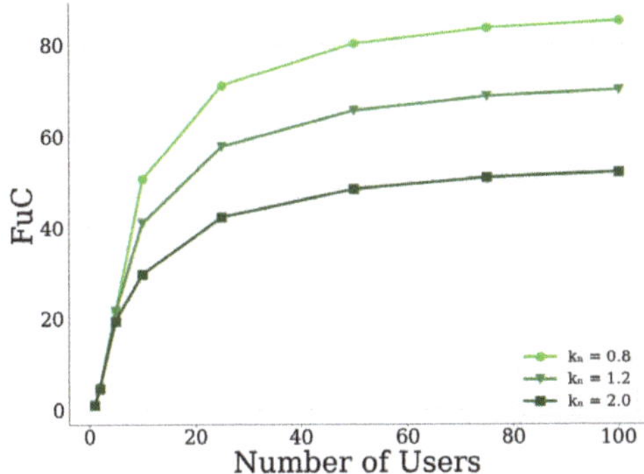

Figure 15. Fragility under Competition vs. no. of users for different loss aversion indices k_n.

7. Conclusions

In this paper, a resource-based pricing and user risk-aware data offloading framework is proposed for UAV-assisted multi-access edge computing systems. In particular, a usage-based pricing mechanism is introduced regarding the exploitation of the MEC server's computing capabilities by the users, and is properly incorporated within the principles and modeling of Prospect Theory, which is used to capture the users' risk-aware behavior in the overall data offloading decision-making. Initially, the user's prospect-theoretic utility function is formulated by quantifying the user's risk seeking and loss aversion behavior, while taking into account the pricing mechanism. Accordingly, the users' pricing and risk-aware data offloading problem is formulated as a distributed maximization problem of each user's expected prospect-theoretic utility function and addressed as a non-cooperative game among the users. The existence of a Pure Nash Equilibrium for the formulated non-cooperative game is shown based on the theory of submodular games. An iterative and distributed algorithm is introduced that converges to the PNE, following the learning rule of the best response dynamics. Detailed numerical results are presented highlighting the operation feature and scalability properties of the proposed framework, while at the same time providing useful insights about the benefits of adopting the usage-based pricing scheme.

Our current and future research work focuses on treating the overall key problem of data offloading in various cloud computing environments, such as fog computing, where a large number of computing devices imposes additional scalability and stability challenges. Moreover, it is noted that, in this work, the data offloading problem was mainly treated from a computing resources perspective. However, depending on the environment assumed, the overall process could be affected by the wireless communication aspects between the UAV and users. The proposed framework could be adapted and extended to treat this aspect, either implicitly through the cost factors and functions considered when using the server resources, or explicitly by modeling the transmission characteristics (e.g., delay, rate, energy) involved in the offloading process.

Author Contributions: All authors contributed extensively to the work presented in this paper. G.M. contributed to the design of the algorithm, developed the code of the overall framework, executed the evaluation experiment, and contributed to the discussions and analysis of the overall theoretical framework. E.E.T. and S.P. were responsible for the overall orchestration of the performance evaluation work and had the overall coordination in the writing of the article. All authors have read and agreed to the published version of the manuscript.

Funding: The research work was supported by the Hellenic Foundation for Research and Innovation (H.F.R.I.) under the "First Call for H.F.R.I. Research Projects to support Faculty members and Researchers and the procurement of high-cost research equipment grant" (Project Number: HFRI-FM17-2436). The research of Eirini Eleni Tsiropoulou was conducted as part of the NSF CRII-1849739.

Conflicts of Interest: The authors declare no conflict of interest.

References

1. Atzori, L.; Iera, A.; Morabito, G. Internet of Things at a Glance. Comput. Netw. **2010**, *54*, 2787–2805. [CrossRef]
2. Ericsson. *Ericsson Mobility Report—Q4 2018*; Ericsson Mobility Report; Ericsson: Stockholm, Sweden, 2019.
3. Porambage, P.; Okwuibe, J.; Liyanage, M.; Ylianttila, M.; Taleb, T. Survey on multi-access edge computing for internet of things realization. *IEEE Commun. Surv. Tutor.* **2018**, *20*, 2961–2991. [CrossRef]
4. Cao, B.; Zhang, L.; Li, Y.; Feng, D.; Cao, W. Intelligent offloading in multi-access edge computing: A state-of-the-art review and framework. *IEEE Commun. Mag.* **2019**, *57*, 56–62. [CrossRef]
5. Vamvakas, P.; Tsiropoulou, E.E.; Papavassiliou, S. Exploiting prospect theory and risk-awareness to protect UAV-assisted network operation. *EURASIP J. Wirel. Commun. Netw.* **2019**, *2019*, 286–306. [CrossRef]
6. Apostolopoulos, P.A.; Torres, M.; Tsiropoulou, E.E. Satisfaction-aware Data Offloading in Surveillance Systems. In Proceedings of the 14th Workshop on Challenged Networks, Los Cabos, Mexcio, 21–25 October 2019; pp. 21–26.
7. He, D.; Qiao, Y.; Chan, S.; Guizani, N. Flight security and safetyof drones in airborne fog computing systems. *IEEE Commun. Mag.* **2018**, *56*, 66–71. [CrossRef]
8. Cheng, N.; Xu, W.; Shi, W.; Zhou, Y.; Lu, N.; Zhou, H.; Shen, X. Air-ground integrated mobile edge networks: Architecture, challenges, and opportunities. *IEEE Commun. Mag.* **2018**, *56*, 26–32. [CrossRef]
9. Luo, F.; Jiang, C.; Yu, S.; Wang, J.; Li, Y.; Ren, Y. Stability of cloud-based uav systems supporting big data acquisition and processing. *IEEE Trans. Cloud Comput.* **2017**, *7*, 866–877. [CrossRef]
10. Valentino, R.; Jung, W.-S.; Ko, Y.-B. Opportunistic computational offloading system for clusters of drones. In Proceedings of the 20th International Conference on Advanced Communications Technology, Chuncheon-si Gangwon-do, Korea, 11–14 February 2018; pp. 303–306.
11. Xiong, J.; Guo, H.; Liu, J. Task offloading in uav-aided edge computing: Bit allocation and trajectory optimization. *IEEE Commun. Lett.* **2019**, *23*, 538–541. [CrossRef]
12. Jeong, S.; Simeone, O.; Kang, J. Mobile cloud computing with an uav-mounted cloudlet: Optimal bit allocation for communication and computation. *IET Commun.* **2016**, *11*, 969–974. [CrossRef]
13. Jeong, S.; Simeone, O.; Kang, J. Mobile edge computing via a uav-mounted cloudlet: Optimization of bit allocation and path planning. *IEEE Trans. Veh. Technol.* **2017**, *67*, 2049–2063. [CrossRef]
14. Zhou, F.; Wu, Y.; Sun, H.; Chu, Z. Uav-enabled mobile edge computing: Offloading optimization and trajectory design. In Proceedings of the IEEE Intern. Conference on Communications, Kansas City, MO, USA, 20–24 May 2018; pp. 1–6.
15. Zhou, F.; Wu, Y.; Hu, R.Q.; Qian, Y. Computation rate maximization in uav-enabled wireless-powered mobile-edge computing systems. *IEEE JSAC* **2018**, *36*, 1927–1941. [CrossRef]

16. Wang, Q.; Guo, S.; Wang, Y.; Yang, Y. Incentive Mechanism for Edge Cloud Profit Maximization in Mobile Edge Computing. In Proceedings of the ICC 2019–2019 IEEE International Conference on Communications (ICC), Shanghai, China, 20–24 May 2019; pp. 1–6.
17. Zhang, T. Data offloading in mobile edge computing: A coalition and pricing based approach. *IEEE Access* **2017**, *6*, 2760–2767. [CrossRef]
18. Vamvakas, P.; Tsiropoulou, E.E.; Papavassiliou, S. Dynamic Spectrum Management in 5G Wireless Networks: A Real-Life Modeling Approach. In Proceedings of the IEEE INFOCOM 2019-IEEE Conference on Computer Communications, Paris, France, 29 April–2 May 2019; pp. 2134–2142.
19. Vamvakas, P.; Tsiropoulou, E.E.; Papavassiliou, S. On the prospect of uav-assisted communications paradigm in public safety networks. In Proceedings of the IEEE INFOCOM WKSHPS: WCNEE 2019: Wireless Communications and Networking in Extreme Environments, Paris, France, 29 April 2019; pp. 1–6.
20. Xiao, L.; Liu, J.; Li, Y.; Mandayam, N.B.; Poor, H.V. Prospect theoretic analysis of anti-jamming communications in cognitive radio networks. In Proceedings of the 2014 IEEE Global Communications Conference, Austin, TX, USA, 8–12 December 2014; pp. 746–751.
21. Thanou, A.; Tsiropoulou, E.E.; Papavassiliou, S. Quality of experience under a prospect theoretic perspective: A cultural heritage space use case. *IEEE Trans. Comput. Soc. Syst.* **2019**, *6*, 135–148. [CrossRef]
22. Hardin, G. The tragedy of the commons. *Science* **1968**, *162*, 1243–1248. [PubMed]
23. Lewandowski, M. Prospect theory versus expected utility theory: Assumptions, predictions, intuition and modelling of risk attitudes. *Cent. Eur. J. Econ. Model. Econom.* **2017**, *9*, 275–321.
24. Kahneman, D.; Tversky, A. Prospect theory: An analysis of decision under risk. In *Handbook of the Fundamentals of Financial Decision Making: Part I*; World Scientific: Singapore, 2013; pp. 99–127.
25. Vamvakas, P.; Tsiropoulou, E.E.; Papavassiliou, S. On controlling spectrum fragility via resource pricing in 5g wireless networks. *IEEE Netw. Lett.* **2019**, *1*, 111–115. [CrossRef]
26. Vamvakas, P.; Tsiropoulou, E.E.; Papavassiliou, S. Risk-Aware Resource Control with Flexible 5G Access Technology Interfaces. In Proceedings of the 2019 IEEE 20th International Symposium on "A World of Wireless, Mobile and Multimedia Networks" (WoWMoM), Washington, DC, USA, 10–12 June 2019; pp. 1–9.
27. National Data Catalog. Available online: https://catalog.data.gov (accessed on 10 April 2020).
28. Vamvakas, P.; Tsiropoulou, E.E.; Papavassiliou, S. Risk-aware resource management in public safety networks. *Sensors* **2019**, *19*, 3853. [CrossRef]
29. Tsiropoulou, E.E.; Vamvakas, P.; Papavassiliou, S. Joint customized price and power control for energy-efficient multi-service wireless networks via S-modular theory. *IEEE Trans. Green Commun. Netw.* **2017**, *1*, 17–28. [CrossRef]
30. Zhang, Y.; Mohsen, G. *Game Theory for Wireless Communications and Networking*; CRC Press: Boca Raton, FL, USA, 2011.
31. Vives, X. Games with strategic complementarities: New applications to industrial organization. *Int. J. Ind. Organ.* **2005**, *23*, 625–637. [CrossRef]
32. Apostol, T.M. *Calculus, Vol. 1: One-Variable Calculus, with an Introduction to Linear Algebra*, 2nd ed.; Blaisdell: Waltham, MA, USA, 1967.
33. Fudenberg, D.; Tirole, J. *Game Theory*; MIT Press: Cambridge, MA, USA, 1993.
34. Milgrom, P.; Roberts, J. Rationalizability, learning, and equilibrium in games with strategic complementarities. *Econom. J. Econom. Soc.* **1990**, *58*, 1255–1277. [CrossRef]
35. Topkis, D.M. *Supermodularity and Complementarity*; Princeton University Press: Princeton, NJ, USA, 1998.
36. Apostolopoulos, P.A.; Tsiropoulou, E.E.; Papavassiliou, S. Cognitive Data Offloading in Mobile Edge Computing for Internet of Things. *IEEE Access* **2020**, *8*, 55736–55749. [CrossRef]
37. Singhal, C.; De, S. (Eds.) *Resource Allocation in Next-Generation Broadband Wireless Access Networks*; IGI Global: Hershey, PA, USA, 2017.
38. Hota, A.R.; Garg, S.; Sundaram, S. Fragility of the commons under prospect-theoretic risk attitudes. In *Games and Economic Behavior*; Elsevier: Amsterdam, The Netherlands, 2016; Volume 98, pp. 135–164.

Article

Edge Computing Resource Allocation for Dynamic Networks: The DRUID-NET Vision and Perspective

Dimitrios Dechouniotis [1,*], Nikolaos Athanasopoulos [2], Aris Leivadeas [3], Nathalie Mitton [4], Raphael Jungers [5] and Symeon Papavassiliou [1]

[1] School of Electrical and Computer Engineering, National Technical University of Athens—NTUA, 15780 Zografou, Greece; papavass@mail.ntua.gr
[2] School of Electronics, Electrical Engineering and Computer Science, Queen's University of Belfast—QUB, Belfast BT7 1NN, UK; n.athanasopoulos@qub.ac.uk
[3] École de Technologie Supérieure (ÉTS Montreal) | Université du Québec, Montreal, QC H3C 1K3, Canada; Aris.Leivadeas@etsmtl.ca
[4] Inria Lille-Nord Europe, Lille, 59650 Villeneuve d'Ascq, France; nathalie.mitton@inria.fr
[5] Institute of Information and Communication Technologies, Electronics and Applied Mathematics, Université Catholique de Louvain—UCLouvain, 1348 Louvain-la-Neuve, Belgium; raphael.jungers@uclouvain.be
* Correspondence: ddechou@netmode.ntua.gr; Tel.: +30-210-772-1449

Received: 23 March 2020; Accepted: 10 April 2020; Published: 14 April 2020

Abstract: The potential offered by the abundance of sensors, actuators, and communications in the Internet of Things (IoT) era is hindered by the limited computational capacity of local nodes. Several key challenges should be addressed to optimally and jointly exploit the network, computing, and storage resources, guaranteeing at the same time feasibility for time-critical and mission-critical tasks. We propose the DRUID-NET framework to take upon these challenges by dynamically distributing resources when the demand is rapidly varying. It includes analytic dynamical modeling of the resources, offered workload, and networking environment, incorporating phenomena typically met in wireless communications and mobile edge computing, together with new estimators of time-varying profiles. Building on this framework, we aim to develop novel resource allocation mechanisms that explicitly include service differentiation and context-awareness, being capable of guaranteeing well-defined Quality of Service (QoS) metrics. DRUID-NET goes beyond the state of the art in the design of control algorithms by incorporating resource allocation mechanisms to the decision strategy itself. To achieve these breakthroughs, we combine tools from Automata and Graph theory, Machine Learning, Modern Control Theory, and Network Theory. DRUID-NET constitutes the first truly holistic, multidisciplinary approach that extends recent, albeit fragmented results from all aforementioned fields, thus bridging the gap between efforts of different communities.

Keywords: edge computing; internet of things; mobile robots; resource allocation; control co-design

1. Introduction

The Internet of Things (IoT) consists of low-cost efficient sensors, actuators, and computing units and provides great benefits to people to synthesize a system of interrelated computing, sensing, and communication devices that facilitates and improves everyday life, in cities and industry. IoT is foreseen to reach 500 billion devices that are connected to the Internet by 2030 [1], while the global mobile traffic is expected to increase sevenfold by 2021 [2]. Though significant improvements have been obtained in terms of hardware advances and processing capabilities at the device level; still, in most cases, IoT devices (e.g., smart devices, sensors, actuators, mobile agents) cannot meet, and more importantly cannot guarantee the required high performance and/or fulfillment of time constraints, for time-critical and mission-critical IoT-enabled applications. Thus, offloading computation and energy intensive tasks to powerful computing infrastructure for further processing becomes of vital importance.

The success of the computation offloading, and consequently the performance of IoT-enabled applications, depends on many contextual parameters, e.g., the user's mobility, various wireless parameters, and the resource availability of the computing resources in the data center. Most of the modern IoT-enabled applications rely on continuously moving people or mobile agents. Regarding the latter, various types of autonomous mobile agents or unmanned vehicles are used. Typical examples of these agents are the unmanned aerial vehicles (UAV), which are widely used in several human activities in the context of smart city, agriculture, area surveillance, rescue missions, and event coverage [3]. UAVs can be used individually or in a swarm, and they are equipped with various sensors in order to complete a mission or to execute their own tasks, such as trajectory planning and positioning. Their limited computing resources and energy reserves do not allow local data processing. Thus, the data offloading seems the only viable solution for using massively UAVs in various daily scenarios. Data are transmitted through wireless links, i.e., cellular or WiFi, and the quality of the wireless connection heavily depends on signal strength, interference, packet dropouts, and other parameters related to the wireless environment, which must be considered in the offloading decision.

The computation offloading aims to save time and energy at the end user's side. Cloud computing seems the natural selection for offloading, as it is the prevalent service delivery model nowadays. However, the high network delay for sending data over public internet counterbalances the benefits of the powerful computing resources that are available at a cloud data center. Accordingly, Multi-Access Edge Computing (MEC) [4] and Fog Computing [5] have arisen as promising approaches to overcome this obstacle and provide the benefits of cloud computing in the proximity of the end-users. Over the last few years, powerful UAVs have been considered as a means to provide computing support to the end-users by acting as UAV-mounted MEC servers [6]. In that respect, the UAV-mounted MEC servers in combination with ground MEC servers collectively create a fog computing system [7], supporting end-users' applications' task offloading. Similarly, the use of clusters of UAV-mounted MEC servers is suggested [8], allowing the opportunistic task offloading to the neighboring UAV clusters with sufficient computing resources. In such a UAV-assisted network, computing intensive tasks are offloaded and executed in a nearby small-size edge data center, either directly connected with a wireless access point, or it is embedded on the UAV itself. The key difference between cloud and edge data center is that the latter has a finite amount of computing resources, which requires fine-grained resource management towards meeting the strict constraints of the deployed time- and mission-critical applications.

The DRUID-NET Perspective and Contributions

This article presents the vision and perspective of the DRUID-NET (eDge computing ResoUrce allocatIon for Dynamic NETworks) framework, along with a detailed description of its main concepts and objectives. While considering the end-user's mobility and the parameters of the wireless connection environment, the DRUID-NET framework aims at developing workload profile holistic and modular dynamic performance models of IoT-enabled applications based on the appropriate theoretical tools. Furthermore, the article aims to outline the control principles of novel resource management systems for these kinds of applications. In particular, the key research threads and topics of this article are summarized as follows:

-   **Workload Profile**: The IoT applications generate time-varying traffic in terms of request size and number of flows. Additionally, the involved wireless communication between the mobile devices and the back-end software components introduces additional uncertainties that considerably affect the offloading decision and the resource scheduling at the edge computing infrastructure. Contrarily to existing average traffic characteristics, building dynamic traffic profiles and prediction mechanisms will enable more accurate, adaptive, and successful data offloading and resource allocation mechanisms.

- **Performance Modelling**: Most of the existing models for describing the performance of IoT-enabled applications are empirical and usually focus on a specific performance metric, e.g., response time, throughput, energy consumption, etc. These models cannot adequately capture the dynamic nature of the emerging applications, which in turn leads to either performance degradation or resource over-provisioning. On the other hand, the DRUID-NET framework proposes formal dynamic multi-input multi-output performance models applicable to various IoT applications. These types of models enable the design of novel controllers for finer regulation of Quality of Service (QoS) metrics.

- **Resource Allocation**: Usually, most studies in the literature combine a static performance model with solving an optimization problem. However, this approach assumes that the workload does not vary significantly, which limits their validity, applicability, and exploitability. In contrast to these approaches, we envisage to design stabilizing controllers in order to guarantee the feasibility of the resource scheduling and the performance requirements.

- **Control co-design** The DRUID-NET framework examines specifically the case where the application is the controller design and its implementation for dynamic processes. In this setting, the performance of the closed-loop system is considered in the overall application performance. The control co-design approach aims to design feedback mechanisms achieving closed-loop system properties such as reachability, stabilization, and other complex specifications, and simultaneously design and implement resource allocation algorithms for the dynamic network.

The rest of the article is organized as follows. In Section 2, the current state of the art is presented. Section 3 demonstrates the conceptual architecture of the DRUID-NET framework, while Section 4 describes three IoT-enabled use cases where the proposed solution is applicable. Finally, Section 5 draws the conclusions and future directives of our research.

2. Related Work and Motivation

This section provides a thorough yet comprehensive presentation of the most relative studies to the DRUID-NET framework, in the recent literature. Aligned with the DRUID-NET objectives, Abdelzaher et al. [9] presented five challenges on IoT applications and Edge Computing. This study focused mostly on deep learning-based application modeling, optimal offloading, closed loop guarantees, and collaborative offloading. Towards these directions, the related work is categorized under three major classes; (i) IoT workload profile, (ii) performance modeling and resource allocation, and (iii) control co-design.

2.1. IoT Workload Profile

The estimation of the workload and communication patterns in IoT-Fog/Edge networks has only been explored a little due the high heterogeneity of co-existing devices. Nevertheless, there is no doubt that the proper estimation of the offered workload and communication patterns could lead to a more efficient utilization of the underlying infrastructure.

Authors in [10] considered a two-tier network architecture consisting of shallow and deep cloudlets, and explored the benefits of hierarchical capacity provisioning based on queuing analysis. Although shown to be efficient in very specific cases, this approach cannot be generalized in principle. Osmotic Computing [11] relied on the deployment of lightweight microservices on resource-constrained IoT platforms at the network edge, coupled with more complex microservices running on large-scale datacenters. MobiQoR [12] introduced a new metric, Quality of Results, to validate the quality of edge resource deployment. Nevertheless, none of these approaches attempted to estimate the IoT workload, which in turn could significantly enhance the corresponding deployments. The authors of [13] and subsequently of [14] analyzed the resource allocation of a three-layer infrastructure (IoT, Edge, Cloud) under dynamic network conditions. However, they took into consideration the dynamic opt-in and out of IoT devices into the network, while ignoring their instantaneous workload generation. To the best of our knowledge, the only attempts to estimate workload are referring to the cloud utilization [15,16], and as such they did not capture the locality of the heterogeneous IoT traffics.

A promising approach to derive workload profile is to use machine learning techniques. Applying deep or machine learning techniques for IoT applications is not new [17], but most of the time they are centralized and do not need any adaptation to fit specific IoT devices limitations. In DRUID-NET, we will rely on existing estimation methods, such as [18], to estimate the workload of hardware constrained devices. In existing works, the focus is placed on one specific resource each time (e.g., energy, memory, computing, etc.) [19]. The DRUID-NET framework aims at extending them to multiple resources, while combining these approaches with predictive methods, which have only been slightly explored for IoT due to resource limitations. Thus far, methods such as ARIMA [20], deadreckonning [21], Kalman filters [22], Thompson sampling [23], or Bayesian approaches [24] have mainly been investigated for navigation and position prediction [20], data reduction [24], link prediction [25], or medium occupation [23]. Our aim is to provide a unique distributed and adaptive multi-resource estimation and prediction suitable for IoT devices. The DRUID-NET goal is to derive some communication patterns, clearly defined in time and size, towards assessing the need in edge resources in time and space. This edge-resource sizing combined with performance modeling, controlled mobility of edge-resource and resource allocation, will enable the adaptive deployment of sufficient resources, on demand, and in an efficient manner.

2.2. Performance Modeling and Resource Allocation in Cloud and Edge Computing

Resource allocation has become one of the most important open research problems in Cloud and Edge computing and IoT. In the cloud computing environment, the computing resources are assumed to be infinite; thus, static or empirical models combined with coarse resource scheduling techniques have been shown sufficient to provide high performance through over-provisioning. However, these approaches are neither optimal nor able to provide QoS guarantees. Regarding application's performance modeling, the empirical or fixed models considered already known request sizes and execution times, which are not only hardware-specific, but generally very difficult to be precisely computed. Furthermore, many studies relied on queuing models [26], e.g., $G/G/1$ or $G/G/n$, which are reliable only for steady state. It is obvious that this kind of modeling cannot capture transient phenomena due to dynamic workload demand. With this capacity, System Theory [27] can provide dynamic modeling methodologies, appropriate for Cloud/IoT-based applications. The interesting reader may refer to survey [28] for an extended analysis of control theoretic approaches on performance modeling and cloud elasticity. Close to DRUID-NET concepts, Dechouniotis et al. [29] proposed Linear Parameter Varying (LPV) modeling of cloud applications combined with set-theoretic controllers to guarantee a feasible solution of the elasticity in cloud data centers, while Leontiou et al. [30] derived fuzzy Takaki–Sugeno models and designed robust controllers to address simultaneously the problems of vertical and horizontal scaling, and load balancing with stability guarantee.

Contrarily to cloud computing, the resources of edge computing are rather limited; thus, static allocation techniques cannot achieve optimal resource utilization. Furthermore, modern time- and mission-critical IoT-enabled applications [31,32] have strict performance requirements that only dynamic modeling and intelligent allocation algorithms can guarantee. Similarly to cloud, in the edge computing context, most of the relative studies proposed static models alongside with the optimization of a single performance criterion, e.g., energy consumption or response time. Towards this direction, Sonmez et al. [33] proposed a two-stage fuzzy mechanism for offloading requests to edge and cloud infrastructure. The set of fuzzy rules are empirically decided and the VM (Virtual Machine) utilization modeling is threshold-based, which is applicable only for specific types of IoT applications. Queec [34] formulated the problem of scheduling multi-user tasks to multiple edge nodes as an optimization problem which minimizes the overall offloading latency of all tasks. Jalali et al. [35] analyzed fixed flow-based and time-based energy consumption models, and they presented a detailed comparison on energy consumption between cloud and edge computing systems under various network settings. Lyu et al. [36] presented a collaborative Cloud-MEC-IoT architecture and proposed a request modeling scheme and an admission control framework to address the scalability problem of these platforms. Although the authors considered heterogeneous edge resources, the computation model was not dynamic. The authors of [37] addressed both the problems of network selection and service placement for MEC infrastructure. Towards the reduction of the complexity of the general problem, they decomposed it into a series of sub-problems and solved them in an iterative fashion. However, the proposed performance model focused only on network related parameters ignoring the processing time of the application.

In the 5G era, Network Functions Virtualization (NFV) and Software Defined Networks (SDN) play key roles for the realization of many type of verticals, which are comprised of several IoT applications. In this context, virtualized and isolated Service Chains (SCs) comprised of a series of Virtualized Network Functions (VNFs) implemented as VMs need to be deployed in the available MEC infrastructure to offer networking services to the IoT traffic. Normally, the objective of this kind of resource allocation mechanism aims to minimize the overall deployment cost (e.g., the computational and communication resources that an SC needs in order to be provisioned) [38]. Another common approach is to minimize the overall delay, since several IoT applications are characterized as mission critical and delay sensitive. Thus, a valid approach is to utilize the MEC resources that are closer to the IoT devices [39]. An alternative approach to minimize the delay is to create resource clusters inside the MEC infrastructure, where the various requested SCs can be deployed [40]. Minimizing the number of clusters and appropriately positioning the VNFs can lead to a reduction of the communication delay. Efforts have also been dedicated to optimize the energy consumption. The authors of [41] modeled the energy dissipation of the resources in the IoT and MEC infrastructures and constructed a Linear Programming algorithm to carefully select the resources to place the SCs. Another objective focuses on the optimal allocation and scheduling of the available edge resources. This objective can be translated into either: (a) minimizing the overall resource usage, to enable multiple heterogeneous SCs, servicing heterogeneous IoT applications, to co-exist in the MEC layer [42], or (b) minimizing the resource idleness of the infrastructure [43]. Load balancing can also be applied by minimizing the maximum link utilization and reducing the bandwidth consumption [44]. This can be achieved by adopting appropriate queuing and QoS modeling during the optimization problem to minimize the resource utilization [45]. Even though all the above solutions target valid and open challenges of resource allocation in the IoT/MEC, they only propose static approaches failing to provide a holistic mechanism that takes into consideration a multi-objective and dynamic solution. Following the performance modeling and control design principles of [9,46], DRUID-NET aspires to provide multi-variable dynamic models and design modern control methodologies that ensure the desired user's performance requirements and optimize the utilization objectives of the infrastructure provider simultaneously.

2.3. Control-Theoretic Resource Allocation and Control Co-Design

In control theory, the effect of a shared, imperfect communication network between the controller and the sensor/actuator network has been studied extensively for almost three decades, generating the separate branch of Networked Control Systems (NCS), Ref. [47,48]. NCS suffer from many non-idealities. For instance, networked induced delays or, even worse, packet dropouts occur, as the information from the sensor to the controller or from the controller to actuator(s) can be lost in a time interval. Moreover, due to the limited energy available at decentralized nodes, bandwidth can be low, so that the effect of quantization in the communication channels may not be neglected. In addition, switching or hybrid phenomena may occur due to the asynchrony between disconnected agents, or due to event-triggered strategies. Finally, the computational problem, to be performed at the nodes, may be part of a global optimization problem, which is split into decentralized subtasks.

Several methods have addressed these non-idealities separately. Time delays, for instance, have been tackled utilizing perturbation theory, Lyapunov stability theory, and hybrid systems analysis, but also probabilistic methods involving Markov chains and stochastic automata [49]. Quantization problems have led to a rich literature, where the controllability of a plant subject to quantized control is ruled by the so-called *entropy* of the system [50]. From the hybrid control point of view, researchers from real-time computing have dealt with the schedulability problem of distributed control settings, leading to the design of several protocols for a stable closed-loop behavior, [51–53]. Decentralised computation/optimization has been another major topic of research in Systems and Control [54]. Here, though the state of the art is rich, the interaction of this constraint with others is not well understood and studied. Let us note however that the consensus problem has been deeply studied, in many settings, e.g., quantized communications [55].

Additionally to the stochastic results [56], recent theoretical work on the controllability and observability properties of the NCS [57] has shown that a more refined modeling of the communication network allows the proper definition and verification of such properties, thus adding new tools to the NCS community. Furthermore, proof-of-concept work has shown that, under a new modeling framework for hybrid systems and specifically constrained switching systems [58], the control performance can be directly associated with the network quality [59].

Rather than designing the control and communication protocol in two steps, co-design methods aim to synthesize simultaneously controllers and the communication patterns (sampling, delays, scheduling protocols). Applied only to networked control systems with constrained communication resources so far, co-design methods have been extensively studied the last decade [60–63]. Perhaps the most relevant breakthrough in this area is the emergence of event-triggered and self-triggered control mechanisms that allow asynchronous sampling, thus reducing the network traffic, while at the same time behaving sub-optimally [64–66].

Nevertheless, there is limited work on the co-design of controllers taking into account simultaneously more than one phenomena (schedulability, network utilization, edge resource utilization, energy consumption, etc.) caused by the distribution of computing and communication resources. It is anticipated that the research developments in the upcoming decades will allow for encapsulating, comparing, and subsequently altering the impact of the several non-idealities, and this in turn will have a significant impact on future control applications, where resources must be used parsimoniously, in balance with the constraints and the overall considered objective. This will require and motivate new paradigms in Systems and Controls, where multi-objective optimization, model-free (data-driven) approaches, approximate optimality (however with firm safety guarantees), reconfigurability, and resilience take a central place.

3. DRUID-NET Conceptual Architecture

Figure 1 illustrates a high-level overview of the overall DRUID-NET framework. The architecture follows the NFV/SDN paradigm and separates the flow of information into control and data planes. At the lowest layer, the IoT applications are deployed, and the generated workload (data flow) can be offloaded for further processing at the upper level of Edge Computing. In this layer, any component of the application is provided as a virtualized service. As it is shown in the figure, a virtualized service corresponds either to IoT specific functionalities, e.g., path planning and image recognition, or control components such as learning algorithms or optimization solvers. The modeling and control framework collects information (control flow) about the status of the computing and network infrastructure at the edge computing level in order to create workload-resource profiles, update the performance model for every application, and realize the feedback control mechanism for the resource allocation, while simultaneously implementing a resource-aware control strategy for the cyber-physical system to be controlled (control flow). This holistic approach allows the application's dynamical modeling taking various contextual information into account. Furthermore, the controller co-design treats the resource allocation algorithms as application components in the virtualized services. Each major component of the modeling and control framework is described in more detail in the following subsections.

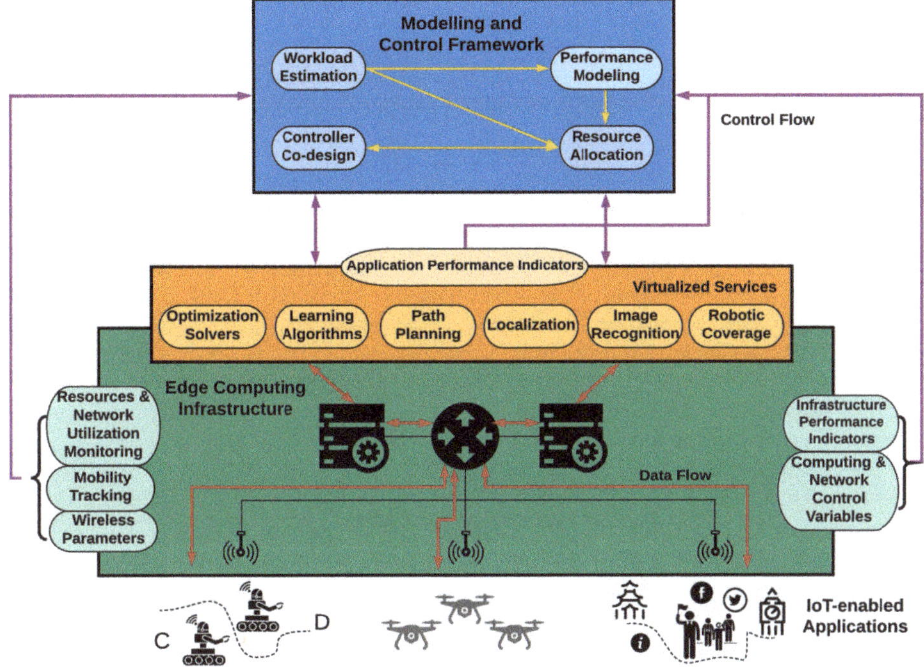

Figure 1. Conceptual architecture.

3.1. IoT Workload Profiling

As mentioned before, a major challenge for solving the resource allocation problem in edge computing settings is to predict the time-varying characteristics of the workload/traffic, as different traffic flows and volatile conditions can influence significantly the resource allocation mechanism. Aspects such as the load generated from an IoT device, latency specifications, the transmitting data frequency, the wireless protocol, the mobility of the devices, and the number of devices associated with the IoT gateway, change the amount of resources requested from the edge, while also influencing the scheduling process. Until now, only generic traffic models have been proposed to estimate the traffic aggregated at the edge layer, while stationary IoT devices are assumed, leading to a static rate model, which however limits its effectiveness and applicability in real scenarios.

Going a step beyond from the pertinent literature, which only considers average and general traffic characteristics of the IoT applications (e.g., Brownian motion as one-fits-all model), DRUID-NET framework aims to differentiate and categorize the requirements of various IoT applications using appropriate data analytic and mathematical models. In particular, we classify and categorize the IoT applications by leveraging the transmission patterns, the spatial and temporal correlation of the traffic, as well as other traffic related characteristics such as the frame size distribution, and the burstiness of the traffic of the IoT applications. The novelty of this approach is that we create prediction mechanisms to treat the dynamics and uncertainty in the corresponding traffic profiles. Each predictive mechanism targets specific categories of IoT applications with similar requirements and characteristics to define the type, the size, as well as the time and the location of the requested resources. Furthermore, with this approach, we can dispose the erroneous assumption that specific tasks are associated with static and pre-specified resource footprints. In contrast, we replace this analogy with an opportunistic association between the requested resources and the IoT traffic dynamicity, thus introducing a holistic mechanism inspired by data analytics, and traffic analysis methods.

3.1.1. IoT Applications Classification

A first classification of the IoT applications can be produced by simply answering yes or no, to questions regarding the involved "things"/devices. Indicative such questions can be identified as follows: (i) Are the devices heterogeneous? (ii) Are they battery-powered? (iii) Are they sending data with high or low frequency? (iv) Are they data rich (e.g., multiple number of sensor measurements)? (v) Are the devices mobile?

The answer to such questions will help us to create a first clustering of the IoT applications. These clusters will contain IoT applications with similar device characteristics and behavior. Nonetheless, this first-phase categorization does not necessarily mean that the IoT applications belonging in the same cluster will present the same exactly resource requirements at the Edge. The reason is that different network access technologies can significantly affect the network requirements of the IoT applications. For example, different access technologies (e.g., LoRaWAN, Wi-Fi, IEEE 802.15.4, cellular, etc.) have different characteristics in terms of packet length, transmission range supported, MAC mechanisms, topological characteristics of the associated IoT devices (e.g., star, mesh, peer-to-peer), number of device connections supported, etc.

Thus, DRUID-NET takes into consideration both the functional and network requirements of the IoT applications in order to provide a complete and realistic IoT application classification.

3.1.2. IoT Applications Workload Prediction

The above categorization will help us to extract the workload generated from each cluster of applications in terms of bandwidth, latency, and other important Key performance Indicators (KPIs) during the offloading of IoT tasks to the Edge. Specifically, through this approach, we can propose appropriate mathematical models to simulate the traffic behavior of the various IoT applications. Nonetheless, even with this modeling, a lot of ambiguity will exist. The reason is that IoT access networks include several uncertainties, usually being wireless, lossy, and unreliable. Hence, the goal of DRUID-NET is not only to categorize and classify IoT applications based on their traffic profiling, but to also apply network analytics to make the communication as deterministic as possible.

Our goal is to replace the so far average estimates of the IoT applications with instantaneous and accurate transmission metrics. To this end, appropriate machine learning algorithms (i.e., Thompson, ARMA, Bayesian) need to be integrated in the traffic profiling in order to learn and predict the network conditions between the IoT and the Edge. This can be decisive in the performance of the subsequent resource allocation at the Edge. The Edge controller will be able to adapt to and predict the changing workload arriving at the Edge infrastructure, creating a holistic and realistic resource allocation approach.

3.2. Performance Modeling

The available resource models are usually single-input single-output. Energy or response time are typically the model's outputs, while computing resources (e.g., CPU, memory), incoming requests, and network bandwidth are the control variables. In most of the current studies, the relation between input and output is fixed and empirically derived. For example, the processing time of a request is proportional to its file size and inversely proportional of the service rate measured in CPU cycles or millions of instructions per second. Although this assumption is reasonable, the actual processing time depends on several time-varying parameters, which are not easily measured. Furthermore, in combination with static resource allocation mechanisms, the offloading decision performs adequately only for specific operating conditions, being unable to guarantee stability under fluctuating workload and heterogeneous IoT communication infrastructure.

Contrary to current approaches that provide empirical static models, we aim to develop formal, realistic, and dynamic traffic and resource models applicable to emulate the generated traffic from various IoT applications. For this purpose, DRUID-NET adopts hybrid dynamical models [67] that have the capacity to include several performance metrics (i.e., state variables) and resources as control parameters (input variables). This type of modeling takes into account in a single formulation the various contributions of the diverse objectives and constraints to the performance/cost. This framework moreover allows for discovering the trade-offs between accuracy, complexity of representation, and real-time feasibility of the resource allocation strategy. Furthermore, the chosen framework will be capable of capturing structural changes interpreted as discrete jumps in the dynamics, e.g., user mobility, change in wireless protocols and topology, and addition/removal of edge servers. Finally, alongside with the dynamic models, the DRUID-NET framework aims to identify the uncertainties of these models and quantify their boundaries in order to facilitate the design of the respective control laws.

3.3. Resource Allocation

The workload profile estimator and the dynamic model of the resources and the overall status of the network/servers provide the foundation upon which the resource allocation algorithm will be developed. Specifically, the objective is to develop a joint communication, computing and storing virtualization paradigm that is updated and adapted dynamically. For this purpose, we consider the problem of simultaneously (i) allocating storage, computing and communication resources, (ii) modifying network topology/ protocol, and (iii) structuring the edge computing data centres (such as VM distribution). Two distinct approaches relating to static and dynamic resource allocation are considered.

3.3.1. Static Resource Allocation

In this approach, we do not take into account the dynamic nature of the processes under study; however, we consider the full resource allocation problem. The method is oriented towards solving multi-objective optimization problems fast that will in turn provide the optimal operating point for the communication network, and the computing and storage allocation in the edge/cloud servers. Our goal is to describe the complex interrelations between the aforementioned resources in an analytic manner, merging available models, e.g., from queuing theory and Markov models. Next, we plan to solve the optimization problems using mixed-integer, linear, and nonlinear programming. Since the complexity of these problems does not allow often exact real-time solutions, our intention is to propose approximate solution algorithms that provide guarantees of the level of suboptimality of the identified solution. Additionally, we will employ machine learning algorithms to relax the complexity of these highly nonlinear/nonconvex problems so they can be solved in real-time, thus respecting hard time constraints. This approach will focus on problems involving complex specifications and mostly static models, aiming to maximize the QoS delivered.

3.3.2. Dynamic Resource Allocation

In this approach, the DRUID-NET framework proposes dynamic control-theoretic resource allocation mechanisms. Utilizing the models established by capturing the performance metrics dynamics in a hybrid dynamical system, our goal is to follow control-theoretic approaches that provide formal guarantees on important properties describing the resource allocation problem. For example, a main objective is to provide guarantees of the speed of convergence of the performance metrics to a pre-determined range, defined by translating the QoS requirements to mathematical statements. Moreover, our goal is to provide decision mechanisms that allow structural changes in some cases (for example, turning on and off edge servers in a cluster, changing the topology in a communication network), together with continuous strategies (such as CPU and memory utilization in a server). The natural, main challenge in this approach is the scalability of the decision algorithm, which will be tackled by proposing smart allocation strategies that allow trade-offs between performance and real-time implementation. Another challenge is to establish resource allocation mechanisms using only partial information, which is the most realistic scenario. This issue will be addressed by proposing distributed control mechanisms that take continuously into account local information and receive only intermittently information about the states of the whole system.

3.4. Co-Design of Controllers

As we have mentioned before, in the broad field of Systems and Control, several different paradigms have emerged in the last few decades, to deal with the control of IoT-enabled cyber-physical systems. Indicative examples include hybrid behavior, quantized control, varying delays, safety-criticality, nonlinear control, etc. Although these challenges are typically met together in IoT environments, the research activities have led to disconnected communities, and likewise very specific and custom control techniques that limit their implementability in a holistic framework. In a real-life IoT control application, these non-idealities take place all together. We argue that the different paradigms separately introduced for each of these non-idealities are hard to reconcile, thus the DRUID-NET framework is devoted to deploying the theoretical results in actual applications.

Modern IoT applications need controllers that address a mixture of these undesired phenomena. Our goal is to establish a formal decision mechanism that will be able to change the provisioning of the resources in real time, adapt its control objective to the available bandwidth, weigh the cost of communication with respect to the advantage of involving decentralized agents, and eventually address a multitude of practical challenges appearing in networked, resource-constrained control applications. Such a new generation of controllers will be made possible by the merging of two sets of hybrid models, namely (a) the performance model having as internal variables performance metrics of the infrastructure and as inputs the resource distribution and utilization, and (b) the process model (having, for example, variables related to position, orientation, velocity and acceleration of mobile agents, lighting conditions, room temperature, mode of operation of sensors, etc). To provide an example of the challenges that will be met in this setting, let us raise the following question: what do traditional data-rate theorems from quantized control (see, e.g., [68]) become in an environment with packet losses and varying delays, and varying computational resources? Another important line of research that will be necessary to follow in a time-varying resources setting is to categorize and model the complexity of the control algorithms. Allowing their dynamic adjustment will eventually provide the coupling between the process/application to be controlled and the control algorithm resource provisioning. In turn, this will enable (i) the establishment of real-time control mechanisms with formal guarantees for the closed-loop system, and (ii) the optimal utilization of resources, either in the network or the edge.

4. IoT-Enabled Applications

The proposed architecture is generic enough offering a holistic paradigm, while its estimation, modeling, and control methodologies are applicable in several categories of IoT applications, such as the ones based on mobile agents (e.g., UAVs) or designed for crowded smart areas or emergency scenarios. The following subsections demonstrate three representative use cases of the DRUID-NET framework.

4.1. Collaborative Robotics

Collaborative robotics is a prerequisite for Industry 4.0, especially in the Industrial Internet of Things setting. The current trend is to produce and program robots that have the capability to work together, or in close proximity, to humans in a shared environment. Removing a physical (or virtual) cage from the robot brings many challenges, the most critical of which is guaranteeing safety/avoiding collision, without leading to an unsatisfactory performance, e.g., the robot working in a non-acceptable speed. The setting can be extended to the case where there are many robotic agents and humans sharing the factory floor, or any other indoor or outdoor environment, e.g., a logistics warehouse, an airport, swarm UAVs networks, etc. In all aforementioned cases, similar challenges appear, namely: (i) intermittent and noisy measurements of the position of the agents either by static sensors or sensors mounted on the robots, (ii) faulty wireless communication networks, (iii) stringent safety specifications as humans and robots move freely in the same environment, and (iv) time-critical specifications. These challenges become harder when the computing/storing/communication resources are limited, or not always available to a control application, which is the typical case. Thus far, a few approaches, aligned with the ones appearing in the cyber-physical systems control problems, take explicitly into account a part of these challenges, e.g., [69]. Controllers, which are co-designed with the resource allocation and computation offloading mechanisms, can be used for human–robot collaboration in the IoT-enabled environment or for real-time, large scale coordination of mobile robots. It should be noted that an additional control challenge in this case, additional to the presence of constraints, is the complex temporal specifications that need to be satisfied. Currently, the control objective has moved away from just ensuring stability or tracking for a prespecified set of reference trajectories, to satisfying statements, for example "robot A and B should collaborate towards a task X and eventually return to their initial positions if these positions are not occupied", as shown in Figure 2. These control applications are often time-critical as well as safety-critical; thus, a very careful co-design procedure should be developed for the controller that leads to formal guarantees without requiring many, possibly idle, resources.

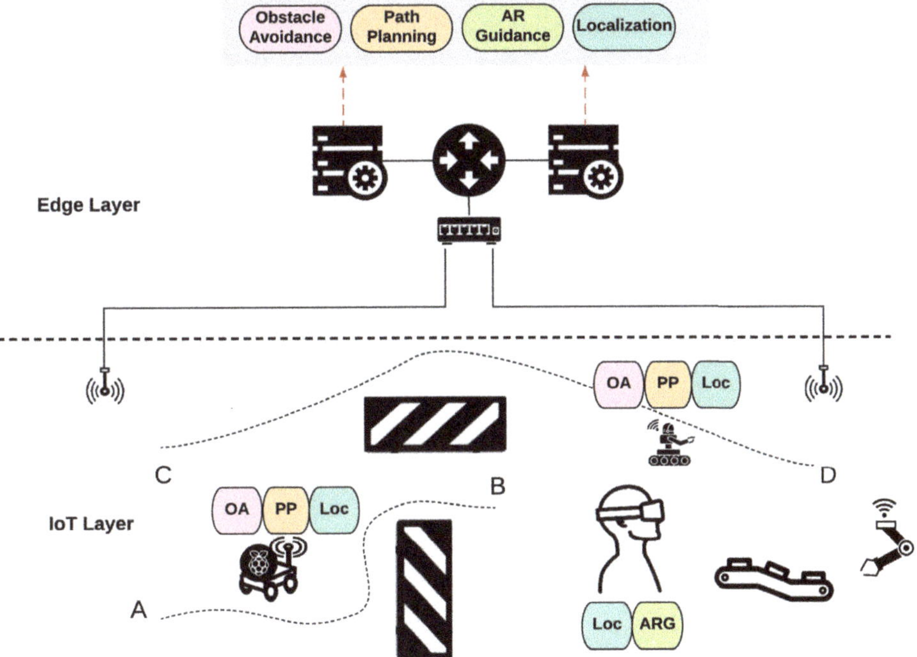

Figure 2. Human–Robot collaboration.

This scenario enables, and is enabled by, a combination of almost all components of the DRUID-NET framework, namely workload and resource estimation, and control co-design of a set of control applications in a platform where resources are shared and their availability is volatile.

4.2. Rapid Resource Deployment for Physical Disaster Scenarios

In the case of a physical disaster, the fixed communication infrastructure could be destroyed or unavailable due to high workload demand. Furthermore, for rescue operations, it could be critical to deploy additional on-demand computing and network resources at the proper place and time, in order to alleviate any remaining network infrastructure and collect data from remaining communicating devices such as mobile phones or sensors, towards helping to locate and rescue survivors. Mobile agents, especially UAVs, are suitable for these kinds of missions and can provide additional edge resources capable of processing the data at low latency and organizing the rescue operation. In order to serve the survivors devices as much as possible, there is a need to predict the kind and amount of resources these devices will request and the location of these resources. Some UAV-mounted edge resources may need to be deployed sporadically and temporarily at different locations based on IoT devices needs and mobility. Thus, there is a need to anticipate the deployment of edge services and to estimate the time they will be required at a given place to decide whether it is worth deploying durable edge resources, or instead mobile temporary resources could suffice. In this latter case, the estimation of the location and quantity of required resources should be anticipated to allow their timely deployment. The deployment of edge resources will be such that a maximum of IoT devices can be served within the required latency, either directly or through multi-hop communications. Direct communications will be favored for devices with very-low latency requirements, while multi-hop communications could be used for weaker latency requirements non-necessary communications. The trajectory of distributed UAV should be consciously planned accordingly, taking into consideration the time restrictions (robots should be deployed at the proper place before we need them).

Figure 3 illustrates the operation of the proposed framework under a physical disaster scenario, such as for example the occurrence of a gas leakage in a large factory. In this case, swarms of mobile robots will be deployed in order to find victims or survivors that require immediate medical assistance. Two types of mobile robots can be deployed, namely, (i) Unmanned ground vehicles (UGVs) and (ii) UAVs. UGVs will cover the ground area (x,y dimensions), while UaVs can provide a certain altitude coverage (z dimension) or coverage in non accessible areas by the UGVs (e.g., upper floors, atriums, etc.). Normally, we expect to find more obstacles in the ground area (e.g., offices, machines, shelfs), which can be translated in a higher number of UGVs in comparison with the UAVs (a ratio of 2:1), as shown in Figure 3. The goal of the interconnected UAVs is to locate living or dead persons, while at the same time send footage of the interior of the factory in order to create a 3D visualization of the area. In this manner, the users of the application (e.g., fire brigade) can immediately detect persons in need and send help to the corresponding location, eliminating the risk of long exposure to harmful gas for the rescuers. For the path planning of the mobile agents, the robots will be capable of detecting the Wi-Fi or LTE preamble and accordingly plan the route towards the source of the signal. The notion behind this behavior is that normally people have in close vicinity their mobile phones or other wireless devices (e.g., smart watches). This will facilitate the path planning and the pointless roaming of the UAVs in space. In order to prioritize the traffic and eliminate the impact of poor wireless communication, the swarm of robots can intensify the load of images/video and increase their quality only in areas with high probability of detecting a person, and send this traffic at the Edge for further processing. UGVs can approach the victims and sense if they are living or unconscious (e.g., detect eye movement, detect sound, etc.). When a UGV finds a survivor, it can communicate with the UAV of the swarm nearby, which in turn can lower its altitude close to the position of the person in need and drop an oxygen mask until help arrives.

In this case, using swarms of mobile robots will assist in eliminating the non-essential communications. Combining service differentiation and smart data offloading to UAVs, there will be reduction of any unnecessary communication between users and overhead due to extensive signaling. Since the number of available robots may still be inadequate to serve all ground services, the prioritization of the applications, flows, and devices is of paramount importance for the success of critical missions. Under these circumstances, the priority will be given to the areas with many victims or of major importance for the completion of the mission; thus, the available swarms of mobile robots should be distributed accordingly. Even in the case of homogeneous UGVs and UAVs with identical computing and networking capabilities, the optimal allocation of UAVs or UGVs formulates a dynamic optimization problem, which depends on the size of the damaged area, the communication ranges of both UGVs and UAVs, the flying altitude of UAVs, the propagation conditions, the data communication requirements (amount of data, frequency of collection, etc.) and the number and type of devices to serve. For example, if the victims are equally spread in different locations, the robot swarms would be equally scattered to deploy their resources in these areas so that the maximum number of devices would be served directly. On the other hand, the mobile robots will be driven to the most damaged area in order to serve the required network traffic.

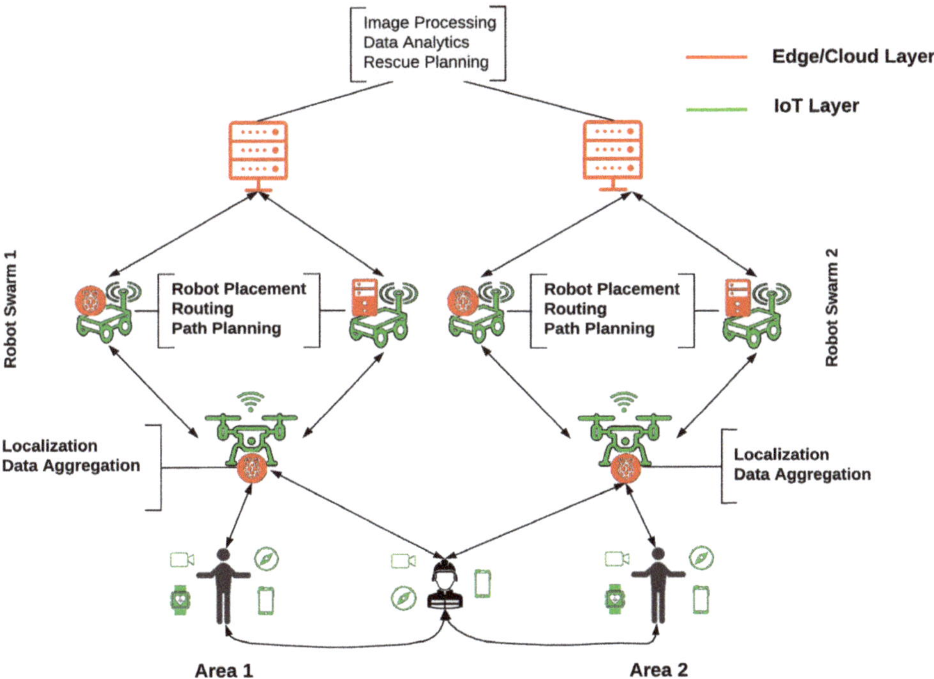

Figure 3. Rapid resource development for physical disasters.

This scenario illustrates the use and combination of the different control components of the DRUID-NET framework, and in particular: (1) workload estimation in quantity, time, and space, (2) resource allocation (tasks assignments to UAV and/or robots) and (3) path trajectory.

4.3. Mobility-Aware Edge Computing

Most of the modern smart city applications rely on mobile end-devices of continuously moving humans. Thus, the user's mobility is a dominant parameter of IoT systems. As shown in Figure 4, in the case of an urban touristic areas, e.g., museums and squares, the visitors collect information about Points of Interests (PoIs) (i.e., exhibits or social events) using their mobile devices. For example, leveraging the augmented or virtual reality technologies, they can retrieve media-enriched information about the surrounding PoIs. However, it is prohibited for the mobile end-devices with limited resources to run these types of applications locally. Thus, the edge computing infrastructure is essential to host the smart applications and meet the user's QoS requirements. Additionally, in crowded touristic areas, the number of visitors varies significantly during short-term (i.e., a day) or long-term periods (summer or winter); therefore, an accurate prediction methodology is important for optimal resource scheduling. Moreover, the offloading decision should be based on both the user's transmission capability and the availability of edge resources. With this capacity, in order to maximize the admittance of users, the main features of the overall generated traffic should be extracted alongside with patterns of the user's mobility. Then, utilizing the dynamic models, effective controllers can be designed towards the horizontal and vertical scaling of resources and the simultaneous guarantee of any QoS and Quality of Experiment (QoE) requirements.

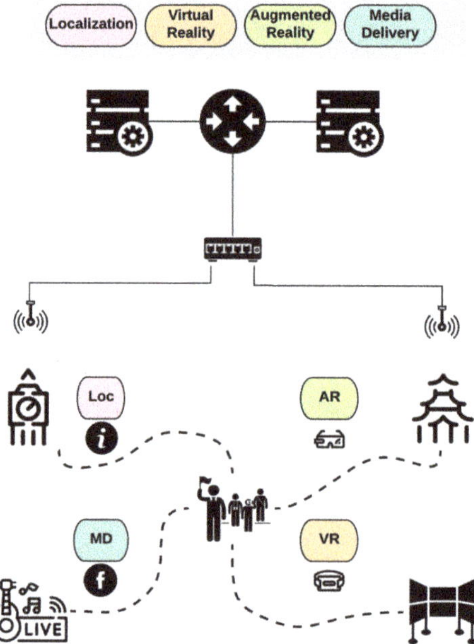

Figure 4. Mobility-aware edge computing.

This use case illustrates the necessity and the collaboration of the involved components of the DRUID-NET framework. Particularly, workload estimation, dynamic performance modeling, and resource allocation components interact to meet the respective requirements and optimize the resource utilization under varying workload conditions.

5. Conclusions

This article presents the most important challenges of IoT-enabled applications, along with the perspective and the basic concepts and objectives of the novel DRUID-NET framework. The corresponding components of the DRUID-NET framework are carefully designed to address several emerging challenges at any level of the IoT/Edge/Cloud system, stemming from mobile end-devices up to powerful cloud data servers.

In particular, the workload estimation aims to create a profile of IoT applications that includes features of the generated data, parameters of the wireless connection, and patterns of the user's mobility. The performance modeling components identify the multi-input multi-output dynamical systems that capture the dynamic operation of the applications, and are utilized to design the resource allocation and the offloading decision strategies. The resource allocation component is in turn responsible for deciding any control action at any level of the hierarchical system. Depending on the objectives of the controller, the resource allocation can be either static or dynamic, providing guarantees on QoS metrics, e.g., response time or energy consumption, and system properties, such as stability. Finally, the co-design of the controllers enables the binding between the IoT application and the resource control algorithm in order to provide guarantees for the closed-loop system.

The DRUID-NET framework aspires to verify its modular architecture and components through different IoT scenarios. These use cases are carefully selected in order to cover all main challenging and emerging aspects of the IoT applications and Edge computing paradigm.

Author Contributions: All authors contributed equally to conceptualization, investigation, and writing—original draft. All authors have read and agreed to the published version of the manuscript.

Funding: Nikolaos Athanasopoulos, Aris Leivadeas, Nathalie Mitton, and Raphael Jungers are partially supported by the CHIST-ERA-2018-DRUID-NET project.

Conflicts of Interest: The authors declare no conflict of interest.

Abbreviations

The following abbreviations are used in this manuscript:

MDPI	Multidisciplinary Digital Publishing Institute
DOAJ	Directory of Open Access Journals
IoT	Internet of Things
UAV	Unmanned Aerial Vehicle
UGV	Unmanned Ground Vehicle
MEC	Multi-Access Edge Computing
DRUID-NET	eDge computing ResoUrce allocatIon for Dynamic NETworks
QoS	Quality of Service
QoE	Quality of Experience
NFV	Network Functions Virtualization
SDN	Software Defined Networks
5G	Fifth Generation
SC	Service Chain
VNF	Virtualized Network Function
VM	Virtual Machine
NCS	Networked Control Systems

References

1. Cisco. White Paper. Internet of Things at a Glance. Available online: https://www.cisco.com/c/dam/en_us/solutions/trends/iot/docs/iot-aag.pdf (accessed on 12 April 2020).
2. Ericsson. Mobility Report—q4, 2018. Available online: https://www.ericsson.com/4932c2/assets/local/mobility-report/documents/2019/emr-q4-update-2018.pdf (accessed on 12 April 2020).
3. Hayat, S.; Yanmaz, E.; Muzaffar, R. Survey on unmanned aerial vehicle networks for civil applications: A communications viewpoint. *IEEE Commun. Surv. Tutor.* **2016**, *18*, 2624–2661. [CrossRef]
4. ETSI. Multi-access Edge Computing (MEC) Group. Available online: https://www.etsi.org/technologies/multi-access-edge-computing (accessed on 12 April 2020).
5. Yousefpour, A.; Fung, C.; Nguyen, T.; Kadiyala, K.; Jalali, F.; Niakanlahiji, A.; Kong, J.; Jue, J. All one needs to know about fog computing and related edge computing paradigms: A complete survey. *J. Syst. Archit.* **2019**, *98*, 289–330. [CrossRef]
6. Jeong, S.; Simeone, O.; Kang, J. Mobile edge computing via a UAV-mounted cloudlet: Optimization of bit allocation and path planning. *IEEE Trans. Vehic. Techn.* **2017**, *67*, 2049–2063. [CrossRef]
7. He, D.; Qiao, Y.; Chan, S.; Guizani, N. Flight security and safety of drones in airborne fog computing systems. *IEEE Commun. Mag.* **2018**, *56*, 66–71. [CrossRef]
8. Valentino, R.; Jung, W.S.; Ko, Y.B. Opportunistic computational offloading system for clusters of drones. In Proceedings of the 20th International IEEE Conference on Advanced Communication Technology, Korea, South, 11–14 February 2018; pp. 303–306.

9. Abdelzaher, T.; Hao, Y.; Jayarajah, K.; Misra, A.; Skarin, P.; Yao, S.; Weerakoon, D.; Undefinedrzén, K. Five Challenges in Cloud-Enabled Intelligence and Control. *ACM Trans. Internet Technol.* **2020**, *20*. [CrossRef]

10. Kiani, A.; Ansari, N.; Khreishah, A. Hierarchical Capacity Provisioning for Fog Computing. *IEEE/ACM Trans. Net.* **2019**, *27*, 962–971. [CrossRef]

11. Villari, M.; Fazio, M.; Dustdar, S.; Rana, O.; Ranjan, R. Osmotic Computing: A New Paradigm for Edge/Cloud Integration. *IEEE Cloud Comp.* **2016**, *3*, 76–83. [CrossRef]

12. Li, Y.; Chen, Y.; Lan, T.; Venkataramani, G. MobiQoR: Pushing the Envelope of Mobile Edge Computing Via Quality-of-Result Optimization. In Proceedings of the 2017 IEEE 37th International Conference on Distributed Computing Systems (ICDCS), Atlanta, GA, USA, 5–8 June 2017; pp. 1261–1270.

13. Leivadeas, A.; Falkner, M.; Lambadaris, I.; Ibnkahla, M.; Kesidis, G. Balancing Delay and Cost in Virtual Network Function Placement and Chaining. In Proceedings of the 2018 IEEE 4th International Conference on Network Softwarization and Workshops (NetSoft), Montreal, QC, Canada, 25–29 June 2018; pp. 1–9.

14. Leivadeas, A.; Kesidis, G.; Ibnkahla, M.; Lambadaris, I. VNF Placement Optimization at the Edge and Cloud. *Future Internet* **2019**, *11*, 1–23. [CrossRef]

15. Nwanganga, F.; Chawla, N. Using Structural Similarity to Predict Future Workload Behavior in the Cloud. In Proceedings of the 2019 IEEE 12th International Conference on Cloud Computing (CLOUD), Milan, Italy, 8–13 July 2019; pp. 132–136.

16. Ra, M.; Lee, H. Fighting with Unknowns: Estimating the Performance of Scalable Distributed Storage Systems with Minimal Measurement Data. In Proceedings of the 2019 35th Symposium on Mass Storage Systems and Technologies (MSST), Santa Clara, CA, USA, 20–24 May 2019; pp. 1–6. [CrossRef]

17. Yao, S.; Zhao, Y.; Hu, S.; Abdelzaher, T. QualityDeepSense: Quality-Aware Deep Learning Framework for Internet of Things Applications with Sensor-Temporal Attention. In Proceedings of the 2nd International Workshop on Embedded and Mobile Deep Learning, Munich, Germany, 15 June 2018.

18. Prakah-Asante, K.O.; Tonshal, B.; Yang, H.; Strumolo, G.; Chen, Y.; Rankin, J.S. Workload estimation for mobile device feature integration. U.S. Patent 9,889,862, 13 February 2018.

19. Sanchez-Alvarez, D.; Linaje, M.; Rodriguez-Pérez, F. A Framework to Design the Computational Load Distribution of Wireless Sensor Networks in Power Consumption Constrained Environments. *Sensors* **2018**, *18*, 954.

20. Pinto Neto, J.B.; Gomes, L.; Ortiz, F.; Almeida, T.; Campista, M.; Kosmalski, M.; Costa, L.; Mitton, N. An Accurate Cooperative Positioning System for Vehicular Safety Applications. *Comp. Electr. Engin.* **2019**, *83*, 1–13. [CrossRef]

21. Laftchiev, E.; Lagoa, C.M.; Brennan, S. Vehicle localization using in-vehicle pitch data and dynamical models. *IEEE Trans. Intellig. Transp. Syst.* **2015**, *16*, 206–220. [CrossRef]

22. Belhajem, I.; Ben Maissa, Y.; Tamtaoui, A. A robust low cost approach for real time car positioning in a smart city using Extended Kalman Filter and evolutionary machine learning. In Proceedings of the 4th IEEE International Colloquium on Information Science and Technology (CiSt), Tangier, Morocco, 24–26 October 2016.

23. Toldov, V.; Clavier, L.; Loscrí, V.; Mitton, N. A Thompson Sampling Approach to Channel Exploration-Exploitation Problem in Multihop Cognitive Radio Networks. In Proceedings of the 27th annual IEEE International Symposium on Personal, Indoor and Mobile Radio Communications (PIMRC), Valencia, Spain, 4–8 September 2016; pp. 1–6.

24. Razafimandimby, C.; Loscrí, V.; Maria Vegni, A.; Aourir, D.; Neri, A. A Bayesian approach for an efficient data reduction in IoT. In Proceedings of the InterIoT 2017—3rd EAI International Conference on Interoperability in IoT, Valencia, Spain, 7 November 2017; pp. 1–7.

25. Li, X.; Mitton, N.; Simplot-Ryl, I.; Simplot-Ryl, D. Dynamic Beacon Mobility Scheduling for Sensor Localization. *IEEE Trans. Parl. Distrib Syst.* **2012**, *23*, 1439–1452. [CrossRef]

26. Ardagna, D.; Panicucci, B.; Trubian, M.; Zhang, L. Energy-aware autonomic resource allocation in multitier virtualized environments. *IEEE Trans. Serv. Comp.* **2010**, *5*, 2–19. [CrossRef]

27. Chen, C.-T. *Linear System Theory and Design*; Oxford University Press, Inc.: New York, NY, USA, 1998.

28. Ullah, A.; Li, J.; Shen, Y.; Hussain, A. A control theoretical view of cloud elasticity: taxonomy, survey and challenges. *Clust. Comp.* **2018**, *21*, 1735–1764. [CrossRef]

29. Dechouniotis, D.; Leontiou, N.; Athanasopoulos, N.; Christakidis, A.; Denazis, S. A control-theoretic approach towards joint admission control and resource allocation of cloud computing services. *Intern. J. Netw. Manag.* **2015**, *25*, 159–180. [CrossRef]

30. Leontiou, N.; Dechouniotis, D.; Denazis, S.; Papavassiliou, S. A hierarchical control framework of load balancing and resource allocation of cloud computing services. *Comp. Elect. Eng.* **2018**, *67*, 235–251. [CrossRef]

31. Zhang, W.; Han, B.; Hui, P. On the networking challenges of mobile augmented reality. In Proceedings of the Workshop on Virtual Reality and Augmented Reality Network, Los Angeles, CA, USA, 11 August 2017; pp. 24–29.

32. Kalatzis, N.; Avgeris, M.; Dechouniotis, D.; Papadakis-Vlachopapadopoulos, K.; Roussaki, I.; Papavassiliou, S. Edge computing in IoT ecosystems for UAV-enabled early fire detection. In Proceedings of the 2018 IEEE International Conference on Smart Computing (SMARTCOMP), Taormina, Italy, 18–20 June 2018; pp. 106–114.

33. Sonmez, C.; Ozgovde, A.; Ersoy, C. Fuzzy workload orchestration for edge computing. *IEEE Trans. Net. Serv. Manag.* **2019**, *16*, 769–782. [CrossRef]

34. Guan, G.; Dong, W.; Zhang, J.; Gao, Y.; Gu, T.; Bu, J. Queec: QoE-aware Edge Computing for Complex IoT Event Processing Under Dynamic Workloads. In Proceedings of the ACM Turing Celebration Conference, Chengdu, China, 17–19 May 2019.

35. Jalali, F.; Hinton, K.; Ayre, R.; Alpcan, T.; Tucker, R.S. Fog computing may help to save energy in cloud computing. *IEEE J. Sel. Areas Comm.* **2016**, *34*, 1728–1739. [CrossRef]

36. Lyu, X.; Tian, H.; Jiang, L.; Vinel, A.; Maharjan, S.; Gjessing, S.; Zhang, Y. Selective offloading in mobile edge computing for the green internet of things. *IEEE Netw.* **2018**, *32*, 54–60. [CrossRef]

37. Gao, B.; Zhou, Z.; Liu, F.; Xu, F. Winning at the starting line: Joint network selection and service placement for mobile edge computing. In Proceedings of the IEEE INFOCOM 2019-IEEE Conference on Computer Communications, Paris, France, 29 April–2 May 2019; pp. 1459–1467.

38. Leivadeas, A.; Kesidis, G.; Falkner, M.; Lambadaris, I. A graph partitioning game theoretical approach for the VNF service chaining problem. *IEEE Trans. Netw. Serv. Manag.* **2017**, *14*, 890–903. [CrossRef]

39. Hegyi, A.; Flinck, H.; Ketyko, I.; Kuure, P.; Nemes, C.; Pinter, L. Application Orchestration in Mobile Edge Cloud Placing of IoT Applications to the Edge. In Proceedings of the IEEE Int. Workshop on Foundations and Applications of Self* Systems, Augsburg, Germany, 12–16 September 2016.

40. Nam, Y.; Song, S.; Chung, J.M. Clustered NFV Service Chaining Optimization in Mobile Edge Clouds. *IEEE Commun. Lett.* **2017**, *21*, 350–353. [CrossRef]

41. Barcelo, M.; Correa, A.; Tulino, A.; Vicario, J.L.; Morell, A. IoT-Cloud Service Optimization in Next, Generation Smart Environments. *IEEE J. Sel. Areas Commun.* **2016**, *34*, 4077–4090. [CrossRef]

42. Zanzi, L.; Giust, F.; Sciancalepore, V. M2EC: A Multi-tenant Resource Orchestration in Multi-Access Edge Computing Systems. In Proceedings of the IEEE International Conference on Wireless Communications and Networking Conference (WCNC), Barcelona, Spain, 15–18 April 2018.

43. Wang, J.; Qi, H.; Li, K.; Zhou, X. PRSFC-IoT: A Performance and Resource Aware Orchestration System of Service Function Chaining for Internet of Things. *IEEE J. Inter. Things* **2018**, *5*, 1400–1410. [CrossRef]

44. Cao, J.; Zhang, Y.; An, W.; Chen, X.; Sun, J.; Han, Y. VNF-FG Design and VNF Placement for 5G Mobile Networks. *Springer J. Inform. Sci.* **2017**, *60*, 040302. [CrossRef]

45. Jemaa, F.; Pujolle, G.; Pariente, M. QoS-aware VNF placement Optimization in Edge-Central Carrier Cloud Architecture. In Proceedings of the IEEE International Conference on Global Communications Conference (GLOBECOM), Washington, DC, USA, 4–8 December 2016.

46. Avgeris, M.; Dechouniotis, D.; Athanasopoulos, N.; Papavassiliou, S. Adaptive resource allocation for computation offloading: A control-theoretic approach. *ACM Trans Inter. Techn.* **2019**, *19*, 1–20. [CrossRef]

47. Gupta, R.; Chow, M. Networked control system: Overview and research trends. *IEEE Trans. Indust. Electron.* **2010**, *57*, 2527–2535. [CrossRef]

48. Zhang, W.; Branicky, M.; Phillips, S. Stability of networked control systems. *IEEE Cont. Syst.* **2001**, *21*, 84–99.

49. Wen, S.; Guo, G. A survey of recent results in networked control systems. *Proc. IEEE* **2007**, *95*, 138–162.

50. Elia, N.; Mitter, S. Stabilization of linear systems with limited information. *IEEE Trans. Autom. Cont.* **2001**, *46*, 1384–1400. [CrossRef]

51. Simon, D.; Robert, D.; Sename, O. Robust control/scheduling co-design: application to robot control. In Proceedings of the 11th IEEE Real Time and Embedded Technology and Applications Symposium, San Francisco, CA, USA, 7–10 March 2005.

52. Xia, F.; Sun, Y. Control-Scheduling Codesign: A Perspective on Integrating Control and Computing. *Dyn. Contin. Discr. Impuls. Syst. Ser. B* **2005**, *13*, 1352–1358.

53. Branicky, M.; Phillips, S.; Zhang, W. Scheduling and feedback co-design for networked control systems. In Proceedings of the 41st IEEE Conference on Decision and Control, Las Vegas, NV, USA, 10–13 December 2002.

54. Bertsekas, D.; Tsitsiklis, J. *Parallel and Distributed Computation: Numerical Methods*; Prentice Hall Englewood: Cliffs, NJ, USA, 1989.

55. Kashyap, A.; Başar, T.; Srikant, R. Quantized consensus. *Automatica* **2007**, *43*, 1192–1203. [CrossRef]

56. Schenato, L.; Sinopoli, B.; Franceschetti, M.; Poolla, K.; Sastry, S. Foundations of control and estimation over lossy networks. *Proc. IEEE* **2007**, *95*, 163–187. [CrossRef]

57. Jungers, R.; Kundu, A.; Heemels, W. Observability and controllability analysis of linear systems subject to data losses. *IEEE Trans. Autom. Cont.* **2018**, *63*, 3361–3376. [CrossRef]

58. Athanasopoulos, N.; Jungers, R. Combinatorial methods for invariance and safety of hybrid systems. *Automatica* **2018**, *98*, 130–140. [CrossRef]

59. Dilip, A.; Athanasopoulos, N.; Jungers, R. The impact of packet dropouts on the reachability energy. In Proceedings of the International Conference on Cyber-Physical Systems (ICCPS), Porto, Portugal, 11–13 April 2018.

60. Zhang, L.; Hristu-Varsakelis, D. Communication and control co-design for networked control systems. *Automatica* **2006**, *42*, 953–958. [CrossRef]

61. Guo, G. A switching system approach to sensor and actuator assignment for stabilisation via limited multi-packet transmitting channels. *Intern. J. Cont.* **2011**, *84*, 78–93. [CrossRef]

62. Wen, S.; Guo, G. Control and resource allocation of cyber-physical systems. *IET Cont. Theory Applic.* **2016**, *10*, 2038–2048. [CrossRef]

63. Martí, P.; Camacho, A.; Velasco, M.; Gaid, M. Runtime allocation of optional control jobs to a set of CAN-based networked control systems. *IEEE Trans. Indust. Inform.* **2010**, *6*, 503–520. [CrossRef]

64. Tabuada, P. Event-triggered real-time scheduling of stabilizing control tasks. *IEEE Trans. Autom. Cont.* **2007**, *52*, 1680–1685. [CrossRef]

65. Abdelrahim, M.; Postoyan, R.; Daafouz, J.; Nei, D.; Heemels, M. Co-design of output feedback laws and event-triggering conditions for the L2-stabilization of linear systems. *Automatica* **2018**, *87*, 337–344. [CrossRef]

66. Donkers, M.; Heemels, W. Co-design of output feedback laws and event-triggering conditions for the L2-stabilization of linear systems. *IEEE Trans. Autom. Cont.* **2012**, *57*, 1362–1376. [CrossRef]

67. Goebel, R.; Sanfelice, R.; Teel, A. Hybrid dynamical systems. *IEEE Cont. Syst. Magaz.* **2009**, *29*, 28–93. [CrossRef]

68. Matveev, A.; Savkin, A. *Estimation and Control over Communication Networks*; Springer Science & Business Media: Boston, MA, USA, 2009.

69. Garoche, P. *Formal Verification of Control System Software*; Princeton University Press: Princeton, NJ, USA, 2019.

 *sensors* 

Article

A Comparison of the Influence of Vegetation Cover on the Precision of an UAV 3D Model and Ground Measurement Data for Archaeological Investigations: A Case Study of the Lepelionys Mound, Middle Lithuania

Algimantas Česnulevičius *, Artūras Bautrėnas, Linas Bevainis and Donatas Ovodas

Department of Cartography and Geoinformatics, Institute of Geosciences, Faculty of Chemistry and Geosciences, Vilnius University, LT-03101 Vilnius, Lithuania; arturas.bautrenas@gf.vu.lt (A.B.); linas.bevainis@gf.vu.lt (L.B.); ovodas@gmail.com (D.O.)
* Correspondence: algimantas.cesnulevicius@gf.vu.lt

Received: 10 October 2019; Accepted: 28 November 2019; Published: 2 December 2019

Abstract: The aim of this research was to conduct a comparative analysis of the precision of ground geodetic data versus the three-dimensional (3D) measurements from unmanned aerial vehicles (UAV), while establishing the impact of herbaceous vegetation on the UAV 3D model. Low (up to 0.5 m high) herbaceous vegetation can impede the establishment of the anthropogenic roughness of the surface. The identification of minor surface alterations, which enables the determination of their anthropogenic origin, is of utmost importance in archaeological investigations. Vegetation cover is regarded as one of the factors influencing the identification of such minor forms of relief. The research was conducted on the Lepelionys Mound (Prienai District Municipality, Lithuania). Ground measurements were obtained using Trimble GPS, and UAV "Inspire 1" was used for taking aerial photographs. Following the data from the ground measurements and aerial photographs, large scale surface maps were drawn and the errors in the measurement of the position of the isolines were compared. The results showed that the largest errors in the positional measurements of fixed objects were conditioned by the height of grass. Grass with a height of up to 0.1 m resulted in discrepancies of up to 0.5 m, whereas grass that was up to 0.5 m high led to discrepancies up to 1.3 m high.

Keywords: GPS measurement; UAV; 3D models; measurement precision

1. Introduction

During the initial stage of an archaeological investigation, one of the most important principles is to identify a potential object, to determine its boundaries and area. Traditionally, large-scale topographic maps and geodetic measurements are widely used during the initial stage of reconstruction.

Aerial photographs were mainly used where archaeological sites coincided with the areas covered by aerial topography and only on fragmentary basis due to their high cost. High-resolution space images have only become possible within the last decade, but they do not cover continuous areas. Moreover, high resolution photos are not always available for academic research or studies. At the beginning of the 21st century unmanned aerial vehicles, better known as drones, were employed to identify and map potential archaeological objects. They have a number of advantages which include the following characteristics: low price, high resolution, large scale, and multispectral. A very important advantage of unmanned aerial vehicles is the creation of 3D models using photogrammetric techniques. These 3D models reveal the small roughness of the surface. Such alterations in the surface serve as identifiers, when searching for potential archaeological sites.

The issues of reliability and accuracy of aerial photographs obtained using unmanned aerial vehicles have already been addressed in studies by many academic researchers [1–21]. The use of unmanned aerial vehicles provide a fast and inexpensive way to explore ground surface and to identify objects of interest [22], however, research on assessing the precision of aerial images from unmanned aerial vehicles is scarce [23–26]. The accuracy of aerial images produced with the help of unmanned aerial vehicles (UAVs) can be affected by a number of factors, for example, altitude of flight, the image quality of the photo camera, the design of the UAV route, the methods of georeferencing, and others. An appropriate design of the UAV route ensures cruise altitude and constant aerial image coverage of the whole territory. An appropriate project for the flight and a high-quality photo camera effect the efficiency of photogrammetric processing of the images obtained. Further investigations are simplified by using the well-tested and broadly applied mathematical and photogrammetric algorithms for image processing. The problems occur while designing a 3D model of the territory captured in aerial images. The initial 3D model is created in the conditional coordinate system, which is later linked to the officially used coordinate system. The coordinates can be connected in one of the following two ways: by direct graphical connection of the position of the object in the aerial image to the coordinate system (less precise) or by linking the GPS measurements of fixed objects to the coordinates of the aerial images. The accuracy of the vertical positioning of given points is a highly important factor in designing the 3D relief models that are used for identification, analysis, and mapping of archaeological objects.

Recent research [18–32] has shown that while aiming for high accuracy of the vertical positioning of the objects, it is not enough to use a global navigation satellite system (GNSS); ground control points (GCPs) have to be applied as well. Such a combined technique allows for the design of a more accurate digital relief model (DRM), where the precision of vertical positioning of points equals 0.7 cm.

The aim of this research is to conduct a comparative analysis of the precision of ground geodetic measurements and aerial photographs from an unmanned aerial vehicle, while establishing the positional accuracy of the identified objects. The archaeological objects of the Middle Ages in the eastern coast of the Baltic sea are often related to natural relief forms, which were modified by people while building fortifications and settlements around them [33–37]. These archaeological objects are now in forests, agricultural lands, and urbanized territories. The surface of archaeological objects in such urban territories has been exposed to significant changes or has been fully destroyed. The use of aerial images from unmanned aerial vehicles for the positional identification of archaeological objects is highly limited. Due to dense vegetation and the foliage of tall trees, the application of aerial imaging in wooded territories is restricted. The surface of archaeological objects in agricultural territories is partially extant. Therefore, aerial images can be rather efficient in seeking to identify positions of archaeological objects in meadows and woodless territories.

The narrow spectral and surface thermal analysis methods are applied for the investigation of the structural diversity of vegetation cover on the basis of UAV aerial images [38–42]. Studies have mainly focused on the influence of big ligneous plants on the mapping of surface elements, whereas the impact of low herbaceous vegetation on low forms of archaeological relief has, so far, not been exhaustively researched [43]. Our research aims to assess the quality of aerial images, ultimately seeking to design accurate digital 3D relief models for the identification of archaeological objects [44–50].

For the identification of small surface irregularities (small archaeological objects) we applied the computer program "Circle_3p", developed by the Department of Cartography and Geoinformatics, Vilnius University, applying the classical Delaunay method (author Artūras Bautrėnas). The results of the study showed that this method is effective in grassy mounds.

2. Research Object, Materials, and Methods

The object of the research is the Lepelionys Mound, which is located in the Prienai Administrative Region of Kaunas County (Figure 1). It dates back to the second half of the first millennium. At the beginning of the second millennium a settlement was established there, covering an area of 9 hectares around the mound. The Lepelionys Mound is on the left side of the road from Vilnius to Prienai (60 km to the west of Vilnius). The territory of the ancient settlement is on both sides of the road, but its bigger part is located on the left side. The Vilnius-Prienai road was built in the second half of the 20th century. While designing the road, the relief of the former ancient settlement was affected but some small and low relief forms of anthropogenic origin still remain, dating back to between the 9th and 12th centuries [51,52]. The main archaeological object, the Lepelionys Mound, was investigated by archaeologists in the second half of the 20th century. During these archaeological investigations the territory boundaries and the protection zone of the ancient settlement were distinguished (Figure 1).

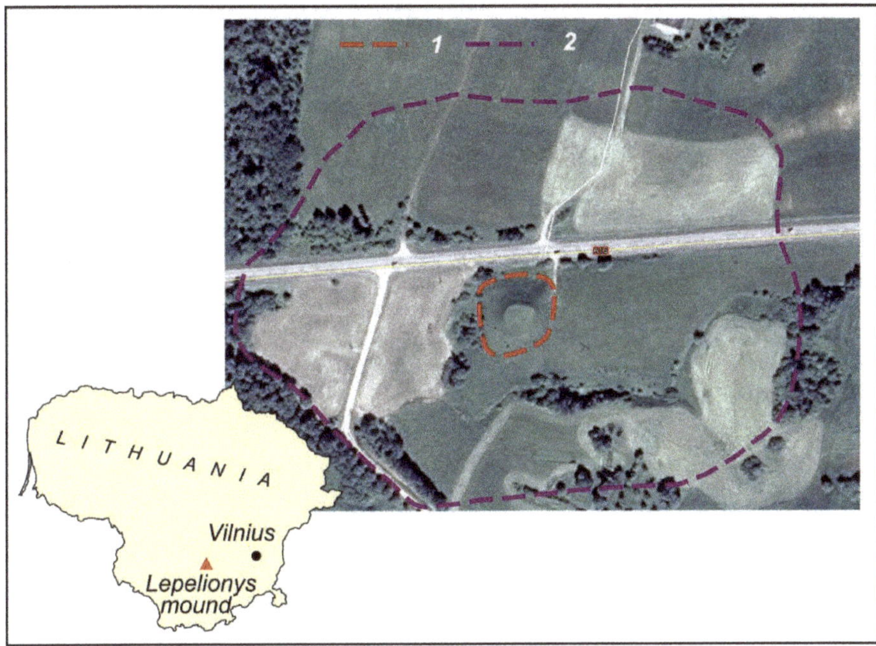

Figure 1. The location of Lepelionys Mound and the ancient settlement: (**1**) mound boundary and (**2**) ancient settlement territory boundary (according to V. Juškaitis [51]).

Ground geodetic measurements and photos taken by the camera on the unmanned aerial vehicles were applied while designing the three-dimensional relief models. Comparisons of accuracy between the UAV 3D model and the ground measurements of the Lepelionys mound were carried out twice, in August 2018 and June 2019.

Ground geodetic measurements were carried out with a Trimble R4 GPS device (measurable accuracy in favorable conditions: X, Y is set to ±8 mm and Z to ±15 mm) on 9 August 2018. Since the mound is in a fully open area and not covered by buildings or greenery (Figure 1), the measurements were collected with maximum accuracy. During the collection of these measurements, the coordinates of 212 characteristic ground-surface points were recorded. After analyzing the accuracy of the measured point coordinates, 179 points were mapped to the LKS-94 coordinate system (Figure 2).

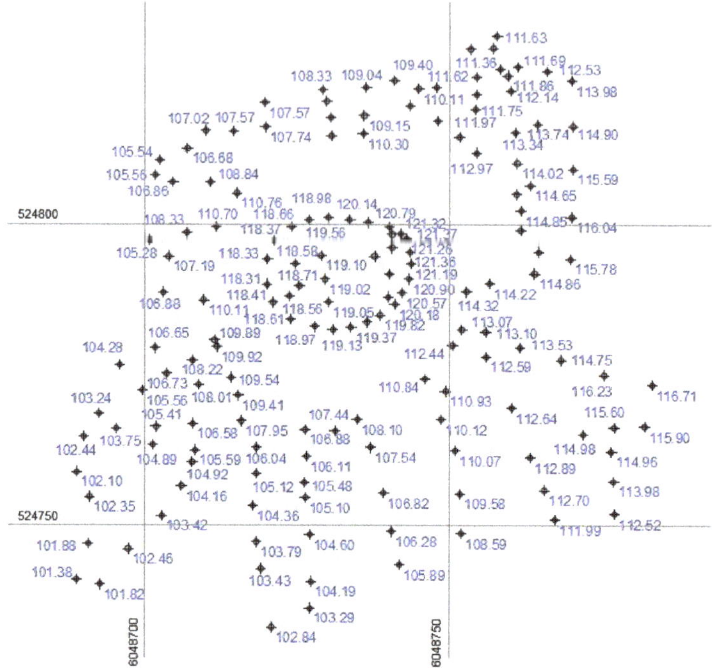

Figure 2. The selected points that were mapped to the LKS-94 coordinate system.

Since the topographic photograph can be used to estimate the accuracy of the aerial photographs, 10 ground control points (GCPs) were measured in parallel to the ground points (Figure 3).

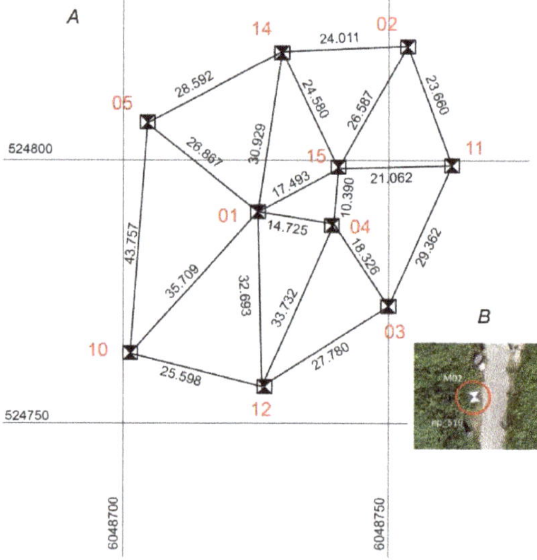

Figure 3. The ground control point marks (**A**) and the diagram of the ground control point (GCP) arrangement (**B**). The red circle defines the location of the ground mark.

Figure 4 shows two objects, the coordinates of which were used for creating the aerial photograph model.

Figure 4. Examples of identified objects. The red circles define the small objects location, whose measured coordinates are used to adjust the 3D model.

The vegetation is one of the most important indicators of archaeological objects. Information on human activities is reflected in the variation of the lushness of vegetation. Homogeneous vegetation is characteristic of the investigated territory, since for several decades most of the surroundings of the mound have been used as pasture. Local differences in herbaceous vegetation in the mound surroundings over a long period of time have been predetermined by changes in the surface relief layer caused by the following human activities:

(i)   Organic waste was thrown at the foot of the mound;
(ii) In the territory of the ancient settlement the ground was excavated for substructures of buildings and the soil (sediment) was poured beside the walls of the building;
(iii) The ancient settlement was surrounded by palisades, the stakes of which were driven into the ground and the excavated soil fortified the foundation of the fence;
(iv) Organic and mineral waste (ceramic fragments, bones of the animals used for food, worn out shoes, and clothes) was thrown over the palisade of the settlement.

All the aforesaid factors resulted in physical differences in the present vegetation, i.e., lusher or sparser vegetation. It is important to point out that currently there is a pasture in the former territory of the settlement, where grazing starts at the end of April and lasts until October. The whole area is grazed in this time and the anthropogenic impact on the surface was equal during photofixation in August.

The picture in Figure 5 provides a visual representation of the camera sensor and the field of view. Using the width of the camera sensor, the focal length, and the drone altitude the ground sample distance (GSD) can be calculated (Figure 5).

The equation we use to calculate the *GSD* is:

$$GSD = \frac{(sensors\ width \times altitude \times 100)}{(focal\ length \times image\ width)} \tag{1}$$

Photofixation of aerial images was conducted using the unmanned aerial vehicle (UAV) INSPIRE 1. Its technical parameters are presented in Table 1.

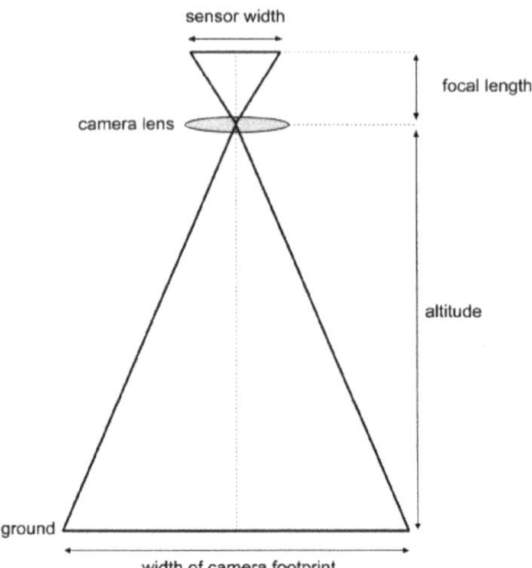

Figure 5. Visual representation of the nadir facing camera on the drone.

Table 1. The specifications of unmanned aerial vehicle (UAV) DJI Inspire1 and the RGB Zenmus X3 camera.

Hovering Accuracy (GPS Mode)	**Vertical: 0.5 m; Horizontal: 2.5 m**
Max angular velocity	Pitch: 300°/s; Yaw: 150°/s
Max tilt angle	35°
Max ascent and descent speed	5 m/s; 4 m/s
Max speed	22 m/s
Max wind speed resistance	10 m/s
Max service ceiling above take-off point	120 m
Type and model	X3; FC350
Total and effective pixels	12.76 M; 12.4 M
Max capacity	64 GB
Maximal image size	4000 × 3000
ISO range	Photo- 100–1600; Video- 100–3200.
The electronic shutter speed	8 s–1/8000 s
A field of view (FOV)	94°
Supported file formats	Photo: JPEG, DNG; Video: MP4/MOV (MPEG-4 AVC/H.264)
Types of electronic media	Micro SD. Maximal capacity 64 GB. Class 10 or UHS-1
Sensor width (mm)	6.17
Focal length (mm)	4.55
Altitude (m)	50
Image width	4000
Image height	3000
GSD (cm/pixel)	1.695054945
Width (m)	67.8021978
Height (m)	50.85164835

The front overlap of the pictures taken is 80% and the side overlap is 70%. The "double grid" mission flight plan was used for a more detailed and accurate 3D model. The flight was made at the height of 50 m, therefore, respectively, the GSD equals 1.7 cm.

There were 199 photos that were processed with special photogrammetric "Pixoprocessing" software. The point cloud, the digital surface model (DSM), and the orthomosaic were obtained during this process.

The study included an assessment of the mismatches between the elevation isoline positions acquired from the ground geodetic measurements and from the aerial images from the UAVs. An associate professor of the Department of Cartography and Geoinformatics, Artūras Bautrėnas, designed the computer program "Circle_3p", which employs the classical method of Delaunay and ensures a consistent systemic selection of points. Using the Delaunay triangulation method, altitude interpolation of the ground measurement points was performed and an isoline view was generated. An analogous method was used for the interpolation of the elevation of surface points and the generation of isolines using the images taken by the camera on the UAV (Figure 6). The following indicators were calculated: $\pm N_i$ which is the sequence number of the analyzed point in the positive or the negative deviation from the base (ground geodetic measurement) isoline, $\pm \Delta S_i$ which is the length of the perpendicular to the positive or the negative side of the analyzed point, $\pm \Delta Z_i$ which is the calculated correction of the overdose to the positive or the negative side, and $\pm D$ which is the distance between the base (ground geodetic measurement) isoline point and the UAV isoline point.

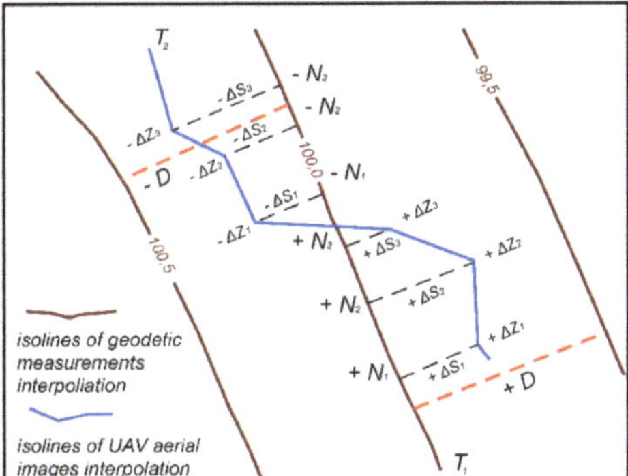

Figure 6. The scheme of elevation isoline position assessment using ground geodetic measurements and the mismatch between it and the aerial images from UAVs.

For the calculation of the deviation of the target position the following formula was used:

$$\pm \Delta S_i = \left(y_{Ni} - y_{Ti}\right)\cos \alpha - (x_{Ni} - x_{Ti})\sin \alpha. \tag{2}$$

where α is the directional angle of the segment N_i–N_{i+1}, T_i is the number of the interpolated UAV measurement point, and i is the number of the point for each fragment of the ground geodetic and UAV isolines.

As we know what the isoline step is (0.5 m), the distance between the horizontal ($\pm D$) at each N_i point can be calculated by geometric interpolation. The difference in height $\pm \Delta Z_i$ is calculated using the formula:

$$\pm \Delta Z_i = \frac{\pm h}{\pm D_i} \times \pm \Delta S_i$$

where h is the isoline step, D is the distance between the horizontal at the calculated point, and the $\pm$ sign depends on the direction of the horizontal deviation.

3. Results

3.1. Creating a Two-Dimensional (2D) Relief Model

In order to confidently state that the 2D relief model is sufficiently precise and can be used as a benchmark for estimating the models, which were made by using aerial photometric methods, the horizontals were drawn automatically in accordance with strict interpolation rules.

In order to perform the automated relief modelling, it was necessary to select pairs of measured points, among which it would be possible to calculate the exact horizontal surfaces of the relief, i.e., interpolate heights. Therefore, the Delaunay triangulation method was chosen to interpolate the heights [53].

3.2. Drawing of a Topographic Plan

First, the Delaunay triangulation (Figure 7) was completed among 179 selected topographic points using the program "Circle_3p". This allowed 321 triangles to be selected, among which the interpolation of the triangle vertices was performed.

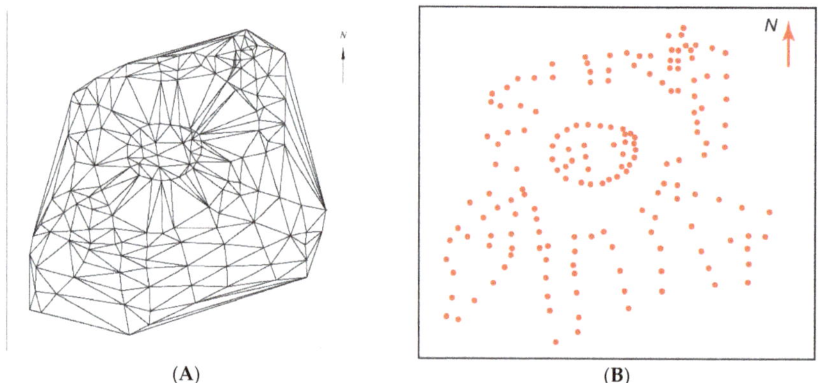

Figure 7. The Delaunay triangulation (**A**) among the 179 selected topographic points (**B**) using the program "Circle_3p".

Among these triangle vertices, horizontal interpolation was performed in the LAS07 height system using a selected step of 0.5 m (Figure 8). The coordinates of 1676 extra points, plotted as horizontals, were calculated during the interpolation.

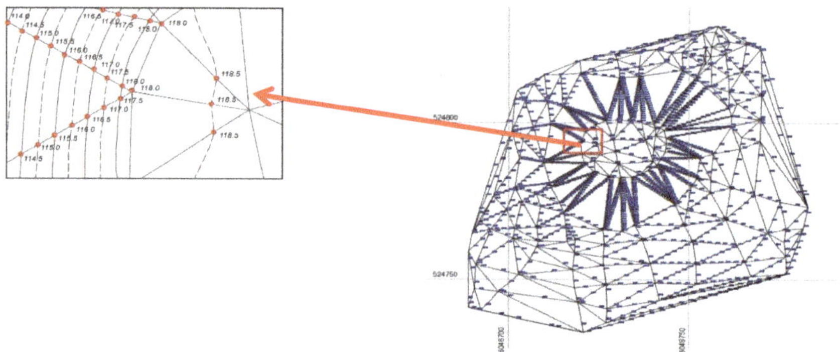

Figure 8. The interpolation among the vertices of selected triangles.

The interpolation points were uploaded to TopoPlan (AutoCAD 2016). The horizontals were plotted using the "Spline" function (Figures 9 and 10).

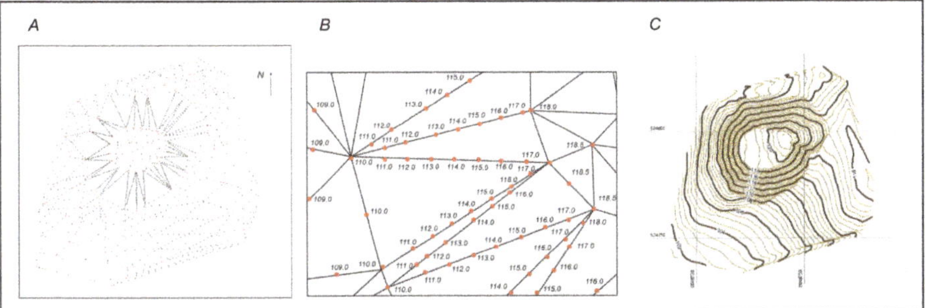

Figure 9. The points of interpolation (**A**), an interpolation sequence (**B**) and the relief isolines (**C**).

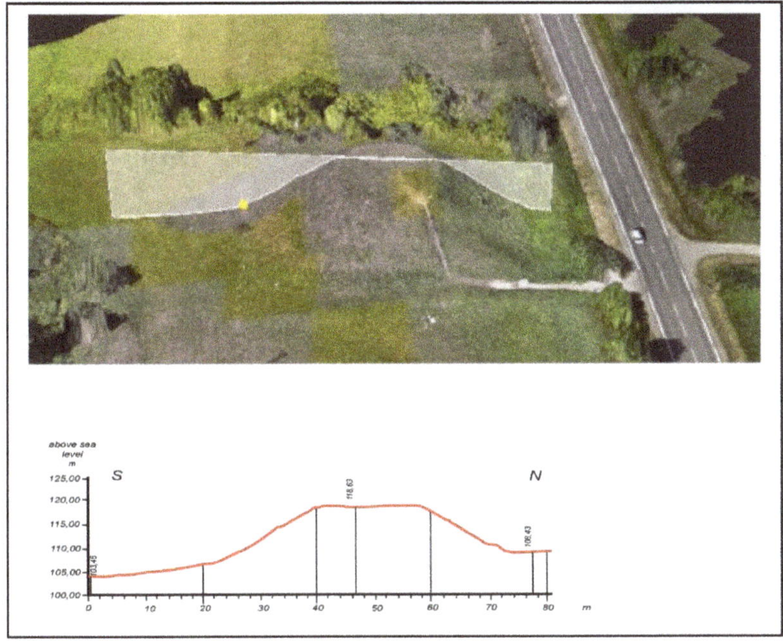

Figure 10. A cross-section of the Lepelionys Mound. The roughness in the red line indicates the remains of the former tree trunk fencing.

The cross-sections of the Lepelionys Mound were created with the help of aerial images taken by the camera on UAV, which highlighted the minor anthropogenic forms of relief on the slope of the mound, i.e., the remains of the former tree trunk wall (Figure 11).

Figure 11. The variance of the equal height isolines in plane position.

4. Discussion

4.1. Evaluation of the Precision of theAaerial Images

One of the most time-consuming tasks in aerial photography is to set out the GCPs and to coordinate them. Therefore, it is necessary to find the optimal number of GCPs in order to minimize the preparatory work. It should also be possible to estimate the feasible use of coordinated stable land objects (Figure 4) instead of bearing marks, which would further simplify the preparatory work. Therefore, ten ground control points (marks) in the study area are used to estimate the accuracy of the coordinates of 10 objects in the study area (Figure 3).

In order to evaluate the accuracy of the 3D model, it was created incorporating all ten marks and the coordinates of all the objects were measured in this model. The differences between the coordinates of objects in the 3D model and the coordinates measured from the topographic image do not exceed the double (Trimble GPS) accuracy (Tables 2 and 3) for those objects that are clearly seen in the 3D model (np-508, -515, -582, and -587). The accuracy of the other objects is poorer due to the vegetation (grass), which complicates their identification.

The random error distribution depends on the accuracy of the object identification, and therefore the graph consists of taking the errors in absolute size in mm.

The error analysis shows that they increase significantly when the orientation marks are fewer than five (Table 5, Figure 5), even for those objects that are visible in the 3D model (np-508, -515, -582, and -587). Therefore, it can be argued that in order to maintain the accuracy of measurements, there should be at least five orientation marks. It has been noticed that the error rate is influenced not only by the vegetation but also by the experience and thoroughness of the operator measuring the 3D model. As the precision of the well-known objects is practically unchanged (from ten to five marks), it can be argued that the 3D model should operate with maximum accuracy with five GCPs and the use of easily visible coordinated objects (Figure 12, Tables 2–5).

Table 2. The parameters of variance of the equal height isolines in plane position.

Line No.	Variance of Equal Height Isolines in Plane Position, m	Slope Inclination, Degree in Brackets (Mean Slope Inclination Values)	Remarks
1	0.91	0	The top of the mound, mown grass
2	0.71	0	The top of the mound, mown grass
3	0.86	0	The top of the mound, mown grass
4	0.94	0	The top of the mound, mown grass
5	0.89	0	The top of the mound, mown grass
6	0.60	30–40	The slope of the mound, short grass (0.1 m)
7	0.36	30–40	The slope of the mound, short grass (0.1 m)
8	0.54	30–40	The slope of the mound, short grass (0.1 m)
9	0.36	30–40	The slope of the mound, short grass (0.1 m)
10	0.34	30–40	The slope of the mound, short grass (0.1 m)
11	0.36	30–40	The slope of the mound, short grass (0.1 m)
12	0.96	21–26	The foot of the mound, high grass (0.5 m)
13	1.26	21–26	The foot of the mound, high grass (0.5 m)
14	1.09	21–26	The foot of the mound, high grass (0.5 m)
15	1.32	21–26	The foot of the mound, high grass (0.5 m)
16	0.91	21–26	The foot of the mound, high grass (0.5 m)
17	0.45	14–19	The foot of the mound, medium-high grass (0.2 m)
18	0.67	14–19	The foot of the mound, medium-high grass (0.2 m)
19	0.71	14–19	The foot of the mound, medium-high grass (0.2 m)
20	0.54	14–19	The foot of the mound, medium-high grass (0.2 m)
21	0.63	14–19	The foot of the mound, medium-high grass (0.2 m)

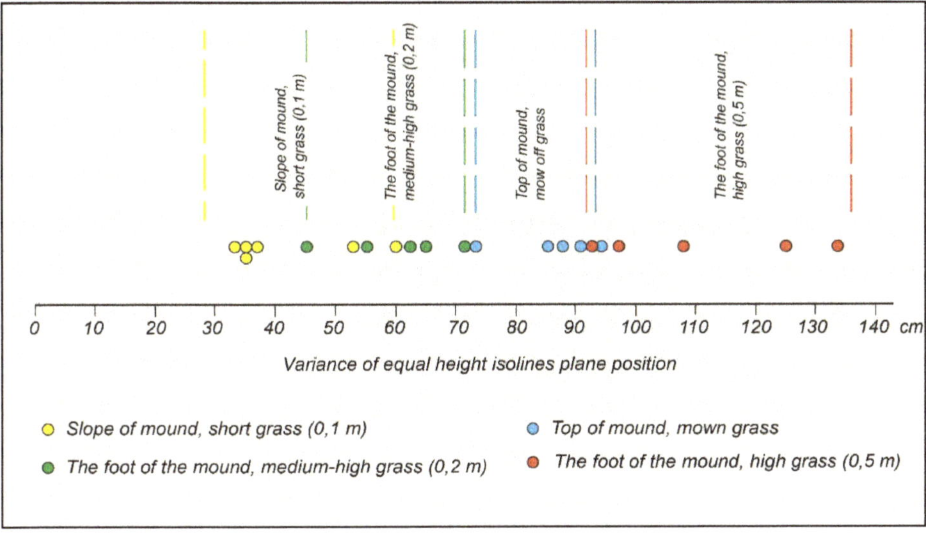

Figure 12. A diagram illustrating the parameters of variance of equal height isolines in plane position.

Table 3. The errors of the measurements of the objects when using ten marks.

No.	GPS	DEM	Error Size (m)	Absolute Error Size (mm)
np-507				
X	6048772.162	6048772.216	−0.054	54
Y	524809.995	524810.072	−0.077	77
Z	112.138	112.212	−0.074	74
np-508				
X	6048771.963	6048771.971	−0.008	8
Y	524810.078	524810.066	0.012	12
Z	112.757	112.785	−0.028	28
np-515				
X	6048774.378	6048774.393	−0.015	15
Y	524804.426	524804.440	−0.014	14
Z	111.620	111.644	−0.024	24
np-522				
X	6048769.132	6048769.230	−0.098	98
Y	524803.655	524803.700	−0.045	45
Z	111.896	111.940	−0.044	44
np-530				
X	6048769.708	6048769.740	−0.032	32
Y	524793.519	524793.562	−0.043	43
Z	110.071	110.140	−0.069	69
np-560				
X	6048716.410	6048716.452	−0.042	42
Y	524751.812	524751.892	−0.080	80
Z	105.409	105.368	0.041	41
np-582				
X	6048712.964	6048712.978	−0.014	14
Y	524768.387	524768.400	−0.013	13
Z	106.038	106.006	0.032	32
np-586				
X	6048692.869	6048692.835	0.034	34
Y	524769.123	524769.100	0.023	23
Z	103.427	103.394	0.033	33
np-587				
X	6048692.729	6048692.734	−0.005	5
Y	524769.143	524769.154	−0.011	11
Z	104.131	104.107	0.024	24
np-596				
X	6048715.725	6048715.685	0.040	40
Y	524781.260	524781.290	−0.030	30
Z	107.443	107.399	0.044	44

Table 4. The absolute errors of objects when a 3D model is made on the basis of 10 marks with, ground control points.

Point No.	Absolute Error (mm)		
	X	**Y**	**Z**
np-507	54	77	74
np-508	8	12	28
np-515	15	14	24
np-522	98	45	44
np-530	32	43	69
np-560	42	80	41
np-582	14	13	32
np-586	34	23	33
np-587	5	11	24
np-596	40	30	44

Table 5. A comparison of the absolute errors when different numbers of ground control points are used for a precise calculation. Values are in mm.

Point No.	3 Marks			4 Marks			5 Marks			7 Marks		
	x	y	z	x	y	z	x	y	z	x	y	z
np-507	402	608	514	309	194	263	56	80	51	48	70	73
np-508	86	105	56	36	42	42	11	12	26	10	9	19
np-515	175	193	64	78	14	10	15	8	25	13	16	26
np-522	95	145	574	48	33	254	95	56	57	88	39	49
np-530	954	657	1419	692	170	164	48	46	52	37	38	55
np-560	400	492	194	387	308	406	45	84	48	43	78	47
np-582	102	266	88	56	32	39	16	11	28	11	15	29
np-586	729	371	803	524	498	527	35	25	31	30	28	27
np-587	109	145	188	64	75	84	17	14	26	9	16	22
np-596	701	398	239	471	480	341	42	32	46	45	33	43

As seen in Table 3, the random error distribution depends on the accuracy of the object identification. Similarly, the absolute errors of objects have been calculated for the 3D models with different number of ground control points (Tables 4 and 5).

4.2. Evaluation of the Influence of Vegetation Covers

In 2019, a comparison of the Lepelionys mound surface isolines obtained using the UAV 3D model or the ground measurements showed that there are significant deviations in the plane and height positions between the two. The comparison was carried out in different vegetation height zones (Figure 13). At the top of the mound, where the grass was mown and its height was only 1 to 2 cm, the maximum discrepancies between the UAV 3D model and the ground measurement isolines were 0.75 m for the plane position and 0.42 m for the height. On the slopes of the mound, where the height of the grass was between 5 and 10 cm, the maximum discrepancies between the plane position of the isolines were up to 0.41 m, and up to 0.42 m for the height. At the foot of the mound, where the height of the unheated grass was 60 to 100 cm, the maximum discrepancies between the plane position of the isolines reached 6.63 m, and up to 0.77 m for the height. The results of the discrepancies between the plane position of the isolines and their height are presented in Table 6.

The sharpness and contrast in aerial images are both becoming important issues for the use of UAV aerial imagery. Aerial image contrast problems occur in areas that fall under the shadow of trees or rough terrain on a sunny day. In this study, the image contrast of the aerial photographs was adjusted and, where necessary, increased. During the 2018 photofixation, the western and southwestern parts of the mound slope were in shadow. To highlight the terrain microforms in parts of the image on the southwest slope we used the brightness/contrast, shadows/highlights, color balance, hue/saturation, and photo tiles tools in Adobe Photoshop software.

The comparison of the large-scale maps of the Lepelionys mound surface created using the UAV 3D model or by using the ground topographic measurements, shows that the plane position of the isolines in the 3D model is highly micro-sinuous. This is due to the methods of isoline interpolation applied in the UAV 3D model, i.e., the calculation of the interfaces between multiple point pairs (about seven million pixel pairs) creates the non-continuous isolines.

Three-dimensional terrain modelling using UAV aerial imagery is currently expanding. The wider application of collaborative mapping initiatives in archaeology [54–56] will lead to an increasing use of nonprofessional UAV aerial imagery to identify undefined and unexplored archaeological sites from the 19th to the early 20th century.

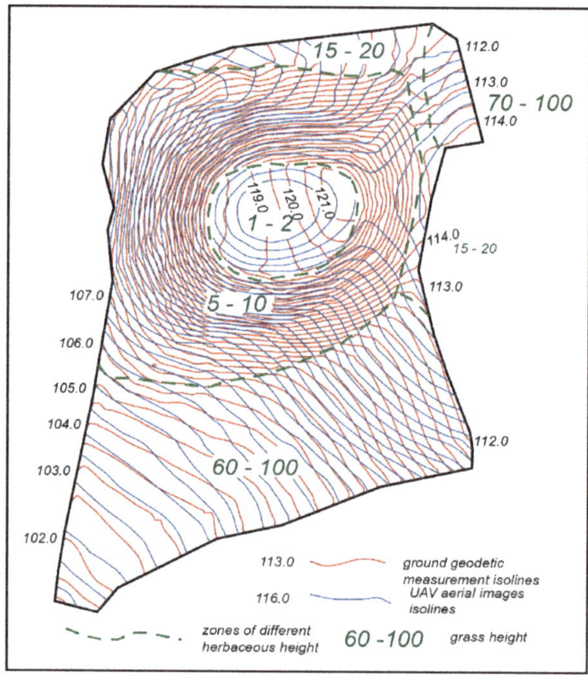

Figure 13. The zones of different herbaceous height on the Lepelionys mound.

Table 6. The deviation between the ground measurement and the UAV isoline plane and height positions (data from June 2019).

Isoline Height	Number of Deviation [a]		Maximum Plane Deviation, m [b]		Average of Plane Deviation, m [c]		Maximum of Height Deviation, m [d]		Average of Height Deviation, m [e]		Proportional Deviation [f]
	−	+	−	+	−	+	−	+	−	+	
The foot of the mound. Height of the grass was 60–100 cm											
102.00	138	232	0.619	1.569	0.269	0.556	−0.109	+0.117	−0.051	+0.038	+0.005
102.50	741	399	6.653	2.527	1.539	0.834	−0.338	+0.214	−0.091	+0.071	−0.034
103.00	449	632	5.791	2.116	0.961	0.864	−0.778	+0.216	−0.110	+0.085	+0.004
103.50	377	431	3.950	1.663	0.569	0.698	−0.480	+0.186	−0.064	+0.079	+0.012
104.00	732	635	1.355	2.162	0.497	0.765	−0.167	+0.213	−0.064	+0.081	+0.003
104.50	882	863	2.301	1.822	0.787	0.569	−0.277	+0.293	−0.086	+0.080	−0.004
105.00	877	608	1.720	1.726	0.577	0.566	−0.190	+0.245	−0.063	+0.091	0.000
105.50	744	631	1.693	1.463	0.539	0.446	−0.199	+0.271	−0.074	+0.075	−0.005
106.00	536	356	1.629	0.902	0.449	0.276	−0.233	+0.139	−0.066	+0.048	−0.021
106.50	532	222	1.534	0.513	0.420	0.197	−0.282	+0.120	−0.077	+0.043	−0.042
107.00	318	251	0.771	0.959	0.310	0.293	−0.197	+0.179	−0.074	+0.059	−0.016
107.50	258	254	0.518	0.818	0.214	0.225	−0.126	+0.176	−0.047	+0.048	+0.000
108.00	203	303	0.681	0.799	0.265	0.253	−0.182	+0.183	−0.066	+0.060	+0.009
108.50	70	382	0.355	0.969	0.127	0.330	−0.109	+0.174	−0.037	+0.062	+0.047
109.00	416	242	1.021	0.508	0.401	0.135	−0.174	+0.147	−0.075	+0.040	−0.033
109.50	374	319	1.736	0.700	0.469	0.254	−0.406	+0.163	−0.115	+0.059	−0.035
110.00	212	305	0.776	0.415	0.250	0.166	−0.167	+0.090	−0.064	+0.036	−0.005
110.50	344	189	1.251	0.449	0.322	0.132	−0.238	+0.103	−0.066	+0.030	−0.032
111.00	449	336	1.401	0.575	0.412	0.211	−0.298	+0.199	−0.093	+0.064	−0.026
111.50	451	361	1.055	1.080	0.438	0.332	−0.251	+0.289	−0.103	+0.081	−0.022
112.00	537	206	1.843	0.779	0.585	0.270	−0.341	+0.141	−0.113	+0.048	−0.069
The slopes of the mound. Height of the grass was 5 to 10 cm											
110.00	138	137	0.422	0.695	0.205	0.325	0.197	+0.424	−0.096	+0.148	+0.027
110.50	112	158	0.349	0.477	0.159	0.195	0.197	+0.272	−0.086	+0.107	+0.027
111.00	162	124	0.326	0.3	0.114	0.109	0.184	+0.184	−0.064	+0.064	−0.006
111.50	169	118	0.355	0.358	0.154	0.133	−0.21	+0.234	−0.089	+0.088	−0.014
112.00	144	144	0.384	0.406	0.159	0.139	−0.247	+0.255	−0.01	+0.091	−0.001
112.50	146	137	0.347	0.381	0.138	0.137	−0.256	+0.246	−0.098	+0.096	−0.004
113.00	125	122	0.297	0.291	0.12	0.12	−0.227	+0.196	−0.09	+0.082	−0.002
113.50	115	113	0.272	0.265	0.124	0.116	−0.182	+0.189	−0.083	+0.081	+0.001
114.00	110	100	0.251	0.213	0.12	0.105	−0.161	+0.161	−0.08	+0.073	−0.007
114.50	103	108	0.277	0.288	0.123	0.132	−0.202	+0.207	−0.008	+0.01	+0.005
115.00	120	125	0.339	0.362	0.141	0.171	−0.199	+0.248	−0.092	+0.116	+0.013
115.50	115	139	0.378	0.329	0.181	0.137	−0.249	+0.219	−0.124	+0.099	−0.001
116.00	120	111	0.34	0.245	0.151	0.109	−0.257	+0.17	−0.108	+0.078	−0.018
116.50	95	124	0.206	0.237	0.095	0.097	−0.159	+0.155	0.072	+0.066	+0.007
117.00	119	88	0.267	0.222	0.119	0.108	−0.185	+0.16	−0.082	+0.078	−0.02
117.50	95	107	0.268	0.285	0.137	0.134	−0.168	+0.207	−0.085	+0.096	+0.009
118.00	91	100	0.271	0.274	0.132	0.131	−0.185	+0.176	−0.096	+0.082	−0.002
118.50	90	99	0.259	0.278	0.115	0.117	−0.193	+0.183	−0.009	+0.075	−0.002
119.00	110	68	0.27	0.242	0.122	0.112	−0.184	+0.158	−0.087	+0.075	−0.026
119.50	612	73	0.236	0.222	0.103	0.088	−0.159	+0.123	−0.079	+0.049	−0.009
120.00	87	92	0.254	0.405	0.094	0.123	−0.217	+0.177	−0.072	+0.063	−0.003
120.50	70	132	0.247	0.253	0.05	0.081	−0.245	+0.114	−0.041	+0.044	+0.014
121.00	57	50	0.116	0.159	0.063	0.069	−0.127	+0.127	−0.041	+0.071	+0.043

Table 6. *Cont.*

Isoline Height	Number of Deviation [a]		Maximum Plane Deviation, m [b]		Average of Plane Deviation, m [c]		Maximum of Height Deviation, m [d]		Average of Height Deviation, m [e]		Proportional Deviation [f]
	−	+	−	+	−	+	−	+	−	+	
The top of the mound. Height of the grass was 1 to 2 cm											
121.00	71	151	0.468	0.443	0.260	0.240	−0.297	+0.055	−0.107	+0.031	−0.013
120.50	153	183	0.396	0.754	0.111	0.310	−0.070	+0.101	−0.021	+0.048	+0.017
120.00	86	251	0.373	0.466	0.148	0.269	−0.042	+0.091	−0.018	+0.051	+0.033
119.50	151	225	0.274	0.694	0.114	0.238	−0.048	+0.125	−0.018	+0.044	+0.019
119.00	205	395	0.736	0.585	0.256	1.228	−0.197	+0.419	−0.033	+0.114	+0.064
118.50	107	490	0.647	2.373	0.302	0.804	−0.235	+0.196	−0.106	+0.074	+0.042
121.00	71	151	0.468	0.443	0.260	0.240	−0.297	+0.055	−0.107	+0.031	−0.013
120.50	153	183	0.396	0.754	0.111	0.310	−0.070	+0.101	−0.021	+0.048	+0.017
120.00	86	251	0.373	0.466	0.148	0.269	−0.042	+0.091	−0.018	+0.051	+0.033
119.50	151	225	0.274	0.694	0.114	0.238	−0.048	+0.125	−0.018	+0.044	+0.019
119.00	205	395	0.736	0.585	0.256	1.228	−0.197	+0.419	−0.033	+0.114	+0.064
118.50	107	490	0.647	2.373	0.302	0.804	−0.235	+0.196	−0.106	+0.074	+0.042
121.00	71	151	0.468	0.443	0.260	0.240	−0.297	+0.055	−0.107	+0.031	−0.013
120.50	153	183	0.396	0.754	0.111	0.310	−0.070	+0.101	−0.021	+0.048	+0.017
120.00	86	251	0.373	0.466	0.148	0.269	−0.042	+0.091	−0.018	+0.051	+0.033
119.50	151	225	0.274	0.694	0.114	0.238	−0.048	+0.125	−0.018	+0.044	+0.019
119.00	205	395	0.736	0.585	0.256	1.228	−0.197	+0.419	−0.033	+0.114	+0.064
118.50	107	490	0.647	2.373	0.302	0.804	−0.235	+0.196	−0.106	+0.074	+0.042

* Explanation of the superscripts in the table: [a] number of negative (−N) and positive (+N) derivation, [b] maximum of negative (−Δ) and positive (+Δ) plane derivation, [c] average of negative ($\sum-\Delta/-N$) and positive ($\sum+\Delta/+N$) plane derivation, [d] maximum of negative (−Δz max) and positive (+Δz max) height derivation, [e] average of negative ($\sum-\Delta z/N_z$) and positive ($\sum+\Delta z/N_z$) height derivation, and [f] ratio of average positive/negative derivation and number of positive/negative derivation $\left(\frac{+\Delta}{-\Delta}\right)/\left(\frac{N+}{N-}\right)$.

5. Conclusions

1. The "Circle_3p" computer program designed by Artūras Bautrėnas, an associate professor of the Department of Cartography and Geoinformatics, employs the classical method of Delaunay and ensures a consistent systemic selection of points.

2. The use of "Circle 3p" for the analysis of aerial photographs of the Lepelionys Mound has shown that the program needs to be improved by adding elements for the correction of the isolines.

3. A comparison of the results of the geodetic measurements and the UAV images, showed that the best overlaps of surface microform isolines are on steep slopes. On the flat top of the mound surface, microform variance makes up 0.7 to 1.0 m. This is due to the rare density of the interpolation points calculated by the program "Circle_3p". The variance of the isolines at the foot of the mound reaches 0.45 to 0.7 m (medium-height grass) and 0.95 to 1.35 m (high grass). The study has shown that external factors have a significant influence on the identification of the mound relief microforms.

4. The research related to the GCPs position and shows that five GCPs arranged at the edges and the center of the object give the best accuracy as compared with other variations (three on the top, four on the edge, and ten GCPs).

5. The unnatural curvature of isolines in the UAV 3D model, resulting in the abundance of unnatural surface microforms, is due to the interpolation techniques used to determine the isoline position, specifically the calculation of the interfaces between numerous point pairs.

Author Contributions: A.Č.: conceptualization, methodology, visualization, writing—original draft, writing—review & editing. A.B.: data curation, formal analysis, investigation, validation. L.B.: investigation, software, visualization. D.O.: investigation, software, validation.

Funding: This research received no external funding.

Conflicts of Interest: The authors declare no conflict of interest.

References

1. Jones, R.J.A.; Evans, R. Soil and crop marks in the recognition of archaeological sites by air photography. In *Aerial Reconnaissance for Archaeology*; Wilson, D.R., Ed.; The Council for British Archaeology: London, UK, 1975; pp. 1–11.
2. Remondino, F.; El-Hakim, S. Image-based 3D modelling, a review. *Photogramm. Rec.* **2006**, *21*, 269–291. [CrossRef]
3. Parcaks, S.H. *Satellite Remote Sensing for Archaeology*; Routledge: London, UK; New York, NY, USA, 2009; p. 312.
4. Bäumker, M.; Przybilla, H.-J. Investigations on the accuracy of the navigation data of unmanned aerial vehicles using the example of the system microcopter. *Arch. Photogramm. Remote Sens. Spat. Inf. Sci.* **2011**, *38*, 113–118.
5. Deseilligny, M.; De Luca, L.; Remondino, F. Automated image-based procedures for accurate artifacts 3D modeling and orthoimages. In Proceedings of the XXIIIth International CIPA Simposium, Prague, Czech Republic, 12–16 September 2011; pp. 291–299.
6. Chiabrando, F.; Nex, F.; Piatti, D.; Rinaudo, F. UAV and RPV systems for photogrammetric surveys in archaeological areas, two tests in the Piedmont Region (Italy). *J. Archaeol. Sci.* **2011**, *38*, 697–710. [CrossRef]
7. Lasaponara, R.; Masini, N. Satellite remote sensing in archaeology, past, present and future perspectives. *J. Archaeol Sci.* **2011**, *38*, 1995–2002. [CrossRef]
8. Lasaponara, R.; Masini, N. (Eds.) *Satellite Remote Sensing: A New Tool for Archaeology*; Springer: Heidelberg, Germany, 2012; p. 364.
9. Lasaponara, R.; Masini, N. Satellite synthetic aperture radar in archaeology and cultural landscape: An overview. *Archaeol. Prospect.* **2013**, *20*, 71–78. [CrossRef]
10. Nex, F.; Remondino, F. UAV for 3D mapping applications, a review. *Appl. Geomat.* **2013**, *6*, 1–15. [CrossRef]
11. Hugenholtz, C.H.; Whitehead, K.; Brown, O.W.; Barchyn, T.E.; Moorman, B.J.; LeClair, A.J.; Riddell, K.; Hamilton, T. Geomorphological mapping with a small unmanned aircraft system (SUAS), feature detection and accuracy assessment of a photogrammetrically-derived digital terrain model. *Geomorphology* **2013**, *194*, 16–24. [CrossRef]
12. Whitehead, K.; Hugenholtz, C.H. Remote sensing of the environment with small unmanned aircraft systems (UAS): A review of progress and challenges. *J. Unmanned Veh. Syst.* **2014**, *2*, 69–85. [CrossRef]
13. Whitehead, K.; Hugenholtz, C.H.; Myshak, S.; Brown, O.W.; LeClair, A.J.; Tamminga, A.; Barchyn, T.E.; Moorman, B.J.; Eaton, B. Remote sensing of the environment with small unmanned aircraft systems (UAS): Scientific and commercial applications. *J. Unmanned Veh. Syst.* **2014**, *2*, 86–102. [CrossRef]
14. Themistocleous, K.; Agapiou, A.; Cuca, B.; Hadjimitsis, D.G. Unmanned aerial systems and spectroscopy for remote sensing applications in archaeology. *Int. Arch. Photogramm. Remote Sens. Spati. Inf. Sci.* **2015**, *XL-7/W3*, 1419–1423. [CrossRef]
15. Hadjimitsis, D.G.; Themistocleous, K.; Agapiou, A. Monitoring Archaeological Site Landscapes in Cyprus using multi-temporal atmospheric corrected image data. *Int. J. Archit. Comput.* **2009**, *7*, 121–138. [CrossRef]
16. Agapiou, A.; Hadjimitsis, D.G.; Themistocleous, K.; Papadavid, G.; Toulios, L. Detection of archaeological crop marks in Cyprus using vegetation indices from Landsat TM/ETM+ satellite images and field spectroscopy measurements. *Proc. SPIE* **2010**, *7831*, 78310V.
17. Ruzgienė, B.; Berteška, T.; Gečytė, S.; Jakubauskienė, E.; Aksamitauskas, V.Č. The surface modelling based on UAV photogrammetry and qualitative estimation. *Measurement* **2015**, *73*, 619–627. [CrossRef]
18. Sužiedelytė-Visockienė, J.; Bručas, D.; Bagdžiūnaitė, R.; Puzienė, R.; Stanionis, A.; Ragauskas, U. Remotely-piloted aerial system for photogrammetry: Orthoimage generation for mapping applications. *Geografie* **2016**, *121*, 349–367.
19. Campana, S. Drones in archaeology. State-of-the-art and future perspectives. *Archaeol. Prospect.* **2017**, *24*, 275–296. [CrossRef]
20. Traviglia, A.; Torsello, A. Landscape pattern detection in archaeological remote sensing. *Geosciences* **2017**, *7*, 128. [CrossRef]

21. Masini, N.; Marzo, C.; Manzari, P.; Belmonte, A.; Sabia, C.; Lasaponara, R. On the characterization of temporal and spatial patterns of archaeological crop-marks. *J. Cult. Herit.* **2018**, *32*, 124–132. [CrossRef]

22. Cowley, D.C.; Moriarty, C.H.; Geddes, G.; Brown, G.L.; Wade, T.; Nichol, C.J. UAVs in context: Archaeological airborne recording in a national body of survey and record. *Drones* **2018**, *56*, 2. [CrossRef]

23. Tapete, D. Remote sensing and geosciences for archaeology. *Geosciences* **2018**, *7*, 41. [CrossRef]

24. Konstantinos, K.G.; Soura, K.; Koukouvelas, I.K.; Argyropoulos, N.G. UAV vs classical aerial photogrammetry for archaeological studies. *J. Archaeol. Sci. Rep.* **2017**, *14*, 758–773.

25. Sanz-Ablanedo, E.; Chandler, J.H.; Rodríguez-Pérez, J.R.; Ordóñez, C. Accuracy of unmanned aerial vehicle (UAV) and SfM photogrammetry survey as a function of the number and location of ground control points used. *Remote Sens.* **2018**, *10*, 1606. [CrossRef]

26. Shahbazi, M.; Sohn, G.; Théau, J.; Menard, P. Development and evaluation of a UAV-photogrammetry system for precise 3D environmental modeling. *Sensors* **2015**, *15*, 27493–27524. [CrossRef] [PubMed]

27. Turner, D.; Lucieer, A.; Watson, C. An automated technique for generating georectified mosaics from ultra-high resolution unmanned aerial vehicle (UAV) imagery, based on structure from motion (SFM) point clouds. *Remote Sens.* **2012**, *4*, 1392–1410. [CrossRef]

28. Shahbazi, M.; Sohn, G.; Théau, J.; Menard, P. Robust structure-from-motion computation: Application to open-pit mine surveying from unmanned aerial images. *J. Unmanned Veh. Syst.* **2017**, *5*, 126–145. [CrossRef]

29. Mian, O.; Lutes, J.; Lipa, G.; Hutton, J.J.; Gavelle, E.; Borghini, S. Accuracy assessment of direct georeferencing for photogrammetric applications on small unmanned aerial platforms. *Int. Arch. Photogramm. Remote Sens. Spat. Inf. Sci.* **2016**, *40*, 77–83. [CrossRef]

30. Hugenholtz, C.; Brown, O.; Walker, J.; Barchyn, T.; Nesbit, P.; Kucharczyk, M.; Myshak, S. Spatial accuracy of UAV-derived orthoimagery and topography: Comparing photogrammetric models processed with direct geo-referencing and ground control points. *Geomatica* **2016**, *70*, 21–30. [CrossRef]

31. Benassi, F.; Dall'Asta, E.; Diotri, F.; Forlani, G.; Cella, U.M.; Roncella, R.; Santise, M. Testing the accuracy and repeatability of UAV blocks oriented with GNSS-supported aerial triangulation. *Remote Sens.* **2017**, *9*, 172. [CrossRef]

32. Forlani, G.; Dall'Asta, E.; Diotri, F.; Cella, U.M.; di Roncella, R.; Santise, M. Quality assessment of DSMs produced from UAV flights georeferenced with on-board RTK positioning. *Remote Sens.* **2018**, *10*, 311. [CrossRef]

33. Buivydas, U. The Settlement at the foot of Lepelionys Hill Fort. In *Archaeological Research in Lithuania*; Lithuanian Archaeological Society: Vilnius, Lithuania, 2006; pp. 65–66.

34. Banytė-Rowell, R.; Baronas, D.; Kazakevičius, V.; Vaškevičiūtė, I.; Zabiela, G. *History of Lithuania*; Zabiela, G., Ed.; The Lithuanian Institute of History: Vilnius, Lithuania, 2007; Volume 2, p. 518.

35. Jovaiša, E. *The Aestii: Genesis*; Lithuanian University of Educational Sciences: Vilnius, Lithuania, 2013; Volume 1, p. 379.

36. Jovaiša, E. *The Aestii: Evolution*; Lithuanian University of Educational Sciences: Vilnius, Lithuania, 2014; Volume 2, p. 351.

37. Viršilienė, J.; Zabiela, G. *Revived Mounds*; The Lithuanian Institute of History: Vilnius, Lithuania, 2018; p. 219.

38. Berni, J.; Zarco-Tejada, P.J.; Suarez, L.; Fereres, E. Thermal and narrowband multispectral remote sensing for vegetation monitoring from an unmanned aerial vehicle. *IEEE Trans. Geosci. Remote Sens.* **2009**, *47*, 722–738. [CrossRef]

39. Baluja, J.; Diago, M.P.; Balda, P.; Zorer, R.; Meggio, F.; Morales, F.; Tardaguila, J. Assessment of vineyard water status variability by thermal and multispectral imagery using an unmanned aerial vehicle (UAV). *Irrig. Sci.* **2012**, *30*, 511–522. [CrossRef]

40. Bellvert, J.; Zarco-Tejada, P.J.; Girona, J.; Fereres, E. Mapping crop water stress index in a Pinot-Noir vineyard: Comparing ground measurements with thermal remote sensing imagery from an unmanned aerial vehicle. *Precis. Agric.* **2013**, *15*, 361–376. [CrossRef]

41. Chisholm, R.A.; Cui, J.; Lum, S.K.Y.; Chen, B.M. UAV LiDAR for below-canopy forest surveys. *J. Unmanned Veh. Syst.* **2013**, *1*, 61–68. [CrossRef]

42. Huesca, M.; Merino-de-Miguel, S.; González-Alonso, F.; Martinez, S.; Miguel Cuevas, J.; Calle, A. Using AHS hyper-spectral images to study forest vegetation recovery after a fire. *Int. J. Remote Sens.* **2013**, *34*, 4025–4048. [CrossRef]

43. Knoth, C.; Klein, B.; Prinz, T.; Kleinebecker, T. Unmanned aerial vehicles as innovative remote sensing platforms for high-resolution infrared imagery to support restoration monitoring in cut-over bogs. *Appl. Veg. Sci.* **2013**, *16*, 509–517. [CrossRef]

44. Mozas-Calvache, A.T.; Pérez-García, J.L.; Cardernal-Escarcena, F.J.; Delgado, J.; Mata-de-Castro, E. Comparison of low altitude photogrammetric methods for obtaining DEMs and orthoimages of archaeological sites. *Int. Arch. Photogramm. Remote Sens. Spat. Inf. Sci.* **2012**, *39*, 577–581. [CrossRef]

45. Erny, G.; Gutierrez, G.; Friedman, A.; Godsey, M.; Gradoz, M. Archaeological topography: Comparing digital photogrammetry taken with unmanned aerial vehicles (UAVs) versus standard surveys with total stations. In Proceedings of the 80th Annual Meeting of the Society for American Archaeology, San Francisco, CA, USA, 15–19 April 2015; p. 409.

46. Mesas-Carracosa, F.-J.; Notario-Garcia, M.D.; Meroño de Larriva, J.E.; García-Ferre, A. An analysis of the influence of flight parameters in the generation of unmanned aerial vehicle (UAV) orthomosaics to survey archaeological areas. *Sensors* **2016**, *16*, 1838. [CrossRef]

47. Mian, O.; Lutes, J.; Lipa, G.; Hutton, J.J.; Gavelle, E.; Borghini, S. Direct georeferencing on small unmanned aerial platforms for improved reliability and accuracy of mapping without the need for ground control points. In Proceedings of the International Conference on Unmanned Aerial Vehicles in Geomatics, Toronto, ON, Canada, 30 August–2 September 2015; Volume XL-1/W4, pp. 397–402.

48. Agüera–Vega, F.; Carvajal–Ramirez, F.; Martínez–Carricondo, P. Assessment of photogrammetric mapping accuracy based on variation ground control points number using unmanned aerial vehicle. *Measurement* **2017**, *98*, 221–227. [CrossRef]

49. Fryskowska, A.; Kedzierski, M.; Walczykowski, P.; Wierzbicki, D.; Delis, P.; Lada, A. Effective detection of sub-surface archaeological features from laser scanning point clouds and imagery data. *Int. Arch. Photogramm. Remote Sens. Spat. Inf. Sci.* **2017**, *42*, 245–251. [CrossRef]

50. Carvajal-Ramírez, F.; Navarro-Ortega, A.D.; Agüera-Vega, F.; Martínez-Carricondo, P.; Mancini, F. Virtual reconstruction of damaged archaeological sites based on unmanned aerial vehicle photogrammetry and 3D modelling: Study case of a southeastern Iberia production area in the Bronze Age. *J. Int. Meas. Conf.* **2019**, *136*, 225–236. [CrossRef]

51. Juškaitis, V. *Report of Archaeological Research in 2007 of the Ancient Settlement of the Lepelionys Mound (22593, Prienai District)*; Trakai Historical Museum: Trakai, Lithuania, 2009.

52. Juškaitis, V. The settlement at the foot of the lepelionys hill fort site. In *Archaeological Research in Lithuania*; The Lithuanian Institute of History: Vilnius, Lithuania, 2007; pp. 73–74.

53. Fang, T.; Piegl, L. Algorithm for Delaunay triangulation and convex hull computation using a sparse matrix. *Comput. Aided Des.* **1992**, *24*, 425–436. [CrossRef]

54. Hacigüzeller, P. Collaborative mapping in the age of ubiquitous internet: An archaeological perspective. *Digit. Class. Online* **2017**, *3*, 5–16.

55. Perkins, C. Community Mapping. *Cartogr. J.* **2007**, *44*, 127–137. [CrossRef]

56. Lee, D. Map Orkney Month: Imagining Archaeological Mappings. *Livingmaps Rev.* **2016**, *1*, 1–25.

Article

Complex Field Network Coding for Multi-Source Multi-Relay Single-Destination UAV Cooperative Surveillance Networks

Rui Xue [1,*], Lu Han [1] and Huisi Chai [2]

[1]   College of Information & Communication Engineering, Harbin Engineering University, Harbin 150001, China; hanlu@hrbeu.edu.cn
[2]   China Research Institute of Radiowave Propagation, Xinxiang 453000, China; chaihs@crirp.ac.cn
*   Correspondence: xuerui@hrbeu.edu.cn; Tel.: +86-131-0451-5299

Received: 31 January 2020; Accepted: 10 March 2020; Published: 11 March 2020

Abstract: Relay-based cooperative communication for unmanned aerial vehicle (UAV) networks can obtain spatial diversity gains, expand coverage, and potentially increase the network capacity. A multi-source multi-relay single-destination structure is the main topology structure for UAV cooperative surveillance networks, which is similar to the structure of network coding (NC). Compared with conventional NC schemes, complex field network coding (CFNC) can achieve a higher throughput and is introduced to surveillance networks in this paper. According to whether there is a direct communication link between any source drone and the destination, the information transfer mechanism at the downlink is set to one of two modes, either mixed or relay transmission, and two corresponding irregular topology structures for CFNC-based networks are proposed. Theoretical analysis and simulation results with an additive white Gaussian noise (AWGN) channel show that the CFNC obtains a throughput as high as 1/2 symbol per source per channel use. Moreover, the CFNC applied to the proposed irregular structures under the two transmission modes can achieve better reliability due to full diversity gain as compared to that based on the regular structure. Moreover, the reliability of the CFNC scheme can continue to be improved by combining channel coding and modulation techniques at the expense of rate loss.

Keywords: unmanned aerial vehicle (UAV); cooperative communication; topology structure; complex field network coding (CFNC)

---

1. Introduction

Recently, wireless communications aided by unmanned aerial vehicles (UAVs, also known as drones) have drawn a lot of attention from academic and industrial fields, as well as the general public [1]. Due to their ease of deployment, low cost, high mobility, and ability to hover [2] compared to conventional terrestrial infrastructure, UAVs hovering in the air are more likely to set up wireless links with favorable channel conditions and thus are considered as a promising vector of support for wireless communications in a great number of practical applications [3], such as security and surveillance, the real-time monitoring of road traffic, providing wireless coverage, remote sensing, search and rescue operations, the delivery of goods, precision agriculture, and civil infrastructure inspection [4]. However, it is difficult to complete the complex missions with a single UAV because of its limited detection capacity, energy resources, load, and other factors [5]. The solution to such a problem is ad-hoc formation using multiple UAVs [6]. The number of UAVs and their travel distances vary over a wide range for different applications here, as shown in Figure 1 [2]. Multiple small UAVs as a swarm to complete various tasks have gained more interest, as they improve the effectiveness of a single UAV system [7].

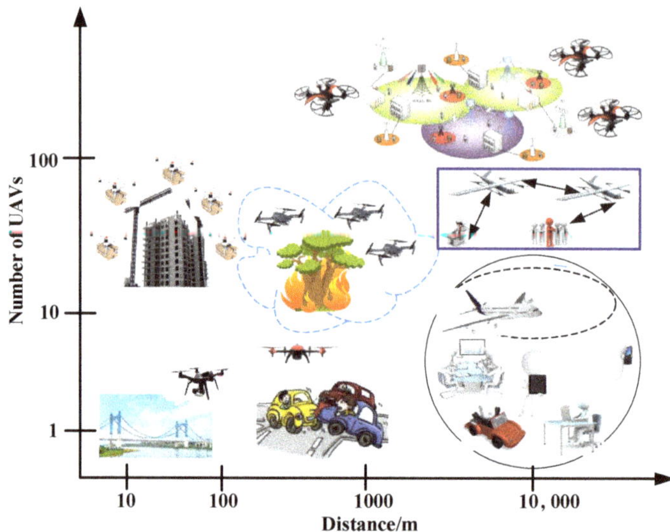

Figure 1. Application areas over a range of distance vs. number of nodes.

An emerging swarm application is the use of small UAVs as source nodes to collect information by their own airborne sensors, and the use of other UAVs as relay nodes to form reliable communication links for ad-hoc ground networks in tactical situations [8–10]. With the application of new sensors (e.g., high-definition aviation digital cameras, airborne imaging spectrometers, aviation imaging radars, etc.) in a single UAV, the information gathered from several source drones is sharply increased. Therefore, determining how to improve the throughput of UAV surveillance networks is a problem worth studying. A multi-source multi-relay single-destination (MSMRSD) structure is the main topology structure of UAV cooperative surveillance networks, and clusters are formed respectively among the source nodes and relay nodes. Effective information sharing among closely spaced intra-cluster nodes (i.e., among source nodes and/or among relay UAVs) is used to facilitate the cooperation [11], which is similar to the structure of network coding (NC) [12]. NC is an effective method to increase network throughput, and a real-time of UAV communication system can be greatly enhanced by introducing network coding principles.

NC is a technique used for effective and secure communication by improving network capacity, throughput, efficiency, and robustness [13]. Its core idea is to employ intermediate nodes to process the received data rather than the traditional forwarding of data, i.e., linear combination or some kind of coding to previously received information. The destination nodes can recover the original data by the part of received data, such that the throughput of the network is efficiently improved and the network's security is increased [14]. Up to now, the main application of network coding in UAV communication networks has been random linear network coding (RLNC) [15,16] or physical-layer network coding (PNC) [17–19]. RLNC can achieve throughput arbitrarily close to the capacity in an unreliable single-hop broadcast network while yielding an acceptable decoding delay [20]. However, the throughput advantage of RLNC in a dynamic UAV network does not seem to be remarkable when the topology of a UAV network is relatively complex [21,22]. Besides, traditional RLNC comes with a sacrifice in service delay because if the users are not able to collect a full size of the encoding packets, the useful information cannot be recovered under the wireless fading channel [23]. Compared with the conventional relay system, PNC can double the throughput of a two-way relay channel (TWRC) by reducing the time slots for the exchange of one packet from four to two [24]. It has been a common belief that PNC requires tight synchronization [25], which is difficult to achieve in UAV networks. Complex field network coding (CFNC), as a generalized version of RLNC, is simple to implement and can facilitate the transmission of 1/2 symbol per source per channel use for multi-source cooperative relay networks [26]. Furthermore, the symbol-level synchronization of CFNC is more convenient to attain than bit-level synchronization [27]. In view of the above advantages, the CFNC is introduced to UAV cooperative surveillance networks in this paper.

The topology structure of NC is also multi-source multi-relay single-destination, as shown in Figure 2 [28]. In the structure, each source node simultaneously connects all relay nodes and the destination node. Moreover, all relay nodes are connected with the destination node. Any source node or relay node links the same numbers of edges, so this structure is called the regular structure by this paper. However, the regular structure is inapplicable to a dynamic time-varying UAV network for two main reasons. One is that not all source drones are always connected with the command and control center (destination node) when the distance between them is beyond communication range, typically for the purpose of expanding the surveillance range or because some obstacles are between them, as illustrated in Figure 3 [29]. It can be seen from Figure 3 that number 4 drone does not have a direct communication link to the command and control center because of a mountain barrier. The other reason for inapplicability is that every source drone cannot be always connected with all relay nodes due to its own mobility or some obstacles between them. In practical applications, any source drone should not always connect with all relay nodes and destination node simultaneously, and the corresponding structure is described as an irregular structure. According to whether there is a direct communication link between any source drone and the command and control center, the information transfer mechanism in downlink is set to one of two modes, either mixed or relay transmission. The specific meaning of mixed and relay transmissions will be expanded upon in Section 2.

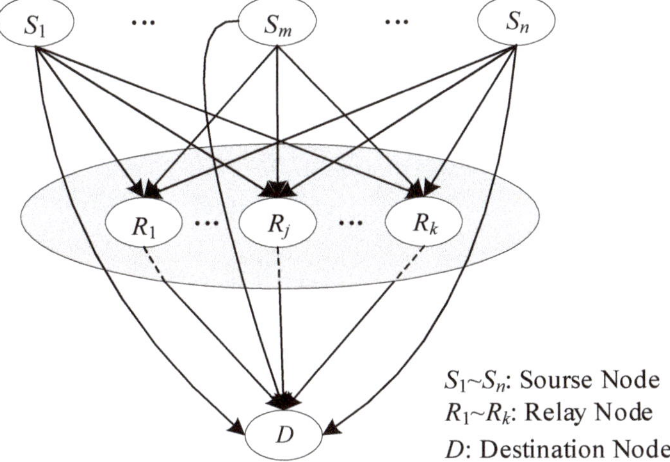

Figure 2. The conventional topology structure of network coding.

Figure 3. An example of an unmanned aerial vehicle (UAV) cooperative surveillance network being applied in a mountainous area.

The rest of this paper is organized as follows: Section 2 presents two irregular topology structures for a CFNC-based network according to the mixed and relay modes. For the different NC schemes, both throughput performance evaluation and the encoding/decoding derivation of CFNC in the two modes are provided by Section 3. Section 4 mainly analyzes the reliability of CFNC combined with the two proposed topology structures over an additive white Gaussian noise (AWGN) channel. Finally, we conclude the paper in Section 5.

2. Design of the Topology Structure

In order to enlarge the coverage area, a UAV cooperative network for surveillance purposes has to employ some drones as relay nodes to transmit messages. A very common topology structure in UAV cooperative networks is multiple surveillance drones, multiple relay drones, and a single command and control center. As shown in Figure 2, a conventional topology structure of NC consists of some source and relay nodes, as well as a destination node. If the source nodes, relay nodes, and the destination node are considered as surveillance drones, relay drones, and the command and control center, respectively, the topology structure of NC is similar to that of the surveillance network. Theoretically, the structure of the former could be applied to the latter.

The prominent feature of a NC structure is that each source node is always connected with all relay drones and the command and control center on the ground. However, this feature is not suitable for the changing dynamics of UAV cooperative networks. On the one hand, some source drones cannot deliver messages to the destination directly because the distance between them exceeds the maximum communication range or because direct communication is blocked by certain obstacles, such as mountains or buildings. On the other hand, it is unreasonable to expect every source drone to connect with all relay drones as obstacle blocking is likely to appear, or the distance among them may be beyond their own individual communication range. From this point of view, the topology structure of NC needs to be appropriately revised before application.

For the multi-source multi-relay single-destination structure expressed as *Ns-Nr-*1, the edges among different types of nodes are the most important factor influencing the total performance of the UAV cooperative surveillance network when the number of source drones (*Ns*) and relay drones (*Nr*) is fixed. The edge refers to a direct communication link between any two different types of nodes in this paper. These edges are divided into three groups, namely, edges between source nodes and the destination node, edges between source nodes and relay nodes, and edges between relay nodes and the destination node. The *Ns-Nr-*1 structure is made up of three types of node and a certain number of edges, so we can consider the structure as a special triple bipartite graph. Based on the characteristics of the bipartite graph, three group edges can be represented by different matrices. A row matrix, **M**, is introduced to express the edges between the source nodes and the destination node. If the ith element of m_i in the matrix is equal to '1', this indicates that the ith source node S_i can deliver messages to the destination node D directly without a relay. Additionally, if $m_i = 0$ this means there is no direct communication link between S_i and D. Likewise, matrix **G** is employed here to represent the edges between the source nodes and relay nodes, and the rows and columns of this matrix indicate the relay and source nodes, respectively. If the element G_{ij} in the matrix is '1', this means that there is a direct communication link between the source node S_j and the relay node R_i. Here, if $G_{ij} = 0$ this represents that S_j cannot send messages to R_i. For convenience, we assume that all relay nodes are always connected to the destination node, which means the edges between them can be expressed as an identity row matrix.

For the conventional topology structure of NC, as illustrated in Figure 2, $\mathbf{M}_{1\times n}$ and $\mathbf{G}_{k\times n}$ are both identity matrices, which is why we call the structure a regular structure. Through the above analysis, we may draw a conclusion that the regular structure of NC is not suitable for UAV cooperative surveillance networks, that is to say that all elements in $\mathbf{M}_{1\times n}$ and $\mathbf{G}_{k\times n}$ cannot always be equal to '1'. The number of edges is variable, even if Ns and Nr are constant, which leads to the diversity in structure. Similar to the characteristics of a check matrix in low-density parity-check (LDPC) codes, the density of '1' in the both matrices is uncertain. The uncertainty results in a large number of irregular structures, even if the values of Ns and Nr are small. According to whether there is a direct communication link between any source drone and the command and control center, the information transfer mechanism at the downlink is set one of two modes, either mixed or relay transmission. In the first mode, the information is transmitted from all source drones to the destination by at least a direct link and multi-relay forwarding, which indicates that **M** is a non-zero matrix. In the other mode, all the source drones deliver messages to relay nodes within their communication range, that is to say, no direct communication link between the source nodes and the destination can be utilized, which means that **M** is a zero matrix.

Based on the two modes, two corresponding irregular topology structures for a CFNC-based network are proposed and Figures 4 and 5 will serve as an example. The matrix **M** is set to $[1\ 0\ \cdots\ 1]_{1\times Ns}$ and $[0\ 0\ \cdots\ 0]_{1\times Ns'}$ in the mixed and relay modes, respectively, and the matrix **G** in the two modes is represented as follows, respectively:

$$
\begin{bmatrix}
1 & 1 & \cdots & 0 \\
1 & 1 & \cdots & 0 \\
\vdots & \vdots & \cdots & \vdots \\
0 & 0 & \cdots & 1
\end{bmatrix}_{Nr\times Ns}
\tag{1}
$$

$$
\begin{bmatrix}
1 & 1 & \cdots & 0 \\
1 & 1 & \cdots & 0 \\
\vdots & \vdots & \cdots & \vdots \\
0 & 1 & \cdots & 1
\end{bmatrix}_{Nr'\times Ns'}
\tag{2}
$$

The process of information transmission in the two topology structures is quite different. For the mixed mode, source drones will transmit information to the destination node via available direct links and the relay nodes within communication range simultaneously in the first time slot. In the second time slot, the relay nodes deliver the demodulated information to the destination node. In the second mode, all the source drones will transmit information to the relay nodes within communication range in the first time slot, then the relay drones deliver the demodulated information to the destination node in the second time slot.

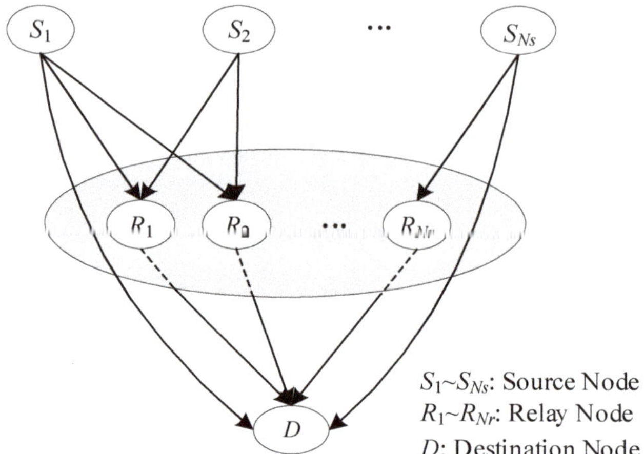

Figure 4. The irregular topology structure for the mixed transmission mode.

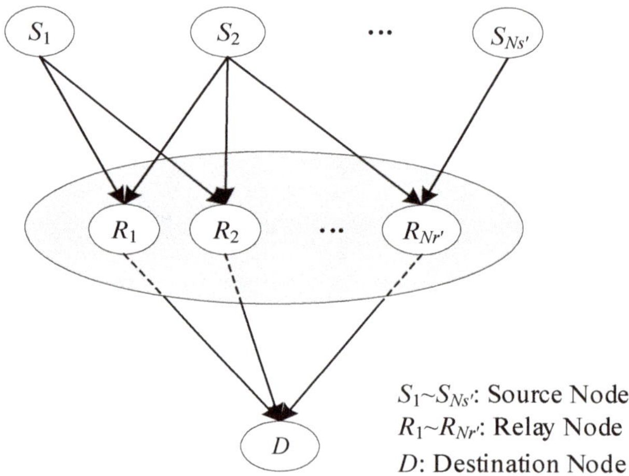

Figure 5. The irregular topology structure for the relay transmission mode.

3. Network Coding

In traditional relay communications, each source node takes advantage of a different time slot to transmit information, and each relay node also successively uses a different time slot to deliver information, which will result in poor real-time performance for information transmission [30]. Network coding can greatly reduce time slots, and the excellent characteristics of this suggest network coding has a very promising future in wireless multicast networks [31,32]. The classification of network coding, different network coding performance evaluations, and the encoding and decoding derivation of CFNC in the two modes are provided by Section 3.

3.1. The Classification of Network Coding

Based on the arithmetic mode, network coding can be divided into several categories, such as the binary field, the Galois field, complex field, and so on. The application of network coding in UAV cluster must consider the characteristics of UAV communication. With the application of new mission payloads, such as large-area and high-resolution digital aerial cameras, synthetic aperture radars, infrared imagers, etc., the information quantity detected by drones is growing exponentially. Saving on the return time of reconnaissance information implies a decrease in discovery probability. Next, we investigate which network coding scheme has the best real-time performance.

In general, network coding designs are based on the Galois field, which implements bit level operations. This coding scheme can improve throughput to some extent, but the advantage is diminished with an increasing number of source and relay nodes. A Ns-source Nr-relay single-destination structure with traditional network coding is depicted in Figure 6. Assuming that each node is equipped with an antenna, Ns sources $(S_1, \cdots, S_{Ns})$ transmit information to the destination (D) directly and via the relays $(R_1, R_2, \cdots, R_{Nr})$. To avoid interference, sources $S_1, \cdots, S_{Ns}$, in the traditional relay format, transmit over orthogonal channels, e.g., via time division multiple access (TDMA) [27]. To start with, source S_1 transmits information symbols x_1 to $R_1, R_2, \cdots, R_{Nr}$ and D simultaneously during channel use (CU) 1. Then, the relay R_1 forwards $\hat{x}_1$ to D in CU 2, and $\hat{x}_1$ is the decoding output of R_1 according to x_1. From CU 3 to CU $(Nr+1)$, the $R_2, \cdots, R_{Nr}$ relays send $\hat{x}_1$ to D successively. The information symbol x_1 takes $(Nr+1)$ CU from source S_1 to the destination D through relays $R_1, R_2, \cdots, R_{Nr}$. For the information symbol sequence $\{x_1, x_2, \cdots, x_{Ns}\}$, a total of $Ns(Nr+1)$ channel uses are needed to deliver Ns symbols with Ns sources, and the throughput of this scheme is $1/(Ns(Nr+1))$ symbol per source per channel use (sym/S/CU).

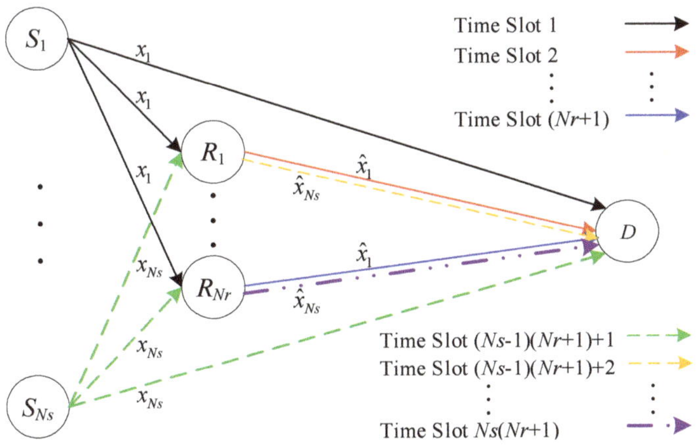

Figure 6. Traditional relay.

The relay scheme based on Galois field network coding (GFNC) is depicted in Figure 7. In CU 1, source S_1 transmits information symbol x_1 to both $R_1, R_2, \cdots, R_{Nr}$ and D, the same as in a traditional relay. From CU 2 to CU Ns, information symbols $x_2, \cdots, x_{Ns}$ are sent to $R_1, R_2, \cdots, R_{Nr}$ and D successively. R_1 forwards the Galois field coded symbol $\hat{x}_1 \oplus \hat{x}_2 \oplus \ldots \oplus \hat{x}_{Ns}$ to D in CU $(Ns+1)$, where $\oplus$ denotes a bitwise exclusive XOR operation. Likewise, R_{Nr} forwards the Galois field coded symbol $\hat{x}_1 \oplus \hat{x}_2 \oplus \ldots \oplus \hat{x}_{Ns}$ to D in CU $(Ns+Nr)$. From the above analysis, we can deduce that $(Ns+Nr)$ channel uses are needed for information symbol sequence $\{x_1, x_2, \cdots, x_{Ns}\}$ transmission from Ns sources to D. Thus, the throughput of a GFNC-based relay is $1/(Ns+Nr)$ sym/S/CU.

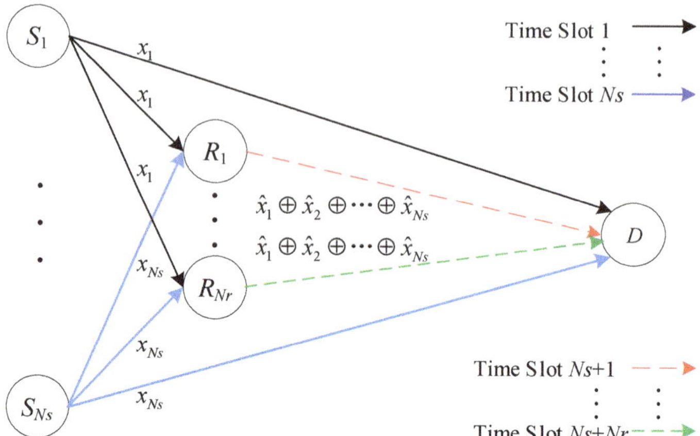

Figure 7. Relay with Galois field network coding (GFNC).

For improving the real-time performance, a CFNC is introduced in this paper. As illustrated in Figure 8, before transmission in time slot 1, the source information x_i from S_i is multiplied by θ_i, which is the ith element of $\theta_S^T = [\theta_1, \theta_2, \cdots, \theta_{Ns}]$. We assume that θ_S^T is available at every node in the network. The choice for a diversity maximizing θ_S^T value is not unique but is available for any Ns. Among the different (parametric/non-parametric) choices for θ_S^T, [28] takes it to be any row of the Vandermonde matrix, i.e.:

$$\theta = \begin{bmatrix} 1 & \delta_1 & \cdots & \delta_1^{Ns-1} \\ 1 & \delta_2 & \cdots & \delta_2^{Ns-1} \\ \vdots & \vdots & \cdots & \vdots \\ 1 & \delta_{Ns} & \cdots & \delta_{Ns}^{Ns-1} \end{bmatrix}_{Ns \times Ns} \tag{3}$$

where the so-called generators, $\{\delta_n\}_{n=1}^{Ns}$, have a unit modulus in complex field C. Relays $R_1, \cdots, R_{Nr}$ simultaneously receive information symbols $\theta_1 x_1, \cdots, \theta_{Ns} x_{Ns}$, transmitted by $S_1, \cdots, S_{Ns}$ in CU 1, and the agreed coefficients $\theta_1, \cdots, \theta_{Ns}$ drawn from C will be specified later. After detecting $x_1, \cdots, x_{Ns}$ as $\hat{x}_1, \cdots, \hat{x}_{Ns}$, $R_1, \cdots, R_{Nr}$ forwards $\theta_1 \hat{x}_1 + \ldots + \theta_{Ns} \hat{x}_{Ns}$ to D in CU 2. Therefore, the throughput of CFNC is 1/2 sym/S/CU. The throughput comparison of the above three schemes is listed in Table 1.

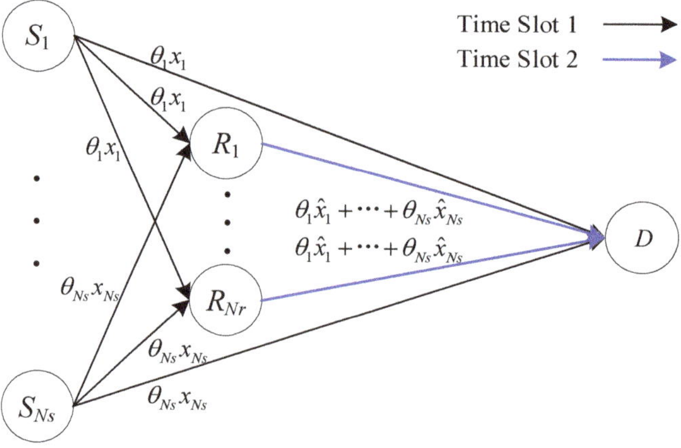

Figure 8. Relay with complex field network coding (CFNC).

Table 1. The throughput performance of various network coding schemes.

Network Coding Scheme	Number of Channels Occupied by the Source Nodes	Number of Channels Occupied by the Relaying Nodes	Throughput (Symbol/Source/Channel Use)
Traditional	Ns	$Ns \times Nr$	$1/(Ns(Nr+1))$
GFNC	Ns	Nr	$1/(Ns+Nr)$
CFNC	1	1	$1/2$

As can be seen from Table 1, GFNC is superior to traditional coding in terms of throughput, and the advantage gradually decreases with the increasing number of source and relay nodes, but CFNC can naturally avoid such a problem. The unique coding method employed by CFNC makes the throughput increase to 1/2 sym/S/CU, which is beneficial to improving the real-time performance. Moreover, the XOR operation is usually adopted by the GFNC, which will cause one-to-one mapping to be impossible between the source information and the received information. By contrast, the received information $\hat{u}$ ($\hat{u} = \theta_1 \hat{x}_1 + \cdots + \theta_{Ns} \hat{x}_{Ns}$) and information symbol sequence $\{x_1, \cdots, x_{Ns}\}$ easily satisfy one-to-one mapping, unless $x_1 = x_2 = \cdots = x_{Ns}$. Meanwhile, the mapping offers a method to detect $\hat{x}_1, \cdots, \hat{x}_{Ns}$ through the received information $\hat{u}$.

3.2. Information Transmission Based on Complex Field Network Coding (CFNC) in Mixed Mode

Based on the theoretical analysis in the previous section, we have deduced that the CFNC obtains overwhelming superiority over other network coding schemes in terms of throughput when the source and relay nodes are of large quantities. Next, the information transmissions based on CFNC applied to the proposed topology structures is derived for the mixed and relay modes, respectively. According to the irregular topology structure Ns-Nr-1 for the mixed mode, as shown in Figure 4, the information symbol transmission based on CFNC merely involves two channel uses. The received symbols at R_j and D after CU 1 are given as follows (see Figure 9):

$$
\begin{aligned}
y_{SR_j}(t) &= h_{S_1 R_j} \theta_1 x_1(t) + \cdots + h_{S_{Ns} R_j} \theta_{Ns} x_{Ns}(t) + n_{SR_j}(t) \\
&= \boldsymbol{\theta}_S^T \mathbf{H}_{SR_j} \mathbf{x}(t) + n_{SR_j}(t)
\end{aligned}
\tag{4}
$$

$$
\begin{aligned}
y_{SD}(t) &= h_{S_1 D} \theta_1 x_1(t) + \cdots + h_{S_{Ns} D} \theta_{Ns} x_{Ns}(t) + n_{SD}(t) \\
&= \boldsymbol{\theta}_S^T \mathbf{H}_{SD} \mathbf{x}(t) + n_{SD}(t)
\end{aligned}
\tag{5}
$$

where for each subscript duplet, $h_{ij} \sim CN(0, \sigma_{ij}^2)$ denotes the channel coefficient and $n_{ij} \sim CN(0, N_0)$ denotes the AWGN term. The instantaneous and average signal-to-noise ratios (SNRs) are given respectively by $_{ij} = |h_{ij}|^2$ and $_{ij} = \sigma_{ij}^2$, where $= P_x/N_0$ and P_x denote the average transmission power of source symbol x, which is assumed to be drawn from a finite alphabet A_x with cardinality $|A_x|$ [27]. Here, $\mathbf{H}_{SR_j} = \text{diag}(h_{S_1 R_j}, h_{S_2 R_j}, \cdots, h_{S_{Ns} R_j})$, $\mathbf{H}_{SD} = \text{diag}(h_{S_1 D}, h_{S_2 D}, \cdots, h_{S_{Ns} D})$, and information symbol vector $\mathbf{x}(t) = [x_1(t), \cdots, x_{Ns}(t)]^T$, where $t = 1, \cdots, Nr$ and $j = 1, \cdots, Nr$.

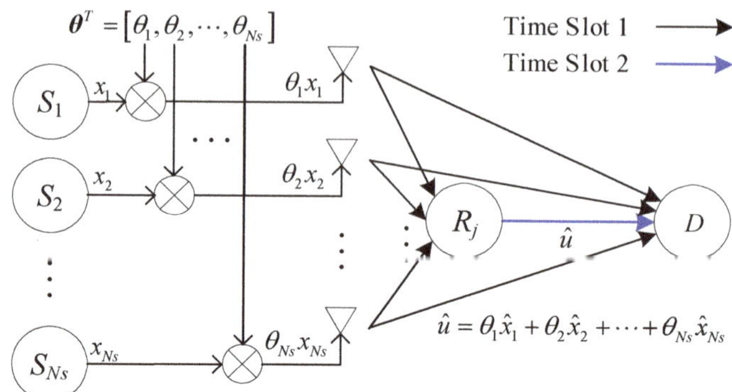

Figure 9. Ns-source setup with CFNC in the mixed mode.

The design of $\boldsymbol{\theta}_S^T$ in Equations (4) and (5) is critical to CFNC. The design relates the linear complex field (LCF) encoder given in [33] for multiple input multiple output (MIMO) systems. Based on the concept of Euler numbers and their properties, two systematic designs of these generators are provided in [34]: $\delta_n = e^{j\pi(4n-1)/2Ns}$ if $Ns = 2^k$ and $\delta_n = e^{j\pi(6n-1)/3Ns}$ if $Ns = 3 \times 2^k$, where n indicates the nth row of Vandermonde matrix. In other words, $\theta_i = e^{j\pi(4n-1)(i-1)/2Ns}$ if $Ns = 2^k$ and $\theta_i = e^{j\pi(6n-1)(i-1)/3Ns}$ if $Ns = 3 \times 2^k$, where $i = 1, \cdots, Ns$. However, the similarities with MIMO-LCF designs stop here. Notice that the coded symbol $u = \theta_1 x_1 + \cdots + \theta_{Ns} x_{Ns}$ in CFNC is transmitted through different nodes (sources) in the network simultaneously, instead of through multiple co-located antennas on one terminal [33]. Therefore, a normalizing factor, as in ([34], Eq. (3.68)), to meet the power constraint on one node is not necessary here [28].

After Nr relay channels, the maximum likelihood (ML) of detection at relay R_j is given as follows:

$$\hat{x}_j(t) = \underset{\mathbf{x}(t)}{\mathrm{argmin}} \| y_{SR_j}(t) - \boldsymbol{\theta}_S^T \mathbf{H}_{SR_j} \mathbf{x}(t) \|, \tag{6}$$

The relaying node R_j re-encodes the demodulation results then sends it to the target node. The input/output (I/O) relationship in CU 2 is expressed as follows:

$$y_{R_jD}(t) = \sqrt{\alpha_j} h_{R_jD} \boldsymbol{\theta}_R^T \hat{x}_j + n_{R_jD}, \quad j = 1, \cdots, Nr, \tag{7}$$

where $\hat{x}_j = \left[\hat{x}_j^T(1), \cdots, \hat{x}_j^T(Nr) \right]^T$, α_j represents a link-adaptive scalar which controls the transmission power at R_j, $\boldsymbol{\theta}_R$ is an $NsNr \times 1$ vector designed as the above, i.e., $\boldsymbol{\theta}_R^T = \left[\theta_1', \theta_2', \cdots, \theta_{Ns\times Nr}' \right]$. For $Nr \times Ns = 2^k$, the entries of $\boldsymbol{\theta}_R$ are given by $\theta_i' = e^{j\pi(4n-1)(i-1)/(2Ns\times Nr)}$ and $i = 1, 2, \cdots, Ns \times Nr$, and for $Nr \times Ns = 3 \times 2^k$, $\theta_i' = e^{j\pi(6n-1)(i-1)/(3Ns\times Nr)}$ for any $n = 1, 2, \cdots, Ns \times Nr$.

The symbol rate is 1/2 sym/S/CU, because Ns sources transmit Ns signals over 2 channels. After passing through 2 channels, the ML detection result at D is given as follows:

$$\hat{x}_D = \underset{\mathbf{x}'}{\mathrm{argmin}} \left\{ \sum_{t=1}^{Nr} \| y_{SD}(t) - \boldsymbol{\theta}_S^T \mathbf{H}_{SD} \mathbf{x}(t) \|^2 + \sum_{j=1}^{Nr} \| y_{R_jD}(t) - \sqrt{\alpha_j} h_{R_jD} \boldsymbol{\theta}_R^T \mathbf{x}' \|^2 \right\}, \tag{8}$$

where $\mathbf{x}' = \left[\mathbf{x}^T(1), \cdots, \mathbf{x}^T(Nr) \right]^T$.

3.3. Information Transmission Based on CFNC in Relay Mode

There are no any direct communication links between the source drones and the command and control center when the source drones move beyond their communication range or the links among them are totally blocked. In such a situation, the conventional topology structure of NC exhibited in Figure 4 is inapplicable for such an application. Thus, an irregular topology structure in the relay mode is proposed by this paper, depicted in Figure 5. The received symbols at R_j after CU 1 (see Figure 10) are the same as in Section 3.2, i.e., $y_{SR_j}(t) = \boldsymbol{\theta}_S^T \mathbf{H}_{SR_j} \mathbf{x}(t) + n_{SR_j}(t)$.

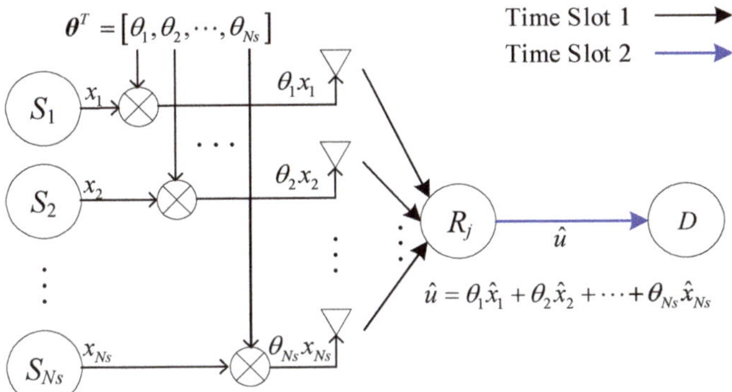

Figure 10. *Ns*-source setup with CFNC in the relay mode.

After *Nr* channel uses, relay R_j detects $\hat{x}_j(t) = \underset{\mathbf{x}(t)}{\text{argmin}} \, \|y_{SR_j}(t) - \boldsymbol{\theta}_S^T \mathbf{H}_{SR_j} \mathbf{x}(t)\|$ and forwards this demodulated symbol with scaling coefficient α_j in next CU. The I/O relationship is $y_{R_j D}(t) = \sqrt{\alpha_j} h_{R_j D} \boldsymbol{\theta}_R^T \hat{x}_j + n_{R_j D}$, where $j = 1, 2, \cdots, Nr$, where $\boldsymbol{\theta}_R$ is the $NsNr \times 1$ vector designed in Section 3.2.

Since *Nr* symbols are transmitted per source over 2*Nr* channel uses, the symbol rate is clearly 1/2 sym/S/CU. After passing through 2 channels, the ML detection result at D is given as follows:

$$\hat{x}_D = \underset{\mathbf{x}'}{\text{argmin}} \left\{ \sum_{t=1}^{Nr} \sum_{j=1}^{Nr} \|y_{R_j D}(t) - \sqrt{\alpha_j} h_{R_j D} \boldsymbol{\theta}_R^T \mathbf{x}'\|^2 \right\}, \tag{9}$$

where the calculation method of $\boldsymbol{\theta}_R$ is referred to the previous section.

4. Simulation Results and Analysis

4.1. Topology Structure Performance Evaluation

The throughput performance of CFNC based on an irregular topology structure in the mixed mode has been assessed in Section 3.1. Compared with CFNC, based on the conventional topology structure, the reliability of CFNC applied in the two proposed topology structures over an AWGN channel has been evaluated by Monte Carlo simulations using MATLAB. In this section, we mainly investigate the influence of the source and relay node numbers to the symbol error probability (SEP) of the two proposed structures. In all simulations, the frame length of information transmitted by each source node was 1000 bits, and the bits in the same position of every information frame constituted a single symbol, i.e., a symbol contained *Ns* bits. The frame number of each source node was fixed at 1500.

We investigated the mixed mode reliability of the proposed irregular topology structure with different numbers of source and relay drones compared to the regular structure. Figures 11 and 12 show the SEP performance of the mixed mode with different numbers of relays in the 6-*Nr*-1 and

8-*Nr*-1 structures, respectively. The edge parameters of the 6-*Nr*-1 and 8-*Nr*-1 structures in the mixed mode are exhibited in Tables 2 and 3 separately, and the other simulation parameters were the same as mentioned above if no special indication is otherwise given. It can be seen from Figure 11 that the SEP performance of the mixed mode increases better with the increasing number of relay drones when the number of source drones is fixed at 6. This is due to the higher diversity gains originating from the increasing number of relay nodes. However, the space for SEP improvement gradually diminishes when increasing the relay drone number when *Nr* is larger than 6. In order to reduce the complexity and cost of UAV cooperative networks, we selected the number of relay drones as 6 for the 6-*Nr*-1 structure. Compared with the regular 6-6-1 structure, the proposed 6-6-1 structure earns gains of at least 3 dB in the region of SEP = 10^{-3}, that is to say, the irregular structure can remarkably improve reliability over the regular structure under the same parameters.

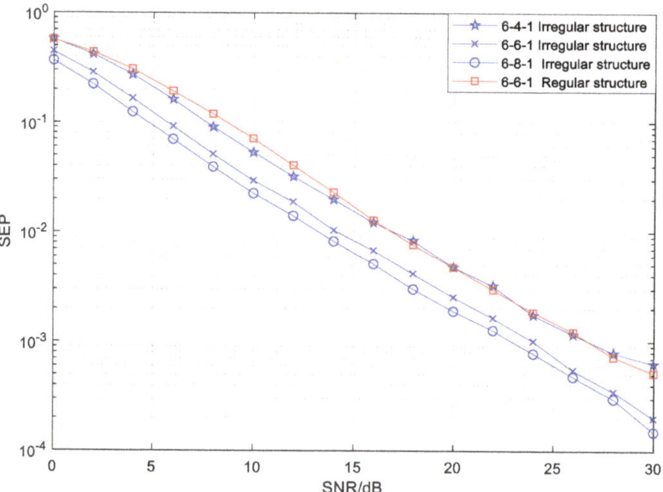

Figure 11. The symbol error probability (SEP) of the mixed mode with different numbers of relays in a 6-*Nr*-1 CFNC-based structure.

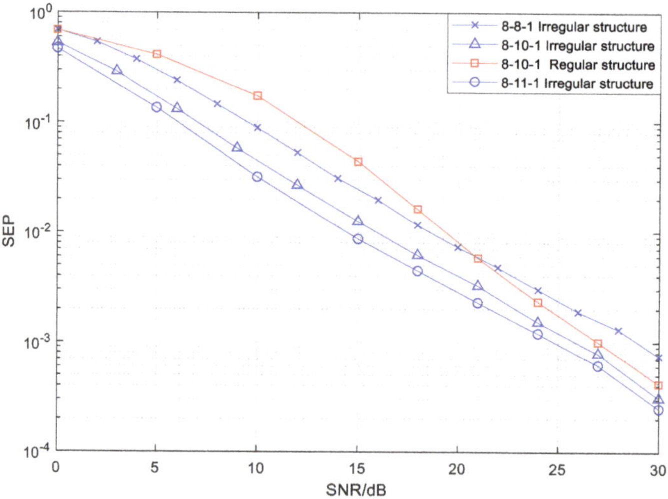

Figure 12. The SEP of the mixed mode with different numbers of relays in a 8-*Nr*-1 CFNC-based structure.

Table 2. The edge parameters of the 6-*Nr*-1 structure in the mixed mode.

Structures	6-4-1	6-6-1	6-8-1
M	$\begin{bmatrix} 1 & 1 & 1 & 0 & 1 & 1 \end{bmatrix}$	$\begin{bmatrix} 1 & 1 & 1 & 1 & 1 & 1 \end{bmatrix}$	$\begin{bmatrix} 1 & 1 & 1 & 1 & 1 & 1 \end{bmatrix}$
G	$\begin{bmatrix} 1 & 1 & 0 & 0 & 0 & 0 \\ 1 & 1 & 1 & 1 & 0 & 0 \\ 0 & 0 & 1 & 1 & 1 & 1 \\ 0 & 0 & 0 & 0 & 1 & 1 \end{bmatrix}$	$\begin{bmatrix} 1 & 1 & 0 & 0 & 0 & 0 \\ 1 & 1 & 1 & 0 & 0 & 0 \\ 1 & 1 & 1 & 1 & 0 & 0 \\ 0 & 1 & 1 & 1 & 1 & 0 \\ 0 & 0 & 1 & 1 & 1 & 1 \\ 0 & 0 & 0 & 1 & 1 & 1 \end{bmatrix}$	$\begin{bmatrix} 1 & 0 & 0 & 0 & 0 & 0 \\ 1 & 1 & 0 & 0 & 0 & 0 \\ 1 & 1 & 1 & 0 & 0 & 0 \\ 1 & 1 & 1 & 1 & 0 & 0 \\ 0 & 1 & 1 & 1 & 1 & 1 \\ 0 & 0 & 1 & 1 & 1 & 1 \\ 0 & 0 & 1 & 1 & 1 & 1 \\ 0 & 0 & 0 & 0 & 1 & 1 \end{bmatrix}$

Table 3. The edge parameters of the 8-*Nr*-1 structure in the mixed mode.

Structures	8-8-1	8-10-1	8-11-1
M	$\begin{bmatrix} 11111111 \end{bmatrix}$	$\begin{bmatrix} 11111110 \end{bmatrix}$	$\begin{bmatrix} 11111110 \end{bmatrix}$
G	$\begin{bmatrix} 11110000 \\ 11110000 \\ 01111000 \\ 01111000 \\ 00111100 \\ 00011110 \\ 00011110 \\ 00001111 \end{bmatrix}$	$\begin{bmatrix} 11110000 \\ 11110000 \\ 01111000 \\ 01111000 \\ 00111100 \\ 00111100 \\ 00011110 \\ 00011110 \\ 00001111 \\ 00001111 \end{bmatrix}$	$\begin{bmatrix} 11110000 \\ 11110000 \\ 01111000 \\ 01111000 \\ 00111100 \\ 00111100 \\ 00011110 \\ 00011110 \\ 00001111 \\ 00001111 \\ 00000111 \end{bmatrix}$

For the 8-*Nr*-1 irregular structure in the mixed mode, the simulation results of the SEP performance shown in Figure 12 are very similar to those in Figure 11. As we see from Figure 12, the SEP decreased with an increasing number of relay drones when the number of source drones was set at 8. It is noteworthy that the improvement on SEP is smaller when the number of relay drones is greater than 10. Too many relay nodes will increase the complexity and cost of a UAV cluster. In view of the reasons given above, the number of relay nodes was selected as 10 for the 8-*Nr*-1 structure. In addition, the reliability of the irregular 8-10-1 structure was superior to that of the regular structure under the same simulation parameters. Through the above analysis, we can deduce that the proposed irregular topology structure in the mixed mode has certain advantages in terms of the reliability when compared with the regular structure under the same conditions.

The effect of the source drone number on the SEP performance of the irregular structure in the mixed mode is illustrated in Figure 13. More details about the edge setting in the *Ns*-6-1 structure are exhibited in Table 4. We can observe from Figure 13 that the SEP performance worsens with an increasing number of source drones when the number of relay drones is fixed at 6. For a single relay node, the more information it receives from the connected source drones, the worse the SEP performance is. The interference among different messages will be intensified when a relay node processes or forwards information, which results in poor SEP performance.

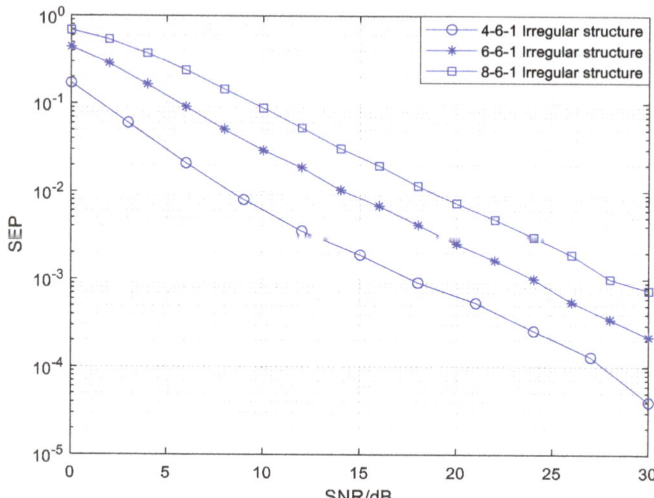

Figure 13. The SEP of the mixed mode with different numbers of sources in the *Ns*-6-1 CFNC-based structure.

Table 4. The edge parameters of the *Ns*-6-1 structure in the mixed mode.

Structures	4-6-1	6-6-1	8-6-1
M	$[1111]$	$[111111]$	$[11111110]$
G	$\begin{bmatrix} 1100 \\ 1110 \\ 1110 \\ 0111 \\ 0111 \\ 0011 \end{bmatrix}$	$\begin{bmatrix} 110000 \\ 111000 \\ 111100 \\ 011110 \\ 001111 \\ 000111 \end{bmatrix}$	$\begin{bmatrix} 11110000 \\ 11110000 \\ 01111000 \\ 00111000 \\ 00011110 \\ 00001111 \end{bmatrix}$

In the relay mode, there is no any connection between the source drones and the command and control center, which indicates that **M** is a zero-row matrix. Figure 14 shows the SEP performance of the relay mode with different numbers of relays in the 2-*Nr*-1 irregular topology structure. More details about the edge setting in the 2-*Nr*-1 structure are exhibited in Table 5. As we see from Figure 14, the SEP performance is gradually improved with an increasing number of relay drones when the number of source drones is fixed at 2. This is because more relay nodes bring more diversity gains, which leads to a better reliability. It is noteworthy that the room for improvement on the SEP performance is limited when the number of relay nodes is larger than a certain value. Moreover, the increasing number of relay nodes will impose a relatively high implementation complexity and cost for cooperative UAV networks. Therefore, the selection of the relay number should take into account reliability, network complexity, system cost, and so on.

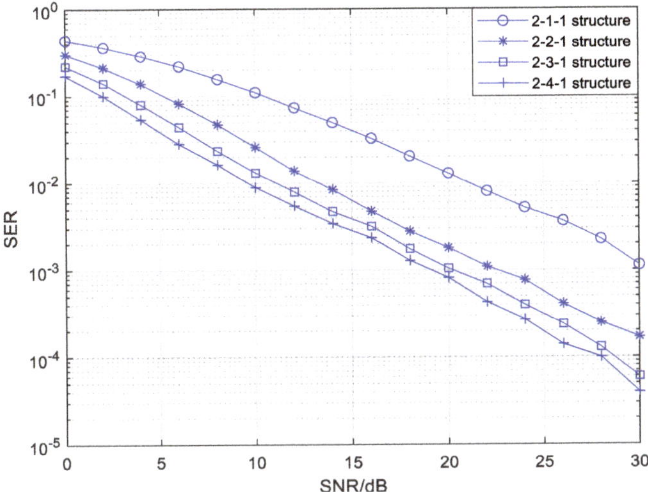

Figure 14. The SEP of the relay mode with different numbers of relays in the 2-*Nr*-1 CFNC-based structure.

Table 5. The edge parameters of the 2-*Nr*-1 structure in the relay mode.

Structures	2-1-1	2-2-1	2-3-1	2-4-1
G	$[1\,1]$	$\begin{bmatrix} 1\,1 \\ 1\,1 \end{bmatrix}$	$\begin{bmatrix} 1\,0 \\ 1\,1 \\ 0\,1 \end{bmatrix}$	$\begin{bmatrix} 1\,0 \\ 1\,1 \\ 1\,1 \\ 0\,1 \end{bmatrix}$

The effect of the source drone number on the SEP performance of the irregular structure in the relay mode is illustrated in Figure 15. The detailed edge parameters in the *Ns*-2-1 structure are exhibited in Table 6. It can observed from Figure 15 that the SEP performance gets worse with an increasing number of source drones. The reason for this is similar to that of the mixed mode. The greater the number of source drones a single relay node links, the more messages it receives. The mutual interference among messages goes against data processing and forwarding, which leads to a considerable decline in reliability.

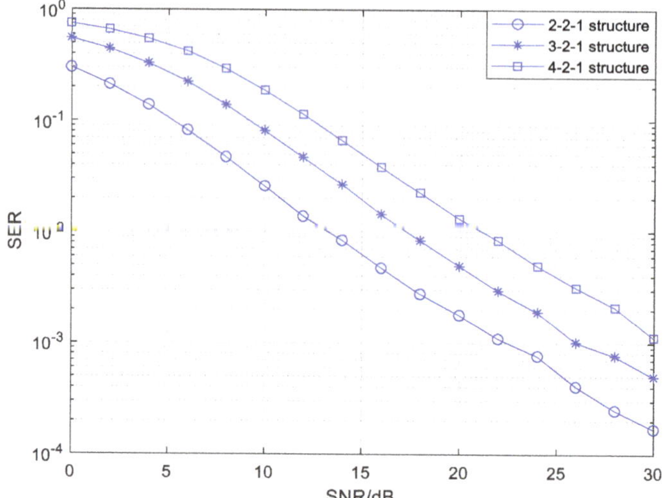

Figure 15. The SEP of the relay mode with different numbers of sources in the Ns-2-1 CFNC-based structure.

Table 6. The edge parameters of the Ns-2-1 structure in the relay mode.

Structures	2-2-1	3-2-1	4-2-1
G	$\begin{bmatrix} 1\,1 \\ 1\,1 \end{bmatrix}$	$\begin{bmatrix} 1\,1\,0 \\ 0\,1\,1 \end{bmatrix}$	$\begin{bmatrix} 1\,1\,1\,0 \\ 0\,1\,1\,1 \end{bmatrix}$

4.2. The Combination of CFNC and Conventional Unmanned Aerial Vehicle (UAV) Datalink

Through the above analysis, we can deduce that the CFNC applied in the proposed irregular structures based on the two transmission modes has a distinct advantage in terms of the reliability and throughput found. Next we discuss the performance of CFNC combined with a UAV datalink signal system and convolutional coded binary phase shift keying (CC-BPSK) modulation, which is a common transmission scheme used in existing UAV datalinks. Figure 16 shows a block diagram of CC-BPSK combined with CFNC (abbreviated as CC-BPSK-CFNC). In this system, the simulation parameters were set as follows: The structure of convolutional code was (2, 1, 3), i.e., one information bit was encoded into a 2-bit codeword each time (code rate was 1/2), and the constraint length was 3; the generator matrix was [1 1 1; 1 0 1]; and the Viterbi algorithm was adopted for decoding. The irregular topology structure of the CFNC in the two modes was chosen as 8-8-1, and the edge parameters in the structure are shown in Table 3.

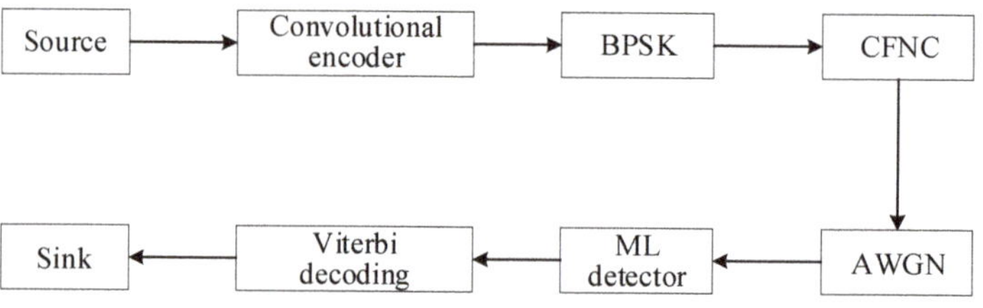

Figure 16. The transmission scheme of coded binary phase shift keying complex field network coding (CC-BPSK-CFNC).

The SEP comparison of CC-BPSK-CFNC, based on the mixed mode in regular and irregular structures, is illustrated in Figure 17. As shown in Figure 17, a SEP value of 10^{-4} is attainable for CC-BPSK-CFNC in the irregular structure at a SNR of around 12 dB, whereas the equivalent SEP performance for CFNC based on the same structure without channel coding and modulation has a SNR of about 30 dB (as shown in Figure 12). Note that the reliability could be improved by invoking a few coded modulation techniques at the expense of rate loss. The transmission scheme, i.e., CC-BPSK-CFNC, in the irregular structure could obtain at least a 14 dB gain at the SEP of 5×10^{-3} compared with the scheme in the regular structure. The SEP comparison of the CC-BPSK-CFNC, based on relay mode in regular and irregular structures, is depicted in Figure 18. We can see that the SEP of 10^{-4} is attainable for CC-BPSK-CFNC in the irregular structure when the SNR is greater than 18 dB. Compared with the regular structure, the scheme based on the irregular one can earn at least a 6.5 dB gain with a SEP of 10^{-3}.

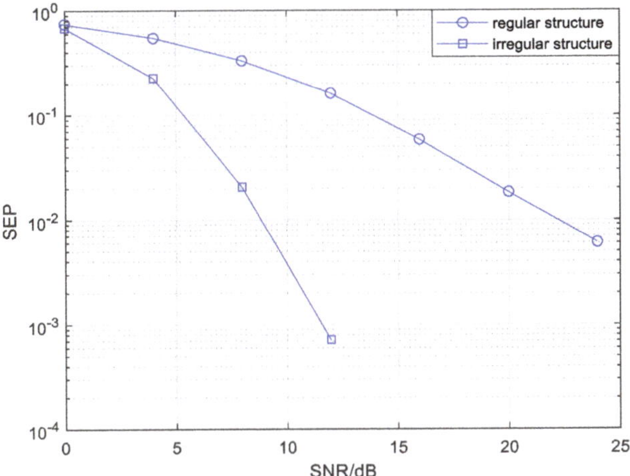

Figure 17. The SEP comparison of CC-BPSK-CFNC based on the mixed mode in regular and irregular structures.

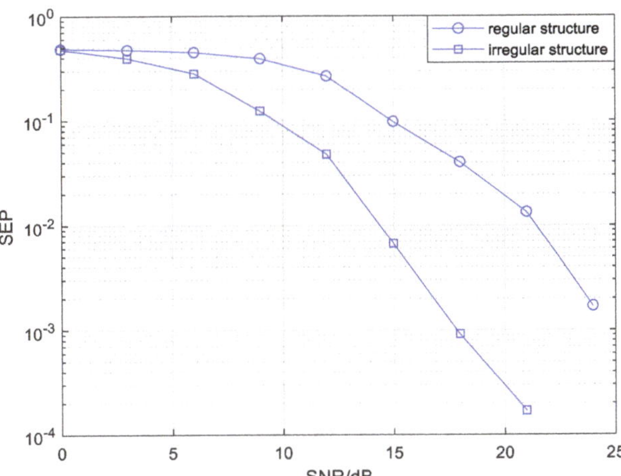

Figure 18. The SEP comparison of CC-BPSK-CFNC based on the relay mode in regular and irregular structures.

5. Conclusions

Using multiple drones to form a collaborative network will become one of the main trends of UAV development in the future. The amount of interactive information among drones in such a collaborative network is expected to increase greatly. Complex field network coding (CFNC) is an effective method to improve network throughput and has been introduced to UAV cooperative surveillance networks in this paper, where the throughput was found to be as high as 1/2 sym/S/CU, which is superior to other network coding schemes. According to whether there is a direct communication link between any source drone and the destination, the information transfer mechanism at the downlink was set to one of two modes, either mixed or relay transmission, and two corresponding irregular topology structures for a CFNC-based network have been proposed, and the information transmissions based on CFNC in the mixed and relay modes were derived. The simulation results over an AWGN channel based on the MATLAB software show that the CFNC applied in the proposed irregular structures under the two transmission modes can remarkably improve reliability using the same parameters when compared with the regular structures. Moreover, the CFNC could easily be combined with the existing channel coding and modulations of UAVs datalinks, such as CC-BPSK, which continues to enhance the SEP performance to a great extent.

Author Contributions: The work presented in this paper was carried out in collaboration with all authors. R.X. conceived the ideas and concept. L.H. implemented the software and carried out the experiments and wrote the manuscript. H.C. critically reviewed and edited the paper. All authors have read and agreed to the published version of the manuscript.

Funding: This research was partially funded by the National Natural Science Foundation of China (Grant No. 61873070), the Technology Development Project of the China Research Institute of Radiowave Propagation (Grant No. JW2019-114), and the Fundamental Research Funds for the Central Universities (Grant No. HEUCFM180803).

Conflicts of Interest: The authors declare no conflict of interest.

Abbreviations

The following abbreviations are used in this manuscript:

UAV	Unmanned Aerial Vehicle
NC	Network Coding
GFNC	Galois Field Network Coding
RLNC	Random Linear Network Coding
PNC	Physical-layer Network Coding
CFNC	Complex Field Network Coding
AWGN	Additive White Gaussian Noise
MSMRSD	Multi-source multi-relay single-destination
LDPC	Low-density Parity-check
LCF	Linear Complex Field
TWRC	Two-way Relay Channel
TDMA	Time Division Multiple Access
CU	Channel Use
MIMO	Multiple Input Multiple Output
ML	Maximum Likelihood
CC	Convolutional Code
BPSK	Binary Phase Shift Keying
SEP	Symbol Error Probability

References

1. Wang, G.; Lee, B.; Ahn, J.; Gihwan, C. A UAV-aided cluster head election framework and applying such to security-driven cluster head election schemes: A survey. *Secur. Commun. Netw.* **2018**, *2018*, 1–18. [CrossRef]
2. Hayat, S.; Yanmaz, E.; Muzaffar, R. Survey on unmanned aerial vehicle networks for civil applications: A communications viewpoint. *IEEE Commun. Surv. Tutor.* **2016**, *18*, 2624–2661. [CrossRef]

3. Yin, S.; Zhao, Y.; Li, L. UAV-assisted cooperative communications with time-sharing SWIPT. In Proceedings of the 2018 IEEE International Conference on Communications, Kansas City, MO, USA, 20–24 May 2018; pp. 1–6.

4. Shakhatreh, H.; Sawalmeh, A.H.; Al-Fuqaha, A.; Dou, Z.; Almaita, E.; Khalil, I.; Othman, N.S.; Khreishah, A.; Guizani, M. Unmanned aerial vehicles (UAVs): A survey on civil applications and key research challenges. *IEEE Access* **2019**, *7*, 48572–48634. [CrossRef]

5. He, D.; Qiao, Y.; Chan, S.; Guizaniet, N. Flight security and safety of drones in airborne fog computing systems. *IEEE Commun. Mag.* **2018**, *56*, 66–71. [CrossRef]

6. Sharma, V.; Kumar, R. Cooperative frameworks and network models for flying ad hoc networks: A survey. *Concurr. Comput. Pract. Exp.* **2017**, *29*, 1–36. [CrossRef]

7. Valentino, R.; Jung, W.S.; Ko, Y. Opportunistic computational offloading system for clusters of drones. In Proceedings of the 2018 20th International Conference on Advanced Communication Technology, Chuncheon-si Gangwon-do, Korea, 11–14 February 2018; pp. 303–306.

8. Zhou, Y.; Cheng, N.; Lu, N.; Shen, X. Multi-UAV-aided networks: Aerial-ground cooperative vehicular networking architecture. *IEEE Veh. Technol. Mag.* **2015**, *10*, 36–44. [CrossRef]

9. Erdelj, M.; Król, M.; Natalizio, E. Wireless sensor networks and multi-UAV systems for natural disaster management. *Comput. Netw.* **2017**, *124*, 72–86. [CrossRef]

10. Kim, B.; Kim, K.; Roh, B.; Choi, H. A new routing protocol for UAV relayed tactical mobile ad hoc networks. In Proceedings of the 2018 Wireless Telecommunications Symposium, Phoenix, AZ, USA, 17–20 April 2018; pp. 1–4.

11. Li, X.; Zhang, Y.D. Multi-source cooperative communications using multiple small relay UAVs. In Proceedings of the 2010 IEEE Globecom Workshops, Miami, FL, USA, 6–10 December 2010; pp. 1805–1810.

12. Ahlswede, R.; Cai, N.; Li, S.R.; Yeung, R.W. Network information flow. *IEEE Trans. Inf. Theory* **2000**, *46*, 1204–1216. [CrossRef]

13. Ayesha, N.; Mubashir, H.R.; Yasir, S.; Imran, R.; Noel, C. Network coding in cognitive radio networks: A comprehensive survey. *IEEE Commun. Surv. Tutor.* **2017**, *19*, 1945–1973.

14. Voskoboynik, N.; Permuter, H.H.; Cohen, A. Network coding schemes for data exchange networks with arbitrary transmission delays. *IEEE Acm Trans. Netw.* **2017**, *25*, 1293–1309. [CrossRef]

15. Lin, C.; Kung, H.T.; Lin, T.; Tarsa, S.J.; Vlah, D. Achieving high throughput ground-to-UAV transport via parallel links. In Proceedings of the 20th International Conference on Computer Communications and Networks, Maui, HI, USA, 31 July–4 August 2011; pp. 1–7.

16. Jiang, S.; Zhang, Q.; Wu, A.; Liu, Q.; Wu, J.; Xia, P. A low-latency reliable transport solution for network-connected UAV. In Proceedings of the 2018 10th International Conference on Communication Software and Networks, Ponta Delgada, Portugal, 6–9 July 2018; pp. 511–515.

17. Gupta, L.; Jain, R.; Vaszkun, G. Survey of important issues in UAV communication networks. *IEEE Commun. Surv. Tutor.* **2015**, *18*, 1123–1152. [CrossRef]

18. Zhang, X.; Miao, L.; Shang, T. Utility analysis of network coding for coordinated formation flight in unmanned aerial vehicles. In Proceedings of the 2016 International Conference on Networking and Network Applications, Hakodate, Japan, 23–25 July 2016; pp. 419–422.

19. Gu, J.; Lamare, R.C.; Huemer, M. Buffer-aided physical-layer network coding with optimal linear code designs for cooperative networks. *IEEE Trans. Commun.* **2018**, *66*, 2560–2575. [CrossRef]

20. Swapna, B.T.; Eryilmaz, A.; Shroff, N.B. Throughput-delay analysis of random linear network coding for wireless broadcasting. *IEEE Trans. Inf. Theory* **2013**, *59*, 6328–6341. [CrossRef]

21. Kusumine, N.; Ishihara, S. Abiding regional data distribution using relay and random network coding on VANETs. In Proceedings of the 2012 IEEE 26th International Conference on Advanced Information Networking and Applications, Fukuoka, Japan, 26–29 March 2012; pp. 105–112.

22. Yang, Z.M.; Hu, Y.J.; Wang, C.L.; Yuan, Q.S. The application of physical layer network coding in the UAVs. *Adv. Mater. Res.* **2014**, *989*, 4147–4151. [CrossRef]

23. Li, B.; Ban, D.; Zhang, R. Efficient scheduling for multicasting multimedia data with adaptive random liner network coding in relay-aided network. In Proceedings of the 2015 IEEE Wireless Communications and Networking Conference, New Orleans, LA, USA, 9–12 March 2015; pp. 1584–1589.

Sensors **2020**, *20*, 1542

24. Zhang, S.; Liew, S.C.; Lam, P.P. Hot topic: Physical-layer network coding. In Proceedings of the 12th annual international conference on Mobile computing and networking, Los Angeles, CA, USA, 23–29 September 2006; pp. 358–365.

25. Liew, S.C.; Zhang, S.; Lu, L. Physical-layer network coding: Tutorial, survey, and beyond. *Phys. Commun.* **2013**, *6*, 4–42. [CrossRef]

26. Rida, K.; Ibrahim, A.; Gunes, K.K. Channel coded complex field network coding in two-way relay networks. In Proceedings of the 2017 IEEE International Black Sea Conference on Communications and Networking, Istanbul, Turkey, 5–8 June 2017; pp. 1–5.

27. Wang, T., Giannakis, G.B. Complex field network coding for multiuser cooperative communications. *IEEE J. Sel. Areas Commun.* **2008**, *26*, 561–571. [CrossRef]

28. Wang, T.; Giannakis, G.B. High-throughput cooperative communications with complex field network coding. In Proceedings of the 2007 41st Annual Conference on Information Sciences and Systems, Baltimore, MD, USA, 14–16 March 2007; pp. 253–258.

29. Sahingoz, O.K. Networking models in flying ad-hoc networks (FANETs): Concepts and challenges. *J. Intell. Robot. Syst.* **2014**, *74*, 513–527. [CrossRef]

30. Liu, X.; Gong, X.; Zheng, Y. Reliable cooperative communications based on random network coding in multi-hop relay WSNs. *IEEE Sens. J.* **2014**, *14*, 2514–2523. [CrossRef]

31. Patil, V.; Gupta, S.; Keshavamurthy, C. S-RLNC based MAC optimization for multimedia data transmission over LTE/LTE-A network. *Int. J. Commun. Netw. Inf. Secur.* **2018**, *10*, 37–49.

32. Li, B.; Li, X.; Zhang, R.; Tang, W.; Li, S. Joint power allocation and adaptive random network coding in wireless multicast networks. *IEEE Trans. Commun.* **2018**, *66*, 1520–1533. [CrossRef]

33. Xin, Y.; Wang, Z.; Giannakis, G.B. Space-time diversity systems based on linear constellation precoding. *IEEE Trans. Wirel. Commun.* **2003**, *2*, 294–309. [CrossRef]

34. Giannakis, G.B.; Liu, Z.; Ma, X.; Zhou, S. *Space-Time Coding for Broadband Wireless Communications*; John Wiley & Sons Inc.: Hoboken, NJ, USA, 2006.

MDPI

St. Alban-Anlage 66

4052 Basel

Switzerland

Tel. +41 61 683 77 34

Fax +41 61 302 89 18

www.mdpi.com

Sensors Editorial Office

E-mail: sensors@mdpi.com

www.mdpi.com/journal/sensors